Correction and Errors

Chapter 1

p. 1 line 6 of para. 3. After "augmented" read: ". . . and even surpassed in numbers . . ."

Chapter 3

p. 36 para. beginning "In the case of", last line: ". . . barrier presented by the velocity of light has never been observed to be broken."

p. 46 end of second line after eq. 3.26 should read: "the four-distance $|\boldsymbol{S}|^2$"

p. 50 middle of page. Reference to eqn (3.35) should be to eq. (3.26). Third line below that should begin "The 3-vector component of $\boldsymbol{P}$, which is $\gamma_u m_0 \boldsymbol{u}$, is . . ."

p. 50 fourth line from bottom: reference should be to (3.30) not (3.42).

Chapter 4

p. 55 Between eqs (4.8) and (4.9), line should read: ". . . which, from (4.1) and (4.6), is . . ."

p. 55 End of second line below eq. (4.9) should read: "Since $k/\omega = 1/c$, . . ."

p. 57 the energy balance equation just below the centre of the page should read:

$$-\frac{\partial}{\partial t}\left(\int U d\tau\right)=\int \boldsymbol{S}\cdot d\boldsymbol{\sigma}+\int \boldsymbol{E}\cdot \boldsymbol{J} d\tau$$

Chapter 5

p. 80 Second line after eq. (5.12). the vector equation should read: $\boldsymbol{n} \times \boldsymbol{n} \times \dot{\boldsymbol{v}} = (\boldsymbol{n} \times \boldsymbol{m})\, \dot{v} \sin\theta$ and in the following line, after "so that" insert $\hat{\boldsymbol{n}} \times \hat{\boldsymbol{v}} = \hat{\boldsymbol{m}} \sin\theta$,

Chapter 6

p. 104 In the penultimate line of text, units are photons/(s rad^2).

Eq. 6.12 should read:

$$\frac{d^2N(\omega,\mathbf{n})}{d\theta d\psi}=\frac{3}{4}\frac{\alpha}{\pi^2}\gamma^2\frac{I}{e}\frac{\Delta\omega}{\omega}\left(\frac{\omega}{\omega_c}\right)^2\left(1+\gamma^2\psi^2\right)^2\left[K_{2/3}^2(\xi)+\frac{\gamma^2\psi^2}{1+\gamma^2\psi^2}K_{1/3}^2(\xi)\right]$$

p. 105 Line after eq. (6.13) should begin: "If we divide eqn. (6.12) by . . ." The units in the next line are photons/(rad^2 s GeV^2).

Figs 6.8–6.12. The y-axis units should be photons/(rad^2 s GeV^2) into unit fractional bandwidth and for one electron.

Chapter 7

p. 122 last sentence should begin. "Values of the integral, multiplied by ω/ω_c, calculated . . ."

p. 123 Table 7.1. Title should read: Values of $\frac{\omega}{\omega_c}\int_{\omega/\omega_c}^{\infty} K_{5/3}(y)\,dy$ as a function of ω/ω_c shown graphically in Fig 7.1

Table 7.1 Second column should be headed $\frac{\omega}{\omega_c}\int_{\omega/\omega_c}^{\infty} K_{5/3}(y)\,dy$. Delete third column.

p. 124 Figure 7.1 Caption should read: Values of $(a)\frac{\omega}{\omega_c}\int_{\omega/\omega_c}^{\infty} K_{5/3}(y)\,dy$ as a function of ω/ω_c

In Figure 7.1. Relabel curve $(a)\frac{\omega}{\omega_c}\int_{\omega/\omega_c}^{\infty} K_{5/3}(y)\,dy$. Delete curve (b).

p. 125 line after eq. 7.44 should begin: ". . . which is identical to eqn. (5.20) obtained . . ."

p. 126 Page ranges in reference 1 should be 1912–1915 and 1551–1554.

Chapter 9

p. 147 para. 3 line 6. Replace "electrons" by "photons". In line 8 of this paragraph, just before eq. (9.3), replace σ_γ by σ'_γ.

Chapter 11

p. 181 Formula on bottom of page should read:

$$x'' = G(s)\frac{\Delta E}{E_0} - [G(s)^2 + K_x(s)]x$$
$$= G(s)\frac{\Delta E}{E_0} - K(s)x$$

p. 182 Replace K(s) with −K(s) in Eqs (11.9a & b), (11.10) and 11.11).

PJD 21 May 2003

OXFORD SERIES ON SYNCHROTRON RADIATION

Series Editors

OXFORD SERIES ON SYNCHROTRON RADIATION

1. S. W. Lovesey, S. P. Collins: *X-ray scattering and absorption by magnetic materials*
2. J. R. Helliwell, P. M. Rentzepis: *Time-resolved diffraction*
3. P. J. Duke: *Synchroton radiation: production and properties*
4. J. A. Clarke: *The science and technology of undulators and wigglers*
5. M. J. Cooper, P. Mijnarends, N. Shiotani, N. Sakai, A. Bansil: *X-ray Compton scattering*
6. D. M. Paganin: *Coherent X-Ray optics*
7. W. Schülke: *Electron dynamics by inelastic X-Ray scattering*

Synchrotron Radiation

Production and Properties

Philip John Duke
Honorary Scientist
Daresbury Laboratory

OXFORD
UNIVERSITY PRESS

OXFORD
UNIVERSITY PRESS

Great Clarendon Street, Oxford OX2 6DP
United Kingdom

Oxford University Press is a department of the University of Oxford.
It furthers the University's objective of excellence in research, scholarship,
and education by publishing worldwide. Oxford is a registered trade mark of
Oxford University Press in the UK and in certain other countries

First published 2000
First published in paperback 2009

British Library Cataloguing in Publication Data
Data available

Library of Congress Cataloging in Publication Data
Data available

ISBN 978-0-19-955909-1

To

Ian Hyslop Munro
Friend, Colleague
and
Synchrotron Radiation Pioneer

A quella luce cotal si diventa,
che volgersi da lei per altro aspetto
è impossibil che mai si consenta;

That light doth so transform a man's whole bent
That never to another sight or thought
Would he surrender, with his own consent.

Dante. The Divine Comedy. 3 Paradise. Canto 33, lines 100–102.
Italian text taken from the Oxford University Press Edition 1939.
English translation by Dorothy L. Sayers and Barbara Reynolds, and published by Penguin Classics.

The European Synchrotron Radiation Facility seen from the air. Photograph supplied by the ESRF and credited to ARTECHNIQUE.

Preface

The purpose of this book is to provide the reader with a thorough introduction to the production of electromagnetic radiation by high energy electron storage rings. This radiation, called synchrotron radiation, has become a research tool of wide application. It is used by physicists, chemists, biologists, geologists, engineers, and other scientific disciplines. It is used as a structural probe for the study of surfaces, bulk materials, crystals, and viruses. The radiation is used for the spectroscopic analysis of solids, liquids, and gases. It provides illumination for the generation of images of special value in the biological and medical sciences.

In order to provide this radiation, centres of excellence have sprung up across the world. At first, in the late 1960s and early 70s the use of synchrotron radiation was a parasitic activity, riding on the back of nuclear and particle physics. The radiation was supplied by electron accelerators such as the synchrotron at Glasgow in Scotland, UK and the electron synchrotron at DESY in Hamburg, which, at that time was in West Germany. In the USA much early work with synchrotron radiation was carried out at three centres. At the high energy electron ring at the Stanford Linear Accelerator Centre in California and at the synchrotron centre at Madison, Wisconsin where a dedicated storage ring provided UV synchrotron radiation and at the National Bureau of Standards, Gaithersburg, Washington DC.

An important step forward was the operation of the world's first centre dedicated to the production of synchrotron radiation in the X-ray region at the Daresbury Laboratory, half way between Liverpool and Manchester in the UK. The electron storage ring, known as the Synchrotron Radiation Source, or SRS, constructed there, began operation in 1981 and replaced a former electron accelerator, NINA which, in its last few years of operation, supplied a small synchrotron radiation laboratory. The SRS was upgraded in 1989 to provide higher beam brightness and, at the time of writing, plans are being prepared for Diamond, a new high brightness source.

Across the world, Europe, the USA, Japan, China, and Russia operate large scale centres for research using synchrotron radiation. Individual countries in Europe, Asia, and South America operate sources of their own.

Alongside these sources there has developed an international interdisciplinary community of synchrotron radiation users and it is to this community that this book is addressed. This book does not lay claim to originality, rather its object is to educate synchrotron radiation users in the fundamental properties and production techniques of the radiation which they are all using. The book brings together a large amount of material which has already been published but which, in many cases, is not easily accessible and which the synchrotron radiation user may not even know about.

The book begins by laying a foundation of electromagnetic theory which is used in later chapters to show how the production of the radiation and the operating principles of the radiation sources are grounded in this theory. The theory itself as based on certain well-known observations in electricity and magnetism so the aim is to relate these simple facts to the rather complex and far reaching conclusions which follow. One key observation is the constancy of the speed of light, independent of the motion of the observer. The consequent methodology, known as the special theory of relativity is essential to an understanding of the synchrotron radiation production process so this is treated in some detail.

The main emphasis of this book is the production of synchrotron radiation from storage ring dipole magnets but additional chapters have been added to introduce the reader to insertion device sources of radiation. A fuller treatment of this important subject will be given in this series of books by other authors.

This text grew out of a series of lectures given at King's College, London, as part of an MSc course entitled 'X-ray science and technology'. I am grateful to Professor R. E. Burge, Dr Alan Michette and the other teachers of this course for their constant encouragement over many years of collaboration.

My interest in the development and uses of high energy particle accelerators began to flourish when I was an undergraduate at Birmingham University, where what was hoped to be the world's first accelerator to reach an energy of 1 GeV was under construction. My experience as a user of these machines continued at Brookhaven National Laboratory and, later, at CERN. It is impossible to mention all those from whom I learned over some 40 years as an accelerator user.

I am particularly grateful to the Director of the Daresbury Laboratory and the Head of the Synchrotron Radiation Division for their hospitality during the writing of this book. I am most grateful to V. P. Suller and the past and present members of the SRS Accelerator Physics Group at Daresbury for answering my questions and providing me with much useful material.

My special gratitude goes to my friend and colleague Ian Munro who was instrumental in bringing synchrotron radiation to the UK and to Daresbury. This book is dedicated to him.

Most of all I would like the readers of this book to share my enthusiasm and delight in the remarkable phenomena of the natural world.

Bibliography

An understanding of the production and properties of synchrotron radiation is based on the classical electromagnetic theory of radiation and the special theory of relativity. There is a massive literature already available on both these vast subjects and it is both impossible and unnecessary to refer to everything. There is an enormous amount of duplication and the books vary not so much in the material covered as in the clarity of presentation. Particularly useful is J. D. Jackson, *Classical electrodynamics* (2nd edition), John Wiley & Sons, New York which covers both electrodynamics and relativity. On relativity alone there is Wolfgang Rindler, *Essential relativity* (Revised 2nd edition), Springer Verlag, New York, Berlin, and Heidelberg.

The literature available on the theory of particle accelerators is much more sparse. The most useful recent book is Helmut Wiedemann, *Particle accelerator physics*, published in 1993 by Springer Verlag. The first volume treats the linear theory of particle accelerators in detail and gives an introduction to synchrotron radiation. The second volume, published more recently, gives a full treatment of synchrotron radiation production from dipole magnets and from insertion devices.

There are a large number of "summer school" and conference proceedings which contain individual contributions on topics connected with storage rings and synchrotron radiation production. Amongst there are:

Synchrotron Radiation Sources and their Applications, Aberdeen, September 1985. Edinburgh University Press, 1989.

CERN Accelerator School, Chester, UK, April 1989. CERN 90-03.

CERN Accelerator School, KFA, Jülich, Germany, September 1990. CERN 91-04.

The mathematical background can be found in

Mary L. Boas, *Mathematical methods in the physical sciences* (2nd edition), John Wiley and Sons, New York.

A world-wide survey of synchrotron radiation sources is available as I. H. Munro, C. A. Boardman, and J. C. Fuggle, *World compendium of synchrotron radiation facilities*, published by The European Synchrotron Radiation Society, c/o Bâtiment 209 D, LURE, Université Paris Sud 91405, Orsay, France.

Acknowledgements

Figures 12.8 and 12.9 were supplied by Lesley Welbourne, formerly of the Daresbury Laboratory. Figure 12.10 was supplied by Rick Fenner, the Advanced Photon Source, Argonne National Laboratory. Figures 13.3 and 13.4 were supplied by James Clarke, Daresbury Laboratory . Figure 15.1 was supplied by Chantal Argoud, and is reproduced by courtesy of the European Synchrotron Radiation Facility.

Grateful thanks are expressed to these people and their organisations for permission to publish this material.

Helsby, Cheshire P. J. D.
May, 1999

Contents

1

Synchrotron radiation and electromagnetic waves

Prelude

Synchrotron radiation was first observed in 1947[1] and began to be used as a research tool in the mid-1960s. The first synchrotron radiation sources were low energy (several hundred MeV) electron accelerators which were also used for nuclear and particle physics. It was quickly realized that the needs of the growing number of people who used synchrotron radiation for their research required dedicated sources of radiation. Many of these research programmes required X-radiation, which pointed to the need for these sources to accelerate electrons to GeV energies in order to provide a useful output of X-rays as well as lower energy radiation.

The first such dedicated X-ray source was the SRS (Synchrotron Radiation Source) at the Daresbury Laboratory in the UK, which operates at 2 GeV electron energy. The SRS did not retain its unique status for long. The NSLS (National Synchrotron Light Source) at Brookhaven, USA and the Photon Factory, Tsukuba, Japan quickly followed. There were similar developments in the then USSR; initially at Akademgorodok, Novosibirsk, and later in Moscow. The most recent advances have included the construction and operation of large scale facilities at Grenoble, France (ESRF, 6 GeV) Argonne, USA (APS, 7 GeV) and Nishi-Harima, Japan (SPring-8, 8 GeV).

In the early days the emphasis in research applications was the exploitation of VUV radiation for atomic and molecular spectroscopy and surface science but the advent of the X-ray sources has stimulated a steady growth of research with harder radiation in such areas as X-ray diffraction, absorption spectroscopy (EXAFS), crystallography, topography, and lithography. The users of the radiation, initially drawn from the physics community, have been augmented by chemists, biologists, geologists, and others so that there is now a world-wide community of people using synchrotron radiation as an essential component of their research.

It is the purpose of this book to provide the basic principles of synchrotron radiation production and properties for all users of synchrotron radiation, whatever their research interest or disciplinary background. The material presented here covers the radiation production process (Chapters 4–7) and the operational principles of the radiation source (Chapters 8–12). Chapters 13 and 14 introduce the user to additional devices which are used to enhance the output of the radiation source. Chapter 15 contains a survey of recent developments and an attempt to scan the horizon for the directions of future progress.

The use of mathematics is essential for a precise description of the radiation and the manner of its production. Fortunately, for our purpose the radiation production process can be described using the concepts of classical electrodynamics although a knowledge of relativity theory will be required. These topics are covered to a sufficient level in Chapters 2 and 3. In this first chapter we consider some basic definitions and properties pertaining to electromagnetic radiation. Readers who are already familiar with all these things should proceed straight to Chapter 4. They will need only to refer to the intervening material if they need to refresh their memory.

Elementary wave properties

Synchrotron radiation is one particular form of energy transmitted from the source to the detector by an electromagnetic wave, known as electromagnetic radiation. Visible light is one example of this radiation. Radio waves and X-rays are others. In practice there is a whole spectrum of this radiation (see Fig. 1.3) which reaches us in the form of wave motion. In order to understand its properties, therefore, we must begin with the mathematics of waves.

The wave model for energy transmission was introduced by Christiaan Huygens (1629–95) following the discovery of the diffraction of light by Francesco Maria Grimaldi (1618–63). The wave theory of light was extended to cover the whole of electromagnetic radiation following the work of James Clerk Maxwell (1831–79). In this theory, the passage of radiation through space generates an electric and a magnetic field at each point traversed by the radiation. An analogous situation occurs when a ripple travels along the surface of a pool of water. Each point on the surface of the water is displaced upwards (or downwards) as the ripple passes across it. The simplest form of wave is a sine wave in which the displacement (or amplitude) of the wave at a certain point is described by

$$y = A_0 \sin 2\pi(\nu t - kz). \tag{1.1}$$

The displacement is a periodic function of both time and position which repeats itself indefinitely in both directions as shown in Fig. 1.1. A snapshot of the wave at a particular moment in time will show peaks and troughs of displacement separated by the wavelength $\lambda = 1/k$. The maximum displacement is A_0. At a particular point along the z-axis the displacement will vary with time in the same way. Clearly, the product $2\pi(\nu t - kz)$, which is the argument of the sine function, has the form of an angle, measured in radians and is called the phase angle, ϕ, or just the phase of the wave. The quantity ν is called the frequency of the wave and is measured in vibrations per second or s^{-1}. The unit of frequency is Hz (hertz, named after Heinrich Rudolf Hertz, 1857–94) so that 50 Hz means 50 vibrations per second. The reciprocal of ν, $1/\nu$, is the time taken for the completion of one vibration of the wave, i.e. for one

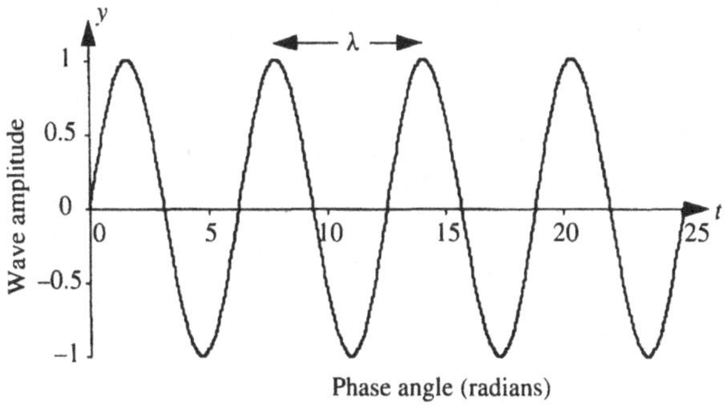

Fig. 1.1 Wave motion (sinusoidal).

crest to be replaced by the next. Often the angular frequency, ω, measured in radians per second, is used instead of ν and $\omega = 2\pi\nu$.

The velocity by which the wave moves forward is just $c = \nu\lambda$. Strictly speaking, this is the phase velocity, the speed with which a particular feature on the wave (a crest for example) appears to move forward. The phase angle, ϕ, changes with position and time according to $\phi = 2\pi(\nu t - kz)$. The rate at which a fixed phase angle is observed to advance along the t-axis is given by $\mathrm{d}z/\mathrm{d}t = \nu/k$ or $\nu\lambda$.

How does the wave transport energy from one place to another? The average displacement at right angles to the direction of the wave, i.e. along the x- (and y-) axes, is zero so we might think that the energy carried by the wave averages to zero as well. However, this is not so because we know from elementary mechanics that energy is proportional to the square of velocity so that at any point the energy stored in the wave is proportional to the square of the transverse velocity $\mathrm{d}y/\mathrm{d}t$, which is, by taking the derivative of eqn (1.1) with respect to time, at a fixed point z_p,

$$\left.\frac{\mathrm{d}y}{\mathrm{d}t}\right|_{z=z_\mathrm{p}} = \dot{y} = 2\pi A_0\nu\cos 2\pi(\nu t - kz_\mathrm{p}).$$

The energy stored is proportional to the square of this velocity and so is proportional to the square of the wave amplitude.

We could equally well have chosen the cosine function instead of the sine function in eqn (1.1) to describe the variation of the amplitude with time. This would produce a wave shifted in phase by 90°. Physically, this alternative choice would make little difference because most detectors are only sensitive to the amplitude of the wave and cannot detect the phase.

Electromagnetic waves

In the case of electromagnetic waves, the wave motion consists of two waves, propagating in phase with the planes of vibration mutually perpendicular (see Chapter 4 and Fig. 1.2).

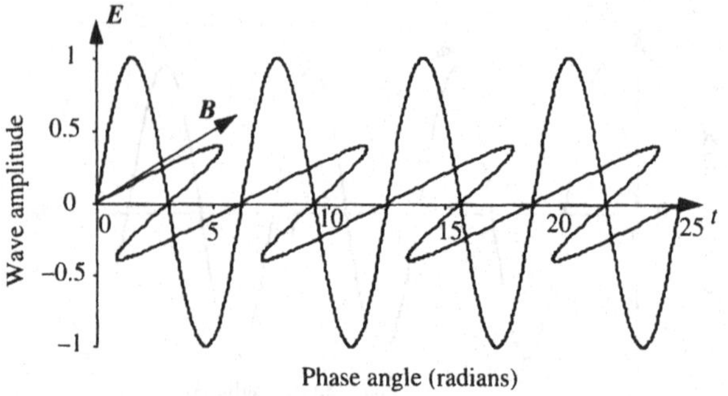

Fig. 1.2 Electromagnetic wave.

The displacement of one wave is proportional to the strength of the electric field (***E***) and the displacement of the other to the magnetic field (***B***) as is shown in Fig. 1.2.

The energy stored in the wave is proportional to the square of the field strength. The phase velocity of the wave is numerically equal to 3×10^8 m/s so that the radiation travels through a distance of 1 m in approximately 3 ns (1ns $= 10^{-9}$ s or one-thousandth of a microsecond). In a vacuum, the velocity does not depend on the wavelength so that all electromagnetic radiation travels at this same speed, often called the speed of light. This is a very high speed by human standards; for example, the time taken for a muscle to contract is about one-tenth of a second and during that time a flash of light or a radio signal will travel thirty thousand (3×10^5) km, which is about the same as the distance to a communication satellite in geostationary orbit above the surface of the Earth.

The electromagnetic spectrum

Although all kinds of electromagnetic radiation (light, X-rays, microwaves, radio, and TV signals) travel at the same speed, they have an extremely wide range of frequencies and wavelengths. For example, the electromagnetic radiation which constitutes red light has a wavelength of about 1 μm. The same quantity for violet light is about one-tenth of that. X-ray wavelengths are shorter still. A standard X-ray tube produces X-rays with a wavelength about 1000 times shorter than the visible light. At the other extreme, a domestic microwave oven operates at a wavelength of 12 cm. TV signal wavelengths occupy a band around 1–10 m and radio wavelengths can be even longer. The product of wavelength and frequency is equal to the speed of the wave so that radio waves oscillate with a frequency of a few megacycles per second and TV signals oscillate at a few gigacycles (10^9 cycles/s or 10^9 Hz).

The shorter the wavelength, the higher the frequency so for visible light and X-rays, the frequencies become so huge as to be hardly useful as a unit of measurement.

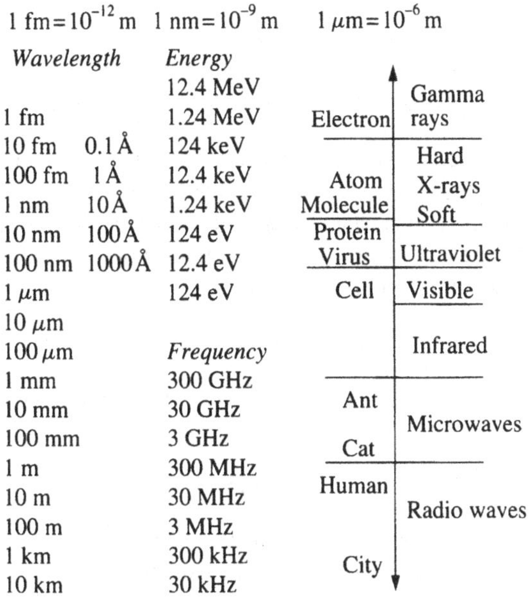

Fig. 1.3 The electromagnetic spectrum.

Figure 1.3 shows (approximately) how the wavelengths of the electromagnetic spectrum relate to human experience of length. You should note at this stage that synchrotron radiation contains all the wavelengths and kinds of electromagnetic radiation just mentioned so it is useful to have a picture in your mind of the wavelengths involved and how these lengths compare with the dimensions of everyday objects.

Electromagnetic waves around us

At any particular point in space and time, the strength of the electric and magnetic fields is the sum of the amplitudes of the waves reaching that point, so that the time variation (or the variation in space as we move from one place to another) can be far removed from the simple expression given in eqn (1.1). If you have the right equipment you will be able to distinguish between the different wavelengths (or frequencies) that are reaching you so that you become aware of the individual frequencies (or the spectrum). One such piece of equipment is a shower of rain which can spread out the visible part of the light that is coming from the Sun into the familiar colours of the rainbow. Sir Isaac Newton (1642–1727), in his book *Opticks* (1703), describes how he used a triangular prism of glass to do the same thing in a controlled way. This simple observation tells us that ordinary daylight, which would be difficult to describe mathematically, can be described as the addition of a series of elementary waves, which are the building blocks of light.

The wave equation

To investigate this in a formal way, we return to the original discussion of wave motion. We can determine the transverse acceleration of the wave at any point by taking the second derivative of eqn (1.1). It is important to be careful about this because we can calculate both the time derivative and the position derivative. In order to allow for this, we must take the two partial derivatives, one derivative in which time is held constant while position along the z-axis changes and one for which position is held constant while the time coordinate changes. These derivatives (together with the original sine wave function) are set out in eqn (1.2) below:

$$
\begin{aligned}
&y = A_0 \sin 2\pi(\nu t - kz); \\
&\frac{\partial y}{\partial z} = -2\pi k A_0 \cos 2\pi(\nu t - kz), \qquad \frac{\partial y}{\partial t} = 2\pi \nu A_0 \cos 2\pi(\nu t - kz); \\
&\frac{\partial^2 y}{\partial z^2} = -4\pi^2 k^2 A_0^2 \sin 2\pi(\nu t - kz), \quad \frac{\partial^2 y}{\partial t^2} = -4\pi^2 \nu^2 A_0^2 \sin 2\pi(\nu t - kz).
\end{aligned} \tag{1.2}
$$

We can use these equations to eliminate the sine and cosine functions to give equations which contain only the first and second partial derivatives and the wave parameters ν and k:

$$\frac{\partial y}{\partial z} = -\frac{k}{\nu}\frac{\partial y}{\partial t} = -\frac{1}{c}\frac{\partial y}{\partial t} \tag{1.3}$$

and

$$\frac{\partial^2 y}{\partial z^2} = -\frac{k^2}{\nu^2}\frac{\partial^2 y}{\partial t^2} = -\frac{1}{c^2}\frac{\partial^2 y}{\partial t^2}. \tag{1.4}$$

Equation (1.3) can be rearranged to tell us that $c = \nu/k = \nu\lambda$ is the velocity of the wave along the z-axis, as we had deduced by examination of the original sine function in eqn (1.1).

Equation (1.4) is very important. It is called the wave equation and is a very general statement about any kind of disturbance. It has a multiplicity of solutions. This is clear from the way we have derived it. When we take the partial derivatives the original functional form of the wave has cancelled out. We could have equally well taken any function, ϕ which is linear in z and t, e.g. $y = \phi(z, t)$. Provided the function is mathematically well-behaved so that it actually has a first and second derivative then it can be a solution of the wave equation. The form of the wave does not have to be a sine function (or a cosine function), almost anything will do.

We can write the wave equation in a more general form to make it clear that a wave disturbance can transport energy throughout space. We write

$$\frac{\partial^2 \phi}{\partial x^2} + \frac{\partial^2 \phi}{\partial y^2} + \frac{\partial^2 \phi}{\partial z^2} = -\frac{1}{c^2}\frac{\partial^2 \phi}{\partial t^2}$$

or

$$\nabla^2 \phi = -\frac{1}{c^2}\frac{\partial^2 \phi}{\partial t^2}. \tag{1.5}$$

The symbol ∇^2 is called 'del squared' (see Appendix) and is shorthand for the sum of the second order partial derivatives along the three spatial axes.

Analysis of the wave spectrum

The discussion above is not merely of theoretical interest. The fact is that light from any kind of source almost never comes to us as a wave with a single unique wavelength, extending over the whole of time and space. The amplitudes of the electromagnetic waves that reach our eyes vary with time in an irregular way and contain many wavelengths. To put this very simply, our eyes detect an electromagnetic disturbance in the visible region of the electromagnetic spectrum. This is the function ϕ in the wave equation, a spatio-temporal electromagnetic disturbance which we can call a wave packet. The optical system in our eyes, the cornea and the lens, analyse the wave packet into an image on our retina. The light sensitive cells in the retina react differently to the optical wavelengths and, in combination with the processing power in the visual cortex of our brain, separates this image into a variety of colours. The process corresponds to an analysis of the original function into a series of spatial and temporal wave trains of the form shown in eqn (1.1).

The idea that the time-varying amplitude of the electromagnetic radiation at any point in space can be resolved into a spectrum of frequency components was first put on a sound mathematical footing by the French engineer Jean Baptiste Joseph Fourier (Baron de Fourier) (1768–1830). He observed that any regular wave motion, no matter how complicated, could be expressed as the sum of a series of cosine and sine waves. This means that if you observe a regular variation of an amplitude $A(t)$ as a function of time, that amplitude can be described as the sum of an infinite series of elementary frequency components in the following way:

$$\begin{aligned} A(t) = \tfrac{1}{2}A_0 + A_1\cos(\omega t) + A_2\cos(2\omega t) + A_3\cos(3\omega t) + \cdots \\ + B_1\sin(\omega t) + B_2\sin(2\omega t) + B_3\sin(3\omega t) + \cdots . \end{aligned} \tag{1.6}$$

Another way of looking at this expression is to say that the shape of the time distribution is being approximated as the sum of a series of cosine and sine functions so that

$$A(t) = \frac{1}{2}A_0 + \sum_{m=1}^{\infty}[A_m\cos(m\omega t) + B_m\sin(m\omega t)]. \tag{1.7}$$

The first coefficient, A_0, is zero so long as $A(t)$ averages to zero. It is important to realize that the Fourier coefficients A_m and B_m give the amplitude of the components of the frequency spectrum of $A(t)$.

The coefficients A and B are calculated by using the following property of the sine and cosine functions:

$$\frac{1}{\pi}\int_{-\pi}^{+\pi} \sin(mu)\sin(nu)\,du = \begin{cases} 1 & \text{when } m = n, \\ 0 & \text{when } m \neq n, \end{cases}$$
$$\frac{1}{\pi}\int_{-\pi}^{+\pi} \cos(mu)\cos(nu)\,du = \begin{cases} 1 & \text{when } m = n, \\ 0 & \text{when } m \neq n, \end{cases} \tag{1.8}$$
$$\frac{1}{\pi}\int_{-\pi}^{+\pi} \sin(mu)\sin(nu)\,du = 0.$$

In eqn (1.8), u is any integration variable and m and n are integers.

To use these identities, take either eqn (1.6) or (1.7), multiply each term by $\cos(\omega t)/\pi$, and integrate the whole thing from $-\pi$ to $+\pi$ (or in fact any interval spanning a total range of 2π) and you will obtain

$$A_1 = \frac{1}{\pi}\int_{-\pi}^{+\pi} A(t)\cos(\omega t)\,dt$$

because [from eqn (1.8)] the integral multiplying the constant A_1 is equal to unity and all the other integrals in the series reduce to zero. You can calculate the constant term A_0 by multiplying eqn (1.6) or (1.7) by unity and integrating as before, remembering that $\cos(\omega t) = 1$ when $\omega = 0$.

All the other coefficients can be obtained in the same way so that, in general, for the nth coefficient:

$$A_n = \frac{1}{\pi}\int_{-\pi}^{+\pi} A(t)\cos(n\omega t)\,dt,$$
$$B_n = \frac{1}{\pi}\int_{-\pi}^{+\pi} B(t)\sin(n\omega t)\,dt \tag{1.9}$$

for $n = 1, 2, 3$, etc.

Once the amplitude function $A(t)$ is known, the Fourier coefficients can be determined. It is worth noticing that, as a consequence of eqn (1.8), each of the coefficients A and B are independent of the others, since the trigonometric functions form what is called an orthogonal set.

It is very important to understand that what is being done here is not an abstruse piece of mathematics but is a way of converting an observed wave amplitude, which may be a very complicated function of time or space coordinates into a sum of elementary cosine and sine waves. When the amplitude function depends on time in a regular repetitive way, the components consist of a fundamental frequency ω and a series of harmonics whose frequencies are an integer multiple of the fundamental frequency. If the amplitude function depends on the space coordinates then the Fourier spectrum is a function of wavelength. The operation of obtaining the spectrum from the amplitude function is called the Fourier transformation [eqn (1.9)] and the result

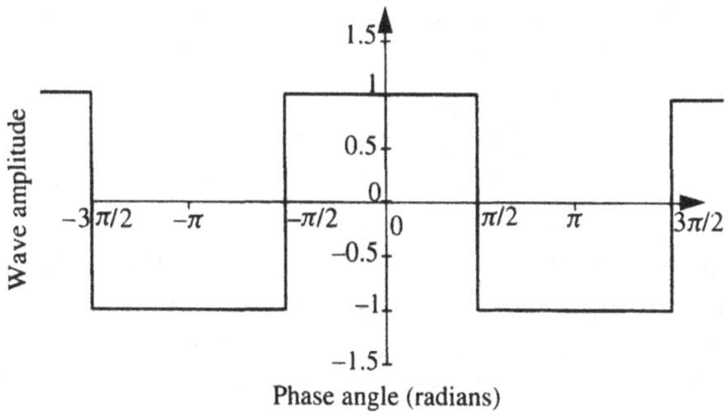

Fig. 1.4 Square wave.

of the operation is called the Fourier transform. The transformation can be carried out in the opposite direction and a wave amplitude can be obtained from the sum of the Fourier components [eqn (1.6) or (1.7)]. Very often, what we observe or calculate is a time-dependent spectrum but what we need to know is the same spectrum as a function of frequency. The process of Fourier transformation enables us to make that connection.

As an example, consider an infinite series of regular square pulses, a brief portion of which is shown in Fig. 1.4. The amplitude function $A(t)$ is equal to 1 at times between 0 and $+\pi/2$, etc., and -1 between $\pi/2$ and $3\pi/2$ and between $-\pi/2$ and $-3\pi/2$ ($\omega = 1$). The pulse train repeats itself indefinitely at times <0 and times >0. It is symmetric about $t = 0$. It averages to zero over one period between $-\pi$ and $+\pi$ so that the constant term A_0 must be zero. The train has the property that $A(t) = A(-t)$ so that the only elementary waves in the Fourier expansion must be cosine waves whose argument is an odd number of periods between 0 and 2π. Cosine waves with an even number of periods and all sine waves have the opposite symmetry, namely $A(t) = (-1)A(-t)$.

The first three terms in the expansion eqn (1.9) are then:

$$\begin{aligned} A_1 &= \frac{1}{\pi}\int_{-\pi}^{+\pi} \cos(t)\,\mathrm{d}t = \frac{4}{\pi}, \\ A_3 &= \frac{1}{\pi}\int_{-\pi}^{+\pi} \cos(3t)\,\mathrm{d}t = -\frac{4}{3\pi}, \\ A_5 &= \frac{1}{\pi}\int_{-\pi}^{+\pi} \cos(5t)\,\mathrm{d}t = \frac{4}{5\pi} \end{aligned} \tag{1.10}$$

and the Fourier transform (or more correctly in this case, the Fourier series), which is the sum of (1.10) and higher order terms, can be written as:

$$A_m = 2\sum_{m=1}^{\infty}\left[\sin\left(m\frac{\pi}{2}\right)\Big/ m\frac{\pi}{2}\right]. \tag{1.11}$$

In this expansion, $A_m = 0$ when m is even.

Examination of the numbered curves in Fig. 1.5 shows how this works in practice. The first order term (curve 1 in the figure), known as the fundamental, is a cosine curve which is in phase with the square wave and represents a first approximation to the shape of the wave train. The addition of further terms (curve 7) improves the match between the square wave and the sum of the spectral components. In particular, the higher order harmonics are needed to generate the sharp corners present in the original wave. The more of these that are present (curves 13 and 19), the better is the fit to the original wave. The high frequency components are needed in the spectrum to generate rapid changes in the behaviour of the amplitude. In the case of the square wave, infinitely high frequencies are needed to generate the zero time duration of the rise and fall of the square wave amplitude.

Figure 1.6 shows the relative importance of the fundamental and the harmonics up to $m = 19$. The spectrum contains only odd numbered harmonics. All even numbered harmonics are zero. We will see in Chapter 14 that the synchrotron radiation output from a magnetic undulator has a frequency spectrum with a similar form.

Although regular pulse trains do occur in nature (from a laser for example), it often happens that the radiation is experienced as a train of brief pulses. Brief means that the time duration of each pulse is long enough to contain many oscillations of the fundamental frequency. In order to describe such a pulse as a series of elementary

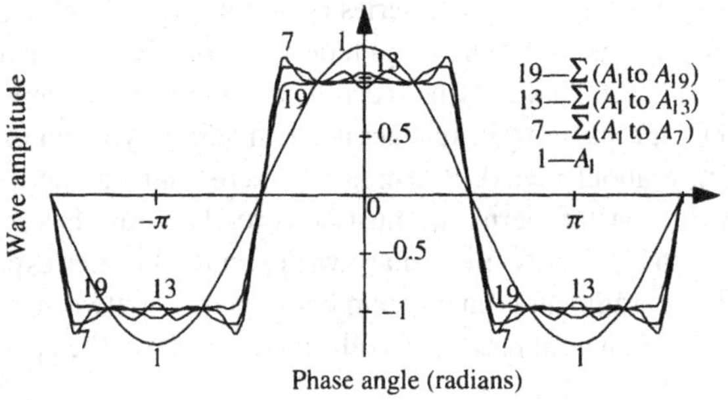

Fig. 1.5 Fourier synthesis of a square-wave train.

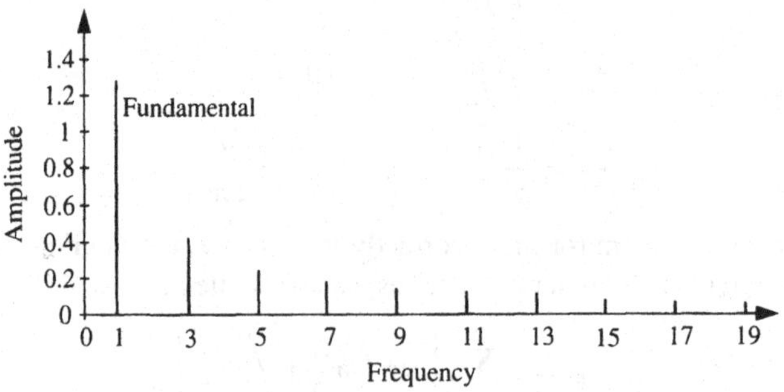

Fig. 1.6 Spectral analysis of a square-wave train.

sine and cosine waves, there must not only be an infinite number of such waves but they must have a continuous frequency spectrum instead of the discrete spectrum described above. In other words, the frequency, which, in eqn (1.7), appeared as a fundamental together with a series of discrete harmonics, must become a continuous variable. In this case, it is best to take the exponential function $\exp(i\omega t)$ (or $e^{i\omega t}$) as the series of functions which are to be used to fit the amplitude. The sine and cosine functions are related to the exponential function through an algebraic identity, called De Moivre's theorem (Abraham De Moivre, 1667–1754) which states that

$$A_0 e^{i\omega t} = A_0 \exp(i\omega t) = A_0 \cos(\omega t) + iA_0 \sin(\omega t).$$

What is the meaning of the symbol i in these expressions? You can understand this if you remember that the sine function lags behind (or leads) the cosine function by a phase angle of 90°. If we plot the amplitudes of the cosine and sine waves (Fig. 1.7), then we can represent the amplitude of the cosine function by an arrow of length $A_0 \cos(\omega t)$ along the x-axis and the amplitude of the sine function by an arrow of length $A_0 \sin(\omega t)$, rotated through 90° relative to the direction of the x-axis which makes it point along the y-axis. Multiplication by i denotes this operation of 90° rotation as can be seen by noting that a further rotation, through the same angle, can be seen as a second multiplication by i ($i \times i = i^2$, in total) which would make the total rotation angle 180° and would cause the arrow to point along the $-x$-direction, as though the amplitude had been multiplied by -1. It follows that i has the algebraic property that $i^2 = -1$ or $i = \pm\sqrt{-1}$. Figure 1.7 is often called the Argand diagram (Jean-Robert Argand, 1768–1822). This diagram is a useful way of representing numbers such as $\exp(i\omega t)$ which are said to be complex numbers with a real part written as $\mathrm{Re}\exp(i\omega t) = \cos(\omega t)$ and an imaginary part written as $\mathrm{Im}\exp(i\omega t) = \sin(\omega t)$. After carrying out the algebraic manipulations, we often keep only the real part of the final expression because this is the portion which has a real physical meaning. In particular, the square of the real part corresponds to the energy

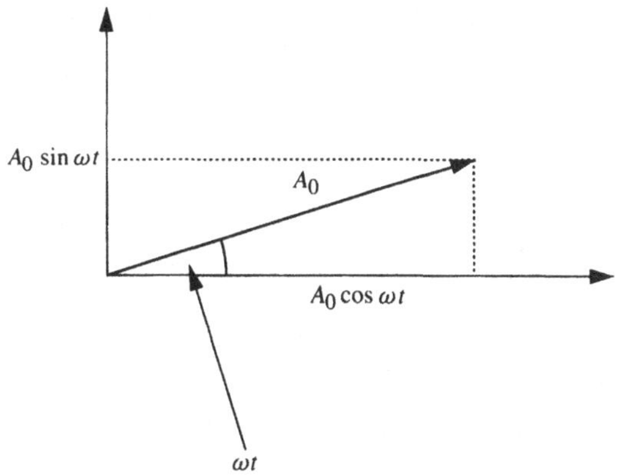

Fig. 1.7 De Moivre's theorem.

stored in the wave. Another approach to finding the square of the wave amplitude, which is, more satisfactory mathematically, is to define the complex conjugate of the wave function A, usually written A^*, as

$$A_0 e^{-i\omega t} = A_0 \exp(-i\omega t) = A_0 \cos(\omega t) - iA_0 \sin(\omega t).$$

Then the stored energy is the product of the wave function with its complex conjugate which is $AA^* = |A_0|^2$ and the phase of the wave (which is undetectable when the energy of the wave is being measured) disappears from the expression.

In order to represent a one-off pulse of radiation, the sum in eqn (1.7) is replaced by an integration so that the time spectrum and the frequency spectrum are related by

$$A(t) = \frac{1}{\sqrt{2\pi}} \int_{-\infty}^{+\infty} f(\omega) e^{i\omega t}\, dt, \qquad f(\omega) = \frac{1}{\sqrt{2\pi}} \int_{-\infty}^{+\infty} A(t) e^{-i\omega t}\, dt. \quad (1.12)$$

Suppose now that we are presented with a pulse of radiation, observed at a time t_0 with an average frequency ω_0 and which has the form of a Gaussian (Carl Friedrich Gauss, 1777–1855) distribution with a standard deviation of Δt so that

$$A(t) = \frac{1}{\Delta t \sqrt{2\pi}} e^{-(t-t_0)^2/2\Delta t^2} \cos \omega_0 t.$$

This pulse is shown in Fig. 1.8 as a snapshot taken at time t_0.

A detector with sufficiently high time resolution would respond to the rapid oscillations of the $\cos(\omega_0 t)$ factor with frequency ω_0. In practice the response time of a typical detector is far too long so the detector output follows the envelope of the rapidly varying field (actually the square of the field as indicated earlier), which is the curve which passes through the peaks of the cosine function. This pulse is neither of infinite duration, nor is it regular and repetitive.

What is the frequency spectrum of this pulse which rises from zero to a maximum and falls again to zero in a finite time? The frequency spectrum $f(\omega)$ of the pulse is obtained by calculating the Fourier transform of $A(t)$ so that, following the

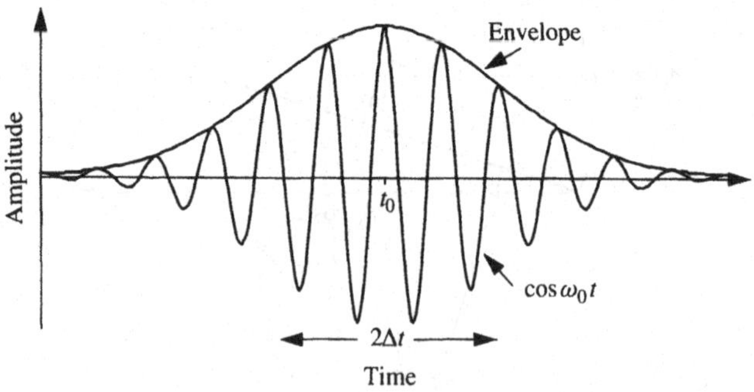

Fig. 1.8 Gaussian pulse (frequency ω_0) modulated by a cosine function.

prescription given in eqn (1.12),

$$f(\omega) = \frac{1}{2\pi\,\Delta t}\int_{-\infty}^{+\infty} \mathrm{e}^{-(t-t_0)^2/2\Delta t^2}\cos\omega_0 t\,\mathrm{e}^{-\mathrm{i}\omega t}\,\mathrm{d}t.$$

Calculation of the integral gives the result

$$f(\omega) = \frac{\Delta t}{\sqrt{2\pi}}\left[\mathrm{e}^{-(\Delta t^2/2)(\omega-\omega_0)^2} + \mathrm{e}^{-(\Delta t^2/2)(\omega+\omega_0)^2}\right]. \tag{1.13}$$

In eqn (1.13), the second term in square brackets is very small compared with the first term, except in the region where ω has a value close to $-\omega_0$, and this region of the spectrum has no physical meaning. We are left with the first term which is also a Gaussian, just like the original pulse with a width $\Delta\omega = (1/\Delta t)$. Just like the square wave which we examined earlier, the sharper the pulse in time, the broader must be the width of the frequency distribution. The reason why a detector with arbitrarily high time resolution detects only the pulse envelope is because all frequencies are present and the detector cannot possibly respond to all of them. To use a more immediate analogy, a lightning flash appears white because it has such a short time duration (a few μs) that all colours are present simultaneously. We shall find later that the broad spectrum of synchrotron radiation from a dipole magnet is related to the shortness of the electromagnetic wave pulse from the electron beam passing through the magnet.

The same discussion could equally well be applied to the spatial extent of the pulse and we would find that if the pulse width were Δx, then the range of wave number present, Δk, would be such that $\Delta x \cdot \Delta k = 1$. The precise relationship between these quantities depends on the shape of the pulse envelope. The exact result of unity for the product $\Delta\omega \cdot \Delta t$ (or $\Delta x \cdot \Delta k$) is reached only for the Gaussian distribution so that in general

$$\Delta\omega\Delta t \geq 1 \quad \text{and} \quad \Delta x\Delta k \geq 1. \tag{1.14}$$

Electromagnetic wave pulses are called photons

These ideas make it possible for us to reconcile two apparently opposite views of electromagnetic radiation. Newton believed that light must consist of particles (which he termed corpuscles). He was sure that this was so because he observed that light travelled in a straight line from one place to another whereas a wave motion would spread out from a point and even after passing through a slit, would spread out further so that light from the slit would spread into the shadow of the slit. It was the eventual observation of this phenomenon (known as diffraction) which made people opt for a wave theory for light. However, in the case of a pulse of electromagnetic radiation, moving through a vacuum, all wavelengths and frequencies move at the same phase velocity, c, so that the pulse stays together as it moves through time and space, getting neither longer nor shorter. This pulse, sometimes called a wave packet, can be imagined as a particle of light (a photon) and in the quantum theory light is imagined to be a stream of such photons, moving with velocity c (3×10^8 m/s). Each photon transports an amount of energy $E = (h/2\pi)\omega$, or $E = h\nu$. The quantity h is called Planck constant (Max Karl Ernst Planck, 1858–1947) and is equal to 6.582×10^{-34} J s. The frequency ν is measured in units of wave vibrations per second and is equal to $\omega/2\pi$ and $h/2\pi$ is very often written as $\hbar$. The Gaussian wave packet carries an energy $E = \hbar\omega_0$, but the energy of the wave packet is subject to an uncertainty in its value of $\Delta E = \hbar\Delta\omega$ and a corresponding uncertainty in its arrival time Δt related by (1.14). This relation, that $\Delta E \cdot \Delta t \geq \hbar$, is known as the Heisenberg uncertainty principle (Werner Karl Heisenberg, 1901–76). It is a fundamental principle of the quantum mechanical description of events at the level of individual photons that it is not possible, at the same moment, to make a precise measurement of both energy and time (or position and momentum). The simple argument, presented here, is meant to show that the idea of the photon wave packet is consistent with this principle.

The practical consequence is that synchrotron radiation can be regarded not only as radiation with a spectrum of wavelengths (or frequencies—see Fig. 1.3), arriving at a detector over a measured time period but also as a flux of photons, each of which carries an energy $\hbar\omega$. The number of photons with a given frequency is the Fourier transform of the time spectrum.

In order to discuss this further, we must sharpen our definitions. We call the photon brightness the number of photons emitted from unit area of the source within a unit solid angle in unit time. Two sources can emit the same number of photons per second but the source with the smaller area (or with the smaller collimation angle) has higher brightness and supplies a larger number of photons (or, equivalently, a greater amount of energy) to a sample target or a radiation detector. Because the amount of energy transported from the source to the detector depends on the wavelength of the radiation (an X-ray photon carries more energy than one in the visible region of the electromagnetic spectrum, see Fig. 1.3), we must also specify the frequency range of the spectrum, often termed the bandwidth. When the bandwidth is included in the definition, the term 'spectral brightness' is often used.

Sometimes, particularly when the source is small, it is useful to integrate the photons emitted per second over the area of the source and just talk about the number emitted in unit time into unit solid angle and unit bandwidth. This quantity is called the spectral brilliance of the source, or just the brilliance if the output is integrated over the wavelength as well. A further integration yields the photon flux emitted in all directions. Unfortunately, there does not appear to be a total agreement about these terms and they are sometimes interchanged. In this book we shall follow the definitions given in Born and Wolf's *Principles of optics*.[2]

Historically, the development of synchrotron radiation sources has tended to follow a pattern of higher and higher brightness. This is illustrated for the X-ray region in Fig. 1.9. What this figure shows is that the brightness of X-ray sources remained essentially constant for 60 years or so (within a factor of 10 or so) ever since the discovery of X-rays by Röntgen in 1895 (Wilhelm Konrad von Röntgen, 1845–1923). The advent of the rotating anode X-ray tube brought about a brightness increase of around two orders of magnitude but this development pales into insignificance

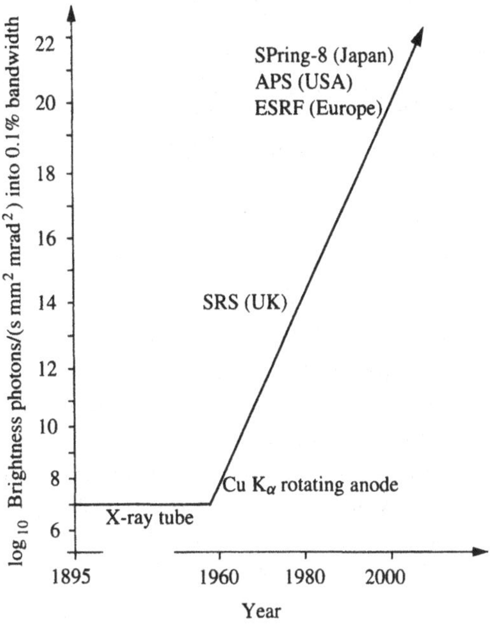

Fig. 1.9 History of X-ray sources.

compared with the radiation now available from the brightest synchrotron radiation sources.

In the subsequent chapters we shall concentrate first on the production of electromagnetic radiation in general and synchrotron radiation in particular. First this will be treated as a theoretical study in electromagnetic theory. Later we will show how a stored beam of electrons can be generated to produce synchrotron radiation in copious quantities. Finally, we shall consider how special devices can be inserted into storage rings to enhance the production of synchrotron radiation and achieve brightness some 14–15 orders of magnitude higher than those available in the first half of this century.

References

1. J. P. Blewett, Synchrotron radiation—Early history. *J. Synchrotron Radiation*, **5**, 135–9 (1998). An historical introduction to synchrotron radiation can be found in Chapter 1 of *Handbook on synchrotron radiation* (ed. Ernst-Eckhard Koch). North Holland Publishing Co. Amsterdam, New York, Oxford (1983).
2. M. Born and E. Wolf, *Principles of optics*. 6th edn. Pergamon Press, Oxford, England(1980).

2

Electromagnetic radiation is produced by electrons

Introduction

In the previous chapter we introduced the idea of the electromagnetic wave as the regular variation of electric and magnetic fields. We explained how these waves transport energy in the form of photons from one place to another. We also described how the photon energy is related to the wavelength and frequency of the wave.

Our aim in this chapter is to begin with the experimental observation that electrons are the source of both electric and magnetic fields and to show how this observation leads to a complete description, in mathematical terms, of the fields themselves and their propagation through space and time. The mathematical description which we are aiming for is known as Maxwell's equations for the electromagnetic field. The argument will be followed at a very elementary level and can be omitted by those who are already familiar with the subject. On the other hand, those who wish for a more detailed and rigorous treatment will need to supplement this chapter with one of the many textbooks on the theory of electromagnetism and electrodynamics.[1]

Electromagnetism cannot be described without using vector quantities. For this reason the properties of vectors and an outline of the basic rules of vector algebra are summarized in the Appendix.

Electrons—source of electric field

Electrons are just as ubiquitous as electromagnetic radiation but they are much more elusive. One can see their effects, but not the things themselves. For example, we probably met electrons for the first time without realizing it when we rubbed a plastic rod on the back of a cat to generate static electricity. The rubbing action transfers carriers of electricity from the cat to the plastic rod, so that the rod picks up tiny pieces of paper. This effect is described by saying that electric charges, stationary on the plastic rod, produce an electric force-field which is able to pass through the space between the rod and the bit of paper and generate a similar force on an electric charge residing on the paper.

We might say that we have 'explained' what is happening by invoking electric charges and fields. On the other hand, I have used the word 'describe' rather than 'explain' because, in my view, the statement about charges and fields does not explain anything. It just replaces one mystery by another. These terms are used because they allow us to write down, as a simple mathematical expression, the experimental result (known as Coulomb's law, after Charles Augustin de Coulomb, 1736–1806) for the force-field (or the electric field), at a distance r from the charge and generated by it, namely,

$$E = \frac{e}{4\pi \varepsilon_0 r^2}. \tag{2.1}$$

In this equation, the electron is said to carry an electric charge e which is equal to -1.6×10^{-19} C. The minus sign is there because, by convention, the electron is said to be negatively charged. The field generated by the electron decreases in strength as the square of the distance from the electron. The symbol ε_0 is called the electric permittivity and indicates the response of the medium (in this case a vacuum) to the electric field around the charge. Its value is chosen so that a charge of 1 C generates a force-field of strength equal to 1 N at a distance of 1 m from the charge. If you ask how an 'empty' vacuum can respond to the presence of an electric charge, the answer of the 19th century scientists would have been that the vacuum is filled with a medium having special properties known as the luminiferous (light-carrying) ether. The present-day answer to the same question demands a knowledge of quantum electrodynamics which is outside the scope of this book.

Electric fields can be described by lines of force

If a second charge of the same numerical value but opposite sign (such as a positron or a proton) is placed at a distance r from the first then the force on each particle (which tends to pull them together) is the product of the charge times the field at that point, or

$$F = eE = \frac{e^2}{4\pi \varepsilon_0 r^2}. \tag{2.2}$$

In this book we often use the word 'particle' for the charge-carrying object.

An electric field can be pictured as lines of force spreading out from an electric charge and ending on a charge of opposite sign. At any point in space, the tangent to the line of force at that point gives the direction of the electric field at that point and the number of lines of force passing through unit area perpendicular to the tangent tells us the strength of the electric field. By convention, the positive charge acts as a source of electric field and the negative charge acts as a sink. The lines of force in the situation described by eqns (2.1) and (2.2) are shown in Fig. 2.1.

Of course, in reality, the lines of force fill the whole of the space and are not just confined to two dimensions as in the figure. The arrows indicate the direction in which a tiny positive charge would tend to move, that is, along the line of force, towards the negative charge. At each point, the electric field has both magnitude and direction and

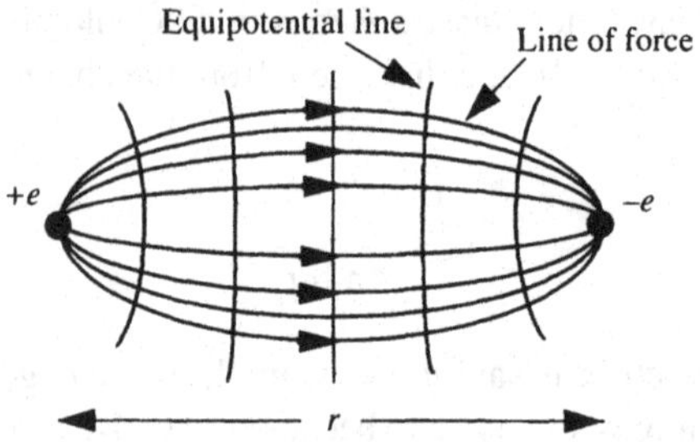

Fig. 2.1 Lines of force between two opposite charges.

is therefore a vector quantity so that, in terms of vector quantities, which are printed in bold italic font to distinguish them from scalar quantities:

$$\boldsymbol{E}(\boldsymbol{r}) = \frac{q\boldsymbol{r}}{4\pi\varepsilon_0|\boldsymbol{r}|^3}. \tag{2.3}$$

Equation (2.3) is very similar to eqn (2.1) but contains more information. By reading eqn (2.3) we understand that the strength of the electric field at a point P (see Fig. 2.2) whose distance and direction from a charge q (composed of many elementary charges, e), is defined by the vector $\boldsymbol{r}$ is obtained by multiplying the vector $\boldsymbol{r}$ by a scalar quantity, $e/4\pi\varepsilon_0|\boldsymbol{r}|^3$, which is a number possessing magnitude but no direction. Here $|\boldsymbol{r}|$ denotes the length (a scalar quantity) of $\boldsymbol{r}$. We also understand, from the same equation, that the direction of the electric field at the point defined by $\boldsymbol{r}$ is the same as the direction of $\boldsymbol{r}$ itself at that point.

Another way of writing the strength of the field is to use the idea of a unit vector, which points in a particular direction but whose magnitude is always equal to unity. Such a unit vector is $\hat{\boldsymbol{r}}$ which is equal to $\boldsymbol{r}/|\boldsymbol{r}|$ and replacement of $\boldsymbol{r}/|\boldsymbol{r}|$ by $\hat{\boldsymbol{r}}$ in eqn (2.3) gives us eqn (2.4), which exhibits the expected inverse-square law, and the force $\boldsymbol{F}$ acting on a charge q at the point $\boldsymbol{r}$ is just $\boldsymbol{F} = q\boldsymbol{E}$:

$$\boldsymbol{E}(\boldsymbol{r}) = \frac{q\hat{\boldsymbol{r}}}{4\pi\varepsilon_0|\boldsymbol{r}|^2}. \tag{2.4}$$

It is important for what comes later to remind you again that q is a scalar, possessing only magnitude, so that multiplication of the electric field strength by e cannot change the direction in which $\boldsymbol{E}$ is pointing, so that $\boldsymbol{F}$ points the same way as $\boldsymbol{E}$. Of course, if q is a negative number (the charge on an electron is negative) $\boldsymbol{F}$ will point in the direction opposite to $\boldsymbol{E}$.

The next thing to notice is that the lines of force begin on a positive charge and end on a negative one. This looks like a new experimental result but it is, in fact, directly related to the law of force in eqns (2.1)–(2.4). We describe this result in mathematical terms by saying that the divergence of the electric field (the number of lines of force streaming out from a point and written div $\boldsymbol{E}$) is equal to the amount of electric charge at that point divided by the electric permittivity. Where there are charges, div $\boldsymbol{E} = \rho/\varepsilon_0$ and, where there are no charges, div $\boldsymbol{E} = 0$. In the statement

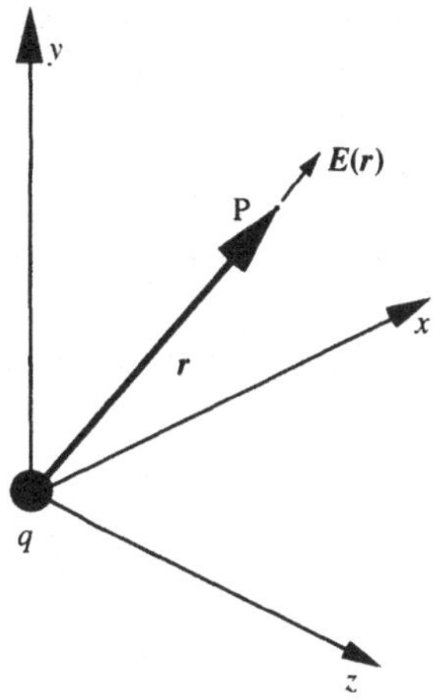

Fig. 2.2 Electric field at distance r from charge q.

we have just made, ρ is the volume density of electric charge at the point in question so that the amount of charge dq in a small volume $d\tau$ around that point is equal to $\rho\, d\tau$. With this definition of charge density, the law of force becomes

$$\boldsymbol{E} = \int_{\tau} \frac{\rho(\boldsymbol{r})\boldsymbol{r}}{4\pi\varepsilon_0|\boldsymbol{r}|^3}\, d\tau,$$

where the integral is calculated over the volume, τ, containing electric charge. We should notice that div $\boldsymbol{E}$ is just a number and is a scalar quantity.

A further observation is that, provided the electric charge is stationary, the lines of force of the electric field cannot form closed loops. The lines of force must begin and end on separate charges. The corresponding equation which expresses this is that curl $\boldsymbol{E} = 0$. In some places 'curl' is known as 'rot', short for rotation. These two statements, that div $\boldsymbol{E} = \rho/\varepsilon_0$ and curl $\boldsymbol{E} = 0$ are sufficient, for all practical purposes, to describe the fields generated by any arrangement of stationary electric charges and we will return to them later.

Moving electrons—source of magnetic field

The situation which I have just described generates what is called static electricity but a static situation does not last for long. If one of the electrons is free to move, it does so once it feels the field generated by the other. Movement of electrons generates an electric current such that if 1 C of electric charge moves past a certain point in a time of 1 s then a suitable measuring device registers an electric current of 1 ampere (A) (named after André Marie Ampère, 1775–1836). In other words, the current I is equal to dq/dt, the rate of change of electric charge with time. To be precise, current, $\boldsymbol{I}$, is a vector quantity and $\boldsymbol{I} = q\boldsymbol{v} = q\, d\boldsymbol{s}/dt$, where $\boldsymbol{s}$ denotes a generalized position vector such as the distance along a trajectory which the charge is following.

Moving electric charges generate magnetic fields but the rule for calculating the strength of the magnetic field is more complicated than that for the electric field. That is because there are no isolated magnetic charges (no magnetic monopoles have ever been detected) so that whereas the lines of force which one can use to visualize the electric field diverge outwards from the charge, the lines of magnetic force circle around the moving charge so that the magnetic field is at right angles to the electric field. This is illustrated in Fig. 2.3.

In Fig. 2.3, a charge q at the origin O is moving with velocity $\boldsymbol{v}$. At a point P defined by the position vector $\boldsymbol{r}$ the magnetic field strength, more precisely, the magnetic flux density $\boldsymbol{B}$ is given by

$$\boldsymbol{B} = \frac{\mu_0}{4\pi} q \frac{\boldsymbol{v} \times \boldsymbol{r}}{|\boldsymbol{r}|^3}$$

or

$$\boldsymbol{B} = \frac{\mu_0}{4\pi} q \frac{\boldsymbol{v} \times \hat{\boldsymbol{r}}}{|\boldsymbol{r}|^2}.$$

The quantity $|\boldsymbol{r}| \sin\theta$ in the expression $\boldsymbol{v} \times \boldsymbol{r} = |\boldsymbol{v}||\boldsymbol{r}| \sin\theta$ is the perpendicular distance from the point P (where $\boldsymbol{B}$ is being determined) to the vector $\boldsymbol{v}$, so that $|\boldsymbol{B}|$ is proportional to the moment (or torque) of $\boldsymbol{v}$ about a line at right angles to the plane containing $\boldsymbol{r}$ and $\boldsymbol{v}$. The direction of $\boldsymbol{B}$ is perpendicular to this plane, and because $\boldsymbol{E}$ points along the direction $\boldsymbol{r}$, $\boldsymbol{B}$ and $\boldsymbol{E}$ are perpendicular to each other. If we imagine $\boldsymbol{r}$ rotating about the axis $\boldsymbol{v}$ then P describes a circle linking points with the same value of $\boldsymbol{B}$, which is a line of magnetic force around the moving charge.

The magnetic field unit is the weber/square metre (Wb/m^2), usually called tesla (T) (Wilhelm Eduard Weber, 1804–91; Nicola Tesla, 1856–1943). A steady current of 1 A in an infinite straight conductor generates a magnetic field of 1 T at a distance of 1 m from the wire. The quantity μ_0, in eqns (2.9) and (2.10) is called the magnetic permeability of the vacuum, which has a value of $4\pi \times 10^{-7}$ Wb/(A m) by definition. The unit Wb/A is named the henry (Joseph Henry, 1797–1878).

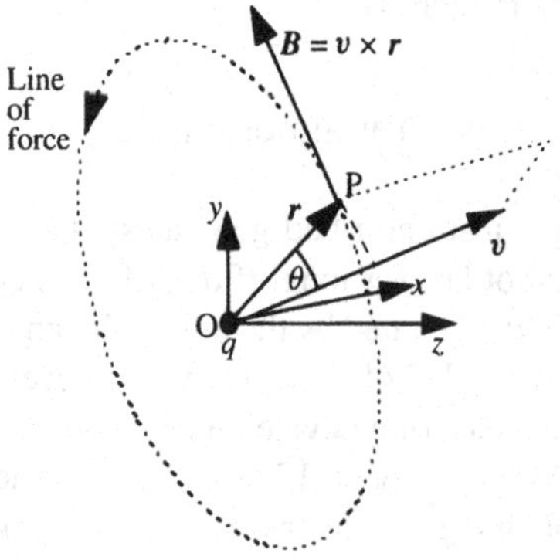

Fig. 2.3 Lines of magnetic force generated by an electric current.

Like the expression for the electric field, the expression for the magnetic field also reduces to an inverse-square law with the current $\boldsymbol{I} = q\boldsymbol{v}$ playing a role similar to the electric charge. By this token the force acting on an elementary charge e in a magnetic field of strength $\boldsymbol{B}$ is given by

$$\boldsymbol{F} = e\boldsymbol{v} \times \boldsymbol{B} \tag{2.5}$$

and acts on the charge in a direction at right angles to both the velocity and the magnetic field. It follows from eqns (2.2) and (2.5) that the total force on the charge e in the presence of both an electric and a magnetic field is the vector sum of the electric and magnetic forces:

$$\boldsymbol{F} = e(\boldsymbol{E} + \boldsymbol{v} \times \boldsymbol{B}). \tag{2.6}$$

This important result is often called the Lorentz formula and the force is called the Lorentz force, after the Dutch scientist Hendrik Antoon Lorentz (1853–1928).

Because there are no free magnetic poles at which the lines of force can begin or end, div $\boldsymbol{B} = 0$ everywhere but, because the magnetic field forms closed loops around the direction of the electric current, curl $\boldsymbol{B} = \mu_0 \boldsymbol{J}$, where $\boldsymbol{J} = \rho \boldsymbol{v}$ is called the current density and is measured in A/m^2.

Fields described by potential functions

An alternative way of describing electric and magnetic fields is to use the idea of potential instead of force. Let us consider this for an electric field generated by a static electric charge distribution, where there are no magnetic fields to confuse the issue. Imagine that a small electric charge q is being held at a fixed point P where the field strength is $\boldsymbol{E}$. In order to prevent the electric charge from moving you will have to exert a force, $\boldsymbol{F}$, on the charge which is equal in strength and opposite in direction to the force produced by the electric field so that $\boldsymbol{F} = -q\boldsymbol{E}$. How does the charge q get there in the first place? Suppose the charge is brought along some arbitrary trajectory from an infinite distance to P. At each step $\mathrm{d}\boldsymbol{s}$ along the trajectory a force $-q\boldsymbol{E}(\boldsymbol{s})$ must be applied. The amount of work done along this step is equal to the component of the force along $\mathrm{d}\boldsymbol{s}$ times the distance travelled, or the component of $\mathrm{d}\boldsymbol{s}$ along the direction of the force times the force, which is the same thing. This work done, or expended energy, is $-q|\boldsymbol{E}(\boldsymbol{s})||\mathrm{d}\boldsymbol{s}|\cos\theta$, or the scalar product $-q\boldsymbol{E}(\boldsymbol{s}) \cdot \mathrm{d}\boldsymbol{s}$ in vector notation as shown in Fig. 2.4.

As you move the charge, you measure $\boldsymbol{s}$ as the distance along the path and because $\boldsymbol{E}(\boldsymbol{s})$ is always pointing parallel to the lines of force, only the component of $\mathrm{d}\boldsymbol{s}$ in this direction contributes to the work done or the energy expended in moving the charge. The component of $\mathrm{d}\boldsymbol{s}$ at right angles to the lines of force is always zero. Integration along the complete path gives V, the total work done in transporting the charge from the distant point to P so that

$$qV = \int_s -q\boldsymbol{E}(s) \cdot \mathrm{d}\boldsymbol{s}.$$

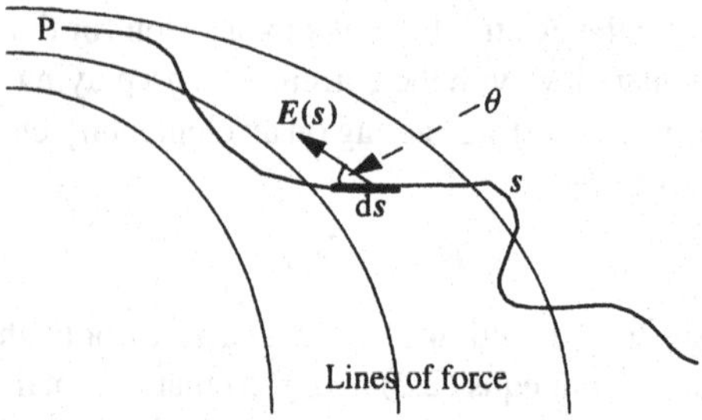

Fig. 2.4 Definition of potential.

The only contribution to the integral along the path comes from moving the charge parallel to the lines of force which means that we can associate a scalar quantity V with each point in space. V is called the electrostatic potential, which is just the energy per unit charge and

$$V = \int_s -\boldsymbol{E}(\boldsymbol{s}) \cdot \mathrm{d}\boldsymbol{s}. \tag{2.7}$$

We can also write this expression in differential form using the gradient operator ∇, and write

$$\boldsymbol{E} = -\,\mathrm{grad}\, V = -\nabla V. \tag{2.8}$$

The expression for the potential in terms of distance $\boldsymbol{r}$ from a charge distribution which is equivalent to the expression for the force [eqn (2.2)] is just

$$V = \frac{q}{4\pi\varepsilon_0 |\boldsymbol{r}|}. \tag{2.9}$$

The unit of potential is the volt (V) (Count Alessandro Giuseppe Anastasio Volta, 1745–1827). One volt is the potential difference between two points A and B when 1 J of work is done to transport a charge of 1 C from A to B. Comparison of eqns (2.7) and (2.8) with (2.9) shows that a constant term which is independent of position can be added to the potential function $V(\boldsymbol{r})$ without affecting the electric field at P. This constant term can be set to zero without any loss of generality.

Like the electric field which can be represented as lines of force, points at the same potential can be linked by equipotential lines, or, in three dimensions, equipotential surfaces as shown in Fig. 2.1.

The divergence equation

We have already stated that the number of lines of force emerging from an electric charge is proportional to the strength of the charge. Let us consider this now in more detail.

In Fig. 2.5, the surface denoted by σ encloses a volume τ. The vector $\boldsymbol{E}$, denoting the field strength at a point P on the surface, emerges from the surface through the small area $\mathrm{d}\sigma$ surrounding P. If $\boldsymbol{n}$ is a unit vector defining a direction at right angles to $\mathrm{d}\sigma$ at P then the scalar product $\boldsymbol{n} \cdot \boldsymbol{E}$ is called the flux of $\boldsymbol{E}$ through the area $\mathrm{d}\sigma$. In terms of lines of force, the electric field strength is a vector tangential to the line of force at P and the total number of lines of force emerging through $\mathrm{d}\sigma$ is $\boldsymbol{n} \cdot \boldsymbol{E}\,\mathrm{d}\sigma$. The total number of lines of force emerging from the entire volume is

$$\int_\sigma \boldsymbol{n} \cdot \boldsymbol{E}\,\mathrm{d}\sigma.$$

We must be careful, when evaluating this integral, to distinguish between regions of the surface where the flux is ingoing (and therefore carrying a negative sign) and regions where the flux is outgoing and positive as in Fig. 2.5.

Let us apply this idea to a simple case of a rectangular volume whose edges form part of an x, y, z coordinate system. This is shown in Fig. 2.6, where the volume is placed within a region where the electric field is increasing with distance from the origin. The rate of increase is supposed to be linear with components $\partial E_x/\partial x$ along the x-direction and similarly for the y- and z-directions. The contributions to the integral come from each of the six faces. Consider the face in the y–z plane ($x = 0$). The area of this face is $\delta y \delta z$, so the contribution to the integral from this face is $-E_x \delta y \delta z$ (the negative sign is there because E_x is pointing inwards). At the opposite face, E_x has increased by an amount $E_x + (\partial E_x/\partial x)\delta x$ so the contribution to the integral from this face is $+(E_x + (\partial E_x/\partial x)\delta x)\delta y \delta z$. The total contribution from the two faces together is the sum of the contributions from the individual faces, namely $(\partial E_x/\partial x)\delta x \delta y \delta z$. Likewise, we can add up the contributions from the other faces so that the overall total comes to

$$\left(\frac{\partial E_x}{\partial x} + \frac{\partial E_y}{\partial y} + \frac{\partial E_z}{\partial z}\right) \delta x \delta y \delta z.$$

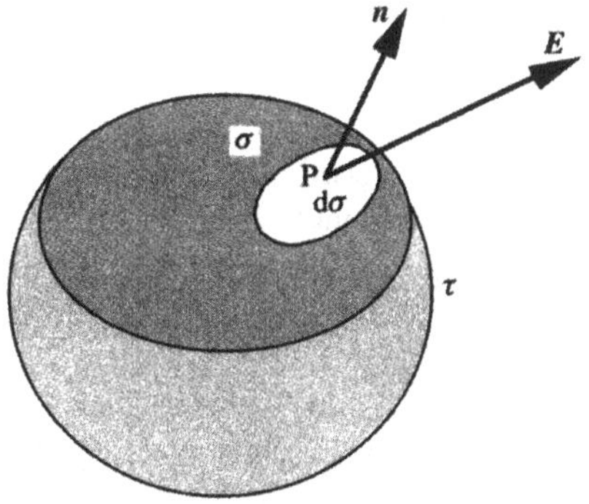

Fig. 2.5 Flux of $\boldsymbol{E}$ from a volume τ through its enclosing surface σ.

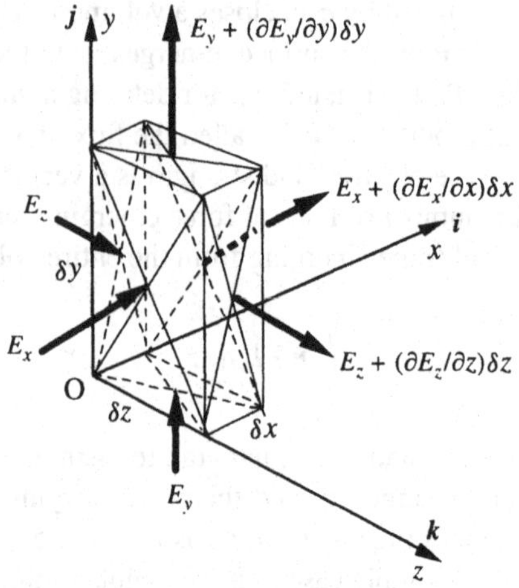

Fig. 2.6 Divergence of a vector $\boldsymbol{E}$.

This is equal to div $\boldsymbol{E}$, multiplied by the volume of the rectangular figure, which means that div $\boldsymbol{E}$ is the increase in the strength of $\boldsymbol{E}$ occurring within unit volume. We can make this volume as small as we please so that div $\boldsymbol{E}$ is the increase in strength of $\boldsymbol{E}$ at a particular point in space and, as stated above, in a region where there is a charge density ρ:

$$\operatorname{div} \boldsymbol{E} = \nabla \cdot \boldsymbol{E} = \frac{\rho}{\varepsilon_0} \tag{2.10}$$

or, where there are no charges

$$\operatorname{div} \boldsymbol{E} = 0. \tag{2.11}$$

The continuity equation

Suppose now that instead of an outward flux of $\boldsymbol{E}$ from the rectangular volume of Fig. 2.6, electric charge was flowing out through the enclosed surface at a rate given by $\partial\rho/\partial t$. This outflow of charge constitutes an electric current density, $\boldsymbol{J}$, defined in Section 2.4. Because we know from experiment that electric charge can be neither created nor destroyed, it follows that

$$\operatorname{div} \boldsymbol{J} + \frac{\partial\rho}{\partial t} = 0. \tag{2.12}$$

Equation (2.12) is known as the equation of continuity.

The rotation or curl equation

In the previous section we examined the divergence of the electric field. Let us now examine the curl or rotation of the magnetic field. We begin by calculating an integral similar to that for the divergence in eqn (2.10) but here the integral is calculated not over a surface enclosing a volume but around a contour enclosing a surface (Fig. 2.7).

In this figure, a quantity related to curl $\boldsymbol{B}$ at the point P is obtained when the integral of $\boldsymbol{B}$ is calculated round the closed loop of length s which encloses P. This is the integral $\int_S \boldsymbol{B} \cdot \mathrm{d}\boldsymbol{s}$.

Figure 2.8 shows a simple case in which the surface is the y–z plane and the contour around which the integral is to be calculated is a rectangle in that plane whose sides are δy and δz.

The integral is obtained by starting from the origin in the bottom left-hand corner of the rectangle and considering each side in turn until one returns to the starting point. The component of the field along the x-axis, B_x makes no contribution to the integral because it is everywhere at right angles to the contour. The contributions are then added as follows; $B_y\delta y + (B_z + (\partial B_z/\partial y)\delta y)\delta z - (B_y + (\partial B_y/\partial z)\delta z)\delta y - B_z\delta z$. The minus signs come about because the integral round the contour is being evaluated in the clockwise direction as viewed along the positive x-axis. This expression for

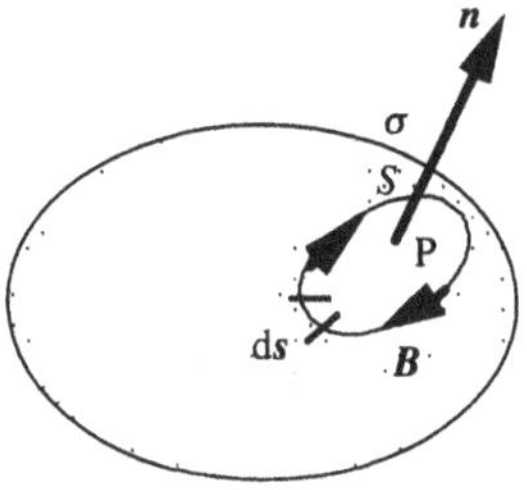

Fig. 2.7 Definition of curl $\boldsymbol{B}$.

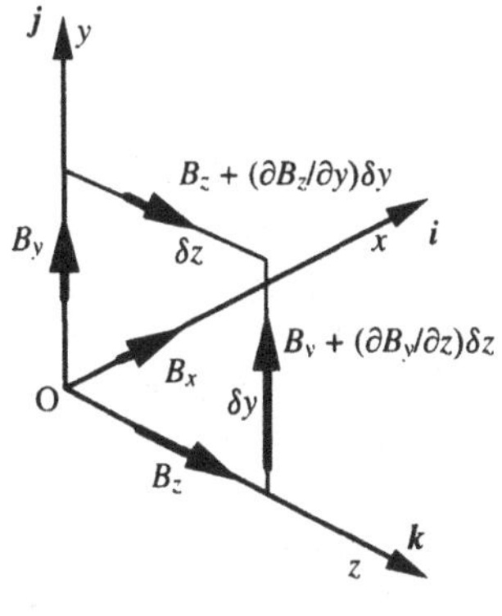

Fig. 2.8 Calculation of curl $\boldsymbol{B}$.

the integral simplifies to

$$\int_S \boldsymbol{B} \cdot \mathrm{d}\boldsymbol{s} = \left(\frac{\partial B_y}{\partial z} - \frac{\partial B_z}{\partial y}\right)\delta y \delta z. \tag{2.13}$$

The right-hand side of eqn (2.13) is the x-component of curl $\boldsymbol{B}$ multiplied by the area enclosed by the contour. This means that curl $\boldsymbol{B}$ is the total change in the value of the field as one goes round the contour, normalized to the area enclosed by the contour. In the case of a static magnetic field this quantity depends on the electric current density passing through the enclosed area so that, as stated previously,

$$\text{curl}\, \boldsymbol{B} = \mu_0 \boldsymbol{J}. \tag{2.14}$$

In the case of a static electric field, there are no loops in the lines of force so that the integral in eqn (2.13), when $\boldsymbol{E}$ is substituted for $\boldsymbol{B}$, is always zero so that curl $\boldsymbol{E} = 0$ everywhere. But the line integral $\boldsymbol{E} \cdot \mathrm{d}\boldsymbol{s}$ is the work done when unit charge is carried around the loop, so it follows that if the loop is split into two sections as in Fig. 2.9, the work done (or the change in potential energy) in moving the unit charge from P to Q round one section of the loop is the same as the energy change in moving from P to Q round the other section (or indeed by any route). This argument shows that for an electrostatic field, the fact that curl $\boldsymbol{E} = 0$ ensures that the field $\boldsymbol{E}$ can be represented by the gradient of a scalar potential which has a single value at each point in the field. In other words, from eqn (2.8),

$$\text{curl}\, \boldsymbol{E} = -\nabla \times \nabla V = 0 \tag{2.15}$$

and, from eqns (2.10) and (2.11),

$$\text{div}\, \boldsymbol{E} = \nabla \cdot \nabla V = \nabla^2 V = \frac{\rho}{\varepsilon_0}, \tag{2.16}$$

which is known as Poisson equation, and, in the absence of electric charge,

$$\nabla^2 V = 0,$$

which is the Laplace equation (Siméon Denis Poisson, 1781–1840; Marquis Pierre Simon Laplace, 1749–1827). These equations, including the relation between field and potential, make it possible to determine the electric field distribution resulting from any arrangement of static electric charges.

In the case of an electric field, the potential is a scalar function which makes it much simpler to use than the field, which is, of course, a vector. Contributions to the potential at a particular point, from charges located at different points, can be added arithmetically. The electric field components can be obtained from the potential by the use of the gradient operator.

When we attempt to apply the concept of potential to the magnetic field we immediately run into the difficulty that the curl of the magnetic field vector $\boldsymbol{B}$ is zero only when there is no current density at the point in question [eqn (2.14)]; so only in that situation can a scalar magnetic potential be defined. However, although, in general, the curl is non-zero, the divergence of the magnetic field must be zero everywhere because free magnetic poles (magnetic monopoles) have never been detected. Assuming therefore that there are no free magnetic poles to act as sources or sinks for the magnetic lines of force we can write

$$\operatorname{div} \boldsymbol{B} = \nabla \cdot \boldsymbol{B} = 0.$$

Thus, although $\boldsymbol{B}$ cannot be written as the gradient of a scalar potential, it can be written as the curl of a vector potential $\boldsymbol{A}$, because

$$\operatorname{div} \operatorname{curl} \boldsymbol{A} = \nabla \cdot (\nabla \times \boldsymbol{A}) = 0$$

and

$$\boldsymbol{B} = \nabla \times \boldsymbol{A} = \operatorname{curl} \boldsymbol{A}.$$

Because the magnetic field can only be described by a vector potential, this concept is not as useful in magnetic field calculations as is the scalar potential in the electrostatic case. However, we shall find that these potentials are very useful when electromagnetic radiation is being considered, so it is to this that we must now turn.

Changing magnetic fields produce electric fields

So far we have thought about fields produced by static charges and steady currents. If the world contained these entities alone, everything would be very simple but much less interesting than the real world in which change is the rule rather than the exception. The rules of the game, developed so far, apply only to an ideal, static, steady situation which can exist only under very exceptional circumstances. Even if one of the charges moves to a new position, or one of the electric currents gets turned off, a new situation exists and the electric and magnetic fields assume a different configuration. The magnitude and direction of the fields at all points in the universe can change, at least in principle, and we would like to know how quickly these changes take effect.

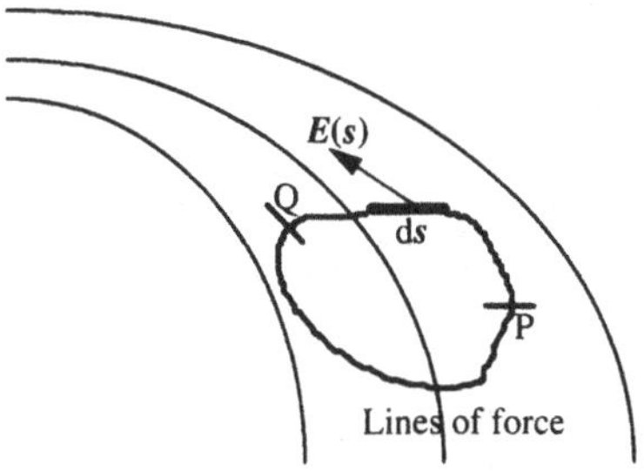

Fig. 2.9 Calculation of $\int \boldsymbol{E} \cdot \mathrm{d}\boldsymbol{s}$ around a loop.

Suppose that a static electric charge is moved over a period of time to a different place where it is again held stationary. During that time of movement, the charge is first accelerated, then moves with constant velocity and finally is decelerated and comes to rest at its new position. Distant observers, with suitable detecting equipment would notice that, during the period of movement of the charge, a magnetic field had been generated, which fell back to zero once the new static situation had been reached. If the observers were themselves some distance apart (though in communication with each other) they would be able to measure the speed with which the new, transient magnetic field travelled from one to the other. If the observers were on other planets, their observations would be complicated by their movement relative to each other and to the moving charge. Indeed, moving observers would not necessarily know if the field had been generated by their own movements or by the movement of the charge or by a combination of both. At the very least, we cannot know if a static magnetic field is a consequence of our own movement at a constant speed past a stationary charge or a movement of the charge past our observation point.

All of these ideas took a long time to develop but the starting point was the discovery of magnetic induction by Michael Faraday (1791–1867) at the Royal institution in Albemarle Street, London, in the summer of 1831. It was already known that a steady current would produce a magnetic field although the mathematical description, as described previously in the chapter, had not been formulated. It followed that if currents could produce fields then fields must produce currents. It was Faraday who first observed and understood that a change in a steady magnetic field would induce an electric current in a nearby electric circuit which was bathed in the lines of force of the magnetic field. The induced current was proportional to the rate of change of the magnetic flux, the direction of the current depended on whether the magnetic flux was increasing or decreasing. The change in flux could be produced by a moving magnet or by changing the current in a nearby electric circuit, or by moving the circuit to a new position. Faraday did not investigate the speed of propagation of the change in the magnetic field and in any case he was not equipped to carry out that kind of measurement. What he did observe was sufficient to start a new industrial revolution with the development of motors, dynamos and transformers and to propel our understanding of the natural world along a direction which would lead to a complete formulation of the theory of electromagnetism and electromagnetic radiation and to a description of relative motion known as the special theory of relativity.

In order to incorporate the observations of Faraday which were confirmed and extended by many other people, all we need to do is to modify eqn (2.15) so that curl $\boldsymbol{E}$ is no longer zero but

$$\text{curl}\,\boldsymbol{E} = -\frac{\partial \boldsymbol{B}}{\partial t}, \tag{2.17}$$

which means that a magnetic field which changes with time at some point in space generates loops of electric lines of force. If there is no magnetic field or if the field does not change, curl $\boldsymbol{E}$ is zero as before. Equation (2.17) is the mathematical statement of the principle of the electrical generator. A rotating magnetic field, whose direction changes with time, produces an electric field which, in turn, generates an electric current in a conductor.

Changing electric fields produce magnetic fields

We can ask the question: is there a reciprocal phenomenon whereby a changing electric field generates a magnetic field loop in addition to the field loop which is always associated with a steady electric current? The answer to this question is 'yes' and that it must be so can be deduced from the equation of continuity [eqn (2.12)] and the statement about magnetic field loops and steady electric currents in eqn (2.14). The second term in eqn (2.12) is the rate of change of electric charge $\partial\rho/\partial t$ which must be equal to $\partial(\text{div}\,\boldsymbol{E})/\partial t$ divided by ε_0, from eqn (2.16). Because $\partial/\partial t$ is a scalar operation, $\partial(\text{div}\,\boldsymbol{E})/\partial t$ is equal to $\text{div}(\partial\boldsymbol{E}/\partial t)$ so that eqn (2.25) may be written as

$$\text{div}\left(\boldsymbol{J} + \frac{1}{\varepsilon_0}\frac{\partial \boldsymbol{E}}{\partial t}\right) = 0.$$

Suppose we now calculate div $\boldsymbol{J}$ from eqn (2.14). It must be equal to div curl $\boldsymbol{B}$, divided by μ_0. But we have already seen that, from the definition of curl, div curl of any vector is always zero so that div $\boldsymbol{J}$ must always be zero. It appears that eqn (2.14) is inconsistent with the continuity equation (2.12) so something must be wrong. The correctness of eqn (2.12) rests, ultimately, on the conservation of electric charge which is not a dogmatic principle but has been demonstrated by experiment to be correct to a high degree of accuracy. Maxwell understood that the charge conservation principle could be saved by modifying eqn (2.14) (which, so far, had only been proved correct for steady electric currents) to read

$$\text{curl}\,\boldsymbol{B} = \mu_0\boldsymbol{J} + \mu_0\varepsilon_0\frac{\partial \boldsymbol{E}}{\partial t}.$$

This tells us that a time-varying electric field generates a contribution to the magnetic field loop which is additional to that generated by the steady current, as was suggested at the start of this section.

Maxwell's equations and electromagnetic radiation

Let us now summarize the set of equations which must be satisfied by electric and magnetic fields. These are statements about the divergence and the rotation (or curl) of the fields at any point in space and are listed below as eqns (2.18a–d). They are known as Maxwell's equations and represent both the culmination of classical physics and the gateway into the modern physics of the 20th century. They make a statement about electricity and magnetism as a universal phenomenon. We include the continuity equation (2.12) because it constrains the movement of charge in space and time. We include the Lorentz equation (2.6) in the list because through it we can relate electricity and magnetism to the mechanical phenomena described by Newton's laws of mechanical motion:

Maxwell's equations:

$$\operatorname{div} \boldsymbol{E} = \frac{\rho}{\varepsilon_0}, \tag{2.18a}$$

$$\operatorname{div} \boldsymbol{B} = 0, \tag{2.18b}$$

$$\operatorname{curl} \boldsymbol{E} = -\frac{\partial \boldsymbol{B}}{\partial t}, \tag{2.18c}$$

$$\operatorname{curl} \boldsymbol{B} = \mu_0 \boldsymbol{J} + \mu_0 \varepsilon_0 \frac{\partial \boldsymbol{E}}{\partial t}, \tag{2.18d}$$

$$\operatorname{div} \boldsymbol{J} + \frac{\partial \rho}{\partial t} = 0, \tag{2.12}$$

$$\boldsymbol{F} = e(\boldsymbol{E} + \boldsymbol{v} \times \boldsymbol{B}). \tag{2.6}$$

Even if there are no sources of electric or magnetic fields at a point where measurements are being made, the fields themselves continue to exist, at least in principle, generated by sources distant from the point of observation. If these distant sources are moving, the fields themselves change in accordance with eqns (2.19). These are identical to eqns (2.18) but with ρ and $\boldsymbol{J}$ both zero:

Maxwell's equations when $\rho = 0$ and $\boldsymbol{J} = 0$:

$$\operatorname{div} \boldsymbol{E} = 0, \tag{2.19a}$$

$$\operatorname{div} \boldsymbol{B} = 0, \tag{2.19b}$$

$$\operatorname{curl} \boldsymbol{E} = -\frac{\partial \boldsymbol{B}}{\partial t}, \tag{2.19c}$$

$$\operatorname{curl} \boldsymbol{B} = \mu_0 \varepsilon_0 \frac{\partial \boldsymbol{E}}{\partial t}. \tag{2.19d}$$

Visual examination of these equations shows that they imply that changes in the electric and magnetic fields propagate through space because a change in the rate of change of $\boldsymbol{B}$, through eqn (2.19c), produces a change in the curl of $\boldsymbol{E}$ and, by implication a change in the value of $\boldsymbol{E}$ as well. This, through eqn (2.19b), changes curl $\boldsymbol{B}$ and hence $\partial \boldsymbol{B}/\partial t$. While this mutual interaction of the fields is taking place, eqns (2.19a) and (2.19b) demand that any change at one point must be accompanied by an equivalent change at adjacent points so that the divergence of both $\boldsymbol{E}$ and $\boldsymbol{B}$ remain zero. In other words, any alteration in $\boldsymbol{E}$ and $\boldsymbol{B}$, once begun not only continues but also propagates throughout the whole of space. This new phenomenon is predicted by the Maxwell equations and is known as electromagnetic radiation.

To express this precisely we must solve eqns (2.19) in order to obtain expressions for $\boldsymbol{E}$ and $\boldsymbol{B}$ separately. Let us apply the curl operation to eqn (2.19c). We write

$$\text{curl curl}\,\boldsymbol{E} = \nabla \times \nabla \times \boldsymbol{E} = -\nabla \times \frac{\partial \boldsymbol{B}}{\partial t}$$

so that, from eqn (2.19d)

$$\nabla \times \nabla \times \boldsymbol{E} = -\mu_0 \varepsilon_0 \frac{\partial^2 \boldsymbol{E}}{\partial t^2}.$$

We can expand the triple vector product to give

$$(\nabla \cdot \boldsymbol{E})\nabla - (\nabla \cdot \nabla)\boldsymbol{E} = -\mu_0 \varepsilon_0 \frac{\partial^2 \boldsymbol{E}}{\partial t^2}.$$

From eqn (2.19a) div $\boldsymbol{E}$, which is $\nabla \cdot \boldsymbol{E}$ is zero so the solution is

$$\nabla^2 \boldsymbol{E} = \mu_0 \varepsilon_0 \frac{\partial^2 \boldsymbol{E}}{\partial t^2} \tag{2.20a}$$

and, by the same argument,

$$\nabla^2 \boldsymbol{B} = \mu_0 \varepsilon_0 \frac{\partial^2 \boldsymbol{B}}{\partial t^2}. \tag{2.20b}$$

Equations (2.20) are examples of the wave equation. We saw in Chapter 1 that an equation of this form describes how a disturbance travels outwards from a source with a phase velocity equal, in this case to $\sqrt{1/\mu_0 \varepsilon_0}$. In this case the equations describe a disturbance which takes the form of changes in the field quantities $\boldsymbol{E}$ and $\boldsymbol{B}$ which propagate in a vacuum with a velocity given by the reciprocal of the product of the electric and magnetic constants. The value of μ_0 has been chosen, in MKSA units [or Systéme Internationale (SI)] to be $4\pi \times 10^{-7}$ H/m and ε_0 is a measured quantity whose best value is 8.85×10^{-12} F/m, in the same units, so that $\sqrt{1/\mu_0 \varepsilon_0}$ is equal to 3×10^8 m/s. This remarkable result, that the wave velocity of an electromagnetic disturbance in a vacuum is equal to the measured value of the velocity of light, allowed Maxwell to infer that light and the electromagnetic disturbance are one and the same thing. What began as the study of the local effects of electric charges and currents has resulted in a description of a phenomenon which fills the whole of time and space and which surrounds us every moment of our lives.

Reference

1. J. D. Jackson, *Classical electrodynamics* (2nd edn). John Wiley & Sons, New York (1975).

3

Electromagnetic radiation—observed and imagined

The special place of the velocity of light

In Chapter 2 it was shown that electromagnetic fields can be propagated through space at the speed of light and that the field in this dynamic form is known as electromagnetic radiation. The identification of electromagnetic radiation with light, combined with the direct observation of radio waves and the measurement of the velocity of these waves by Hertz (Heinrich Rudolf Hertz, 1857–94), generated some serious conceptual problems. The speed at which the light was moving would be expected to depend on the relative movement of the source and the observer. On the other hand, Maxwell's equations would suggest that the speed of light would be independent of this movement because $c = 1/\sqrt{\mu_0\varepsilon_0}$ and in this expression the value of μ_0 is defined by the system of units and ε_0 is defined by the strength of the static electric field. This conclusion is backed by measurements of the speed of light from moving sources, made with increasing accuracy and improved conceptual precision (see ref. 1). The resolution of this problem led, over a period of about 40 years, to the acceptance of the non-intuitive proposition that the speed of light is independent of the motion of the source and the observer. It is an absolute velocity.

An important consequence of the speed of light being an absolute quantity is that the speed of light in vacuum is the maximum that any object can reach and is the same for all observers, whether they are moving or stationary. This is another non-intuitive idea, not something which we expect from our everyday experience, and the question is often asked why the speed of light should be singled out from all other rates of motion for this distinction. The answer to that question has to be that it is an experimental observation whose consequences we must take on board. Under everyday circumstances, we only encounter speeds which are very much less than the speed of light so that the constancy of the speed of light and its importance as the limiting speed is of no practical consequence. In contrast, when we consider the production of electromagnetic radiation by electrons, the fact is that no matter how high the energy of the electrons, their velocity cannot exceed that of the light they are producing. However, there is more to it than this because it is electrical forces which are used whenever information is transferred from one observer to another. This is most obvious when we consider that light enables us to see objects and radio waves (long-wavelength electromagnetic radiation) are used to transmit information over vast distances. But also, on the microscopic scale, it is the electric forces between atoms which are called into action when we touch an object and our brains register that there is something there. Our experience tells us that there appears to be something fundamental about electromagnetism and the appearance of its velocity in the transformation equations, which we will come to in a moment, is just one aspect of its unique place in our perception of the world around us.

Relative motion—classical relativity

It is often necessary to compare experiments done in different places at different times. For example, we shall find that it is useful to calculate the emission of electromagnetic radiation as seen by an observer moving along with an electron (so that the electron appears to be stationary) and then to apply a transformation rule to the result so that we obtain the characteristics of the radiation from a moving electron as seen by a stationary observer. To make this comparison we need a rule which will take us from one situation to the other. Figure 3.1 shows two sets of coordinate axes (two reference frames) in which the X-axis of one is moving with velocity v parallel to the x-axis of the other.

In one system, which we call Σ, the point P has coordinates (x, y, z) and in the frame (Σ') which is seen (by an observer in Σ) to be moving with a steady speed v, parallel to the x-axis, the coordinates of P are (x', y', z'). It then appears to be a simple matter to write down the rule which allows us to compare observations in the two frames:

$$\begin{aligned} x' &= a_0 + x - vt, \\ y' &= b_0 + y, \\ z' &= c_0 + z. \end{aligned} \tag{3.1}$$

In this expression, (a_0, b_0, c_0) are the coordinates of the origin of Σ' in the frame Σ. This expression can be simplified if we suppose that the origins of Σ and Σ' coincide at the time $t = 0$. At that time the observers in the two frames synchronize their

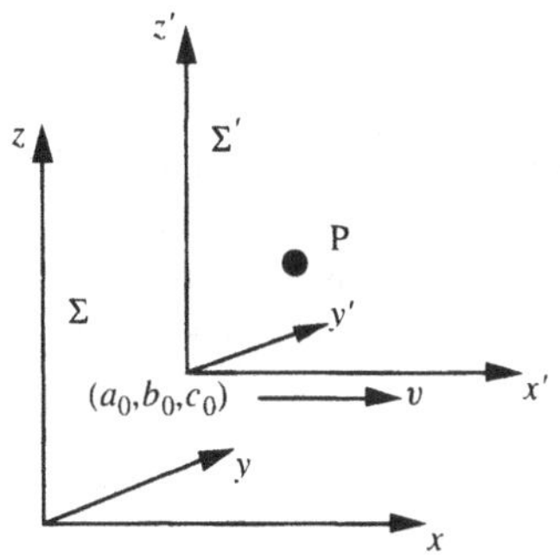

Fig. 3.1 Reference frames.

clocks to the same starting time so that the coordinate transformation rules from one frame to the other become

$$\begin{aligned} x' &= x - vt, \\ y' &= y, \\ z' &= z, \\ t' &= t. \end{aligned} \tag{3.2}$$

These transformation rules are often called the Galilean transformations, after the Italian scientist Galileo Galilei (1564–1642) in recognition of his placing the idea of motion on a firm scientific footing. We find them very easy to write down, with the benefit of 400 years of hindsight, because we know intuitively (or think we do) that in a time t, the origin of the coordinates moves through a distance vt and we must correct the x-coordinate by this amount either by subtraction [as in eqn (3.2)] or by addition depending on whether we are going from Σ' to Σ or from Σ to Σ'. On the other hand, we know (or we think we know) that we do not need to correct the time coordinate because that remains the same for everybody.

The rules also tell us, by taking the first derivative of the equation relating x' and x with respect to time,

$$\frac{\mathrm{d}x'}{\mathrm{d}t'} = \frac{\mathrm{d}x}{\mathrm{d}t} - v. \tag{3.3}$$

In other words, the speed of an object along the x-axis in Σ' is just its speed in Σ minus v. It seems from all this that physicists have a roundabout way of stating the obvious. If someone drops a bottle from the window of a train moving at 100 km/h, the bottle, as seen by a person in the train, has zero velocity in the direction the train is moving but from the point of view of a person standing on the platform that same bottle moves at 100 km/h with, possibly, lethal consequences.

By the same token, the transformation equation for the acceleration of the object can be obtained by taking the first derivative of eqn (3.3) (or the second derivative of eqns (3.2), so that

$$\frac{\mathrm{d}^2x'}{\mathrm{d}t'^2} = \frac{\mathrm{d}^2x}{\mathrm{d}t^2}.$$

Again, this is what we would expect intuitively. The acceleration of an object is the same in both coordinate systems, provided, of course that they are moving with constant speed relative to each other, which is the condition that the second term (in v) in eqns (3.1) and (3.2) vanishes when we take the second derivative. This brief discussion of classical motion has considered movement only along the x-axis but the equations can easily be extended to the y- and z-axes with the same result.

Special Relativity

What does this have to do with electromagnetic radiation? The assertion that the speed of light has a constant value and does not depend on the speed of the source means that emitting light is not like dropping bottles from trains. A different set of rules are required to describe this new situation. Suppose that a light flash is released at the origin of the coordinates in Σ at $t = 0$. This flash moves through Σ with velocity c and is detected by an observer at P, whose space coordinates are (x, y, z) at a time t later, where t is given, using the theorem of Pythagoras, by

$$ct = \sqrt{x^2 + y^2 + z^2}$$

or

$$c^2t^2 - \left(x^2 + y^2 + z^2\right) = 0. \tag{3.4}$$

In Σ', the observer at P sees the source of light moving away with speed v but this does not affect the speed of the light flash which remains unmodified by the movement of the source. In consequence, the observer at P, whose coordinates in Σ' are (x', y', z') detects the light flash at a time t' given by

$$c^2t'^2 - \left(x'^2 + y'^2 + z'^2\right) = 0. \tag{3.5}$$

There is no way that eqns (3.4) and (3.5) can be reconciled with the transformation rules of eqn (3.2) while still retaining the unchanging (or invariant) value of the speed of light. A new set of rules are required which can be derived from eqns (3.4) and (3.5). These rules are known as the Lorentz transformations (Hendrick Antoon Lorentz, 1853–1928) and can be written as

$$\begin{aligned} x' &= \gamma(x - \beta ct), \\ y' &= y, \\ z' &= z, \\ ct' &= \gamma(ct - \beta x), \end{aligned} \tag{3.6}$$

with the definitions of β and γ given by

$$\beta = \frac{v}{c} \quad \text{and} \quad \gamma = \frac{1}{\sqrt{1 - \beta^2}}. \tag{3.7}$$

Insertion of eqns (3.6) and (3.7) in eqns (3.4) and (3.5), followed by a bit of algebra, shows that this transformation of the coordinate system is one which allows both (3.4) and (3.5) to be satisfied and therefore to allow the velocity of light to be the same in both Σ and Σ'. These equations are a statement of the Theory of Special Relativity which was formulated by Albert Einstein (1879–1955) in 1905. It is called the Special Theory because it applies only to comparisons between measurements in reference frames which are moving at a steady speed relative to each another.

The inverse transformation, from Σ' to Σ, is the same as eqn (3.8) but with the sign of v reversed. This is because from the point of view of an observer in Σ', the frame

Σ is moving with the same speed v but in the opposite direction along the x-axis. The transformation equations are written for this case in eqns (3.10):

$$\begin{aligned} x &= \gamma(x' + \beta ct'), \\ y &= y', \\ z &= z', \\ ct &= \gamma(ct' + \beta x'). \end{aligned} \tag{3.8}$$

When $v \ll c$, the binomial theorem can be used to approximate γ [from eqn (3.7)] by

$$\gamma \approx 1 + \frac{1}{2}\left(\frac{v}{c}\right)^2 + \cdots \tag{3.9}$$

so that, when v is small enough, even the second order term in eqn (3.9) can be neglected completely and eqns (3.7) reduce to eqns (3.2). Even for objects moving at quite substantial velocities, the equations of the Galilean transformation are sufficiently accurate for all practical purposes. For example, a rocket, travelling with sufficient speed to leave the earth forever, is only moving at about 9 km/s so that v/c is about 3×10^{-5}. This means that even an astronaut riding in the space shuttle can ignore the equations of the Lorentz transformation and use the Galilean transformation when comparing measurements on the shuttle with a similar measurement at the ground station.

In the case of electrons, things are quite different. An accelerated electron can easily acquire a speed v close to the speed of light so that eqns (3.6) and (3.8) cannot be ignored when its behaviour is being described. It is clear, from the definition of γ in eqn (3.7) that γ gets larger and larger (mathematicians would say that γ tends to infinity) as v gets closer and closer to c. When v is exactly equal to c the value of γ, defined by eqn (3.9), is indeterminate. For $v > c$, the term under the square-root sign becomes negative and γ has no physical meaning so that c, the velocity of light, is the maximum that any object can reach. Whenever electrons (or other sub-atomic particles) are accelerated, this limiting speed is observed as an experimental fact, consistent with the equations of the Lorentz transformation. Unlike the sound barrier which can be easily shattered, by Concorde for example, or by a speeding bullet, the barrier presented by the velocity of light has never been broken.

Understanding the Lorentz transformation

The principal difficulty in understanding the implications of Special Relativity comes from the Lorentz transformation equation relating the time coordinates in two reference frames. Intuitively, we imagine that we can view the reference frames as though we were detached observers for whom time is an absolute quantity about which we can all agree. What the theory says to us is that we need to be quite specific about the state of motion of the observer relative to the reference frame in which the observations are taking place. The reason for this need for precision is because not only the space coordinate but also the time coordinate depends on the reference frame in which it is being measured. This is stated mathematically in the fourth equation of (3.6) and (3.8).

We can illustrate this by considering the phenomenon of time dilation and the corresponding length contraction. The idea that a moving body (such as a ruler or some more sophisticated equipment for measuring distance) can be reduced in length (as viewed by a stationary observer) was originally proposed to explain the null result of the Michelson–Morley experiment (Albert Abraham Michelson, 1852–1931; Edward Williams Morley, 1838–1923). In that experiment the speed of light was compared along two directions at right angles to each other. Light was understood to be a disturbance of the electric and magnetic fields, which was propagating itself through some kind of elastic medium known as the ether (or the luminiferous ether). It appeared to follow from this picture that at any given moment the Earth must be moving through this ether so that a comparison of the velocity of light at the surface of the Earth along the two directions at right angles should be a measurement of the velocity of the Earth through the medium [according to eqns (3.2)]. No difference in the two light velocities could be detected, which implied that the Earth was stationary in the medium, no matter what time of the year the experiment was carried out. This result was scarcely credible. To account for it, Fitzgerald (George Francis Fitzgerald, 1851–1901) and, independently, Lorentz suggested that motion through the electromagnetic medium could contract the length of any body in the direction of motion by a factor equal to $1/\sqrt{1-(v/c)^2}$. The effect would be very small but would be sufficient to account for the null result.

The Special Theory of Relativity accounts for this contraction in a satisfactory way, as a consequence of the velocity of light having the same value in all reference frames even though these may be moving (at a steady speed) relative to each other. To be specific, let us consider an electron which is moving with a velocity βc past a series of fixed alternating magnetic poles (this anticipates the theory of a magnetic undulator which we will encounter in more detail in Chapter 14). These poles, as shown in Fig. 3.2 are a distance λ_0 apart in the reference frame Σ. In Fig. 3.2(a), the electron, which is at rest at a point x'_0 in its own reference frame (Σ') 'sees' the first pole passing it at a time t'_1 and the second [Fig. 3.2(b)] at a time t'_2 at the same point x'_0. An observer in the laboratory tells a different story. That observer notes that the electron is moving (with the same velocity βc) past a series of stationary poles and that the time interval in the laboratory frame (Σ) is $t_1 - t_2$. These times are related by the equations for the inverse Lorentz transformation [eqns (3.8)] so that

$$\begin{aligned} ct_1 &= \gamma(ct'_1 + \beta x'_0), \\ ct_2 &= \gamma(ct'_2 + \beta x'_0), \\ t_1 - t_2 &= \gamma(t'_1 - t'_2). \end{aligned} \tag{3.10}$$

This equation expresses the phenomenon of time dilation. A time interval in the rest frame of the electron (Σ') (often called the proper time interval) transforms into a longer time interval in the frame Σ which is moving relative to Σ'. This surprising

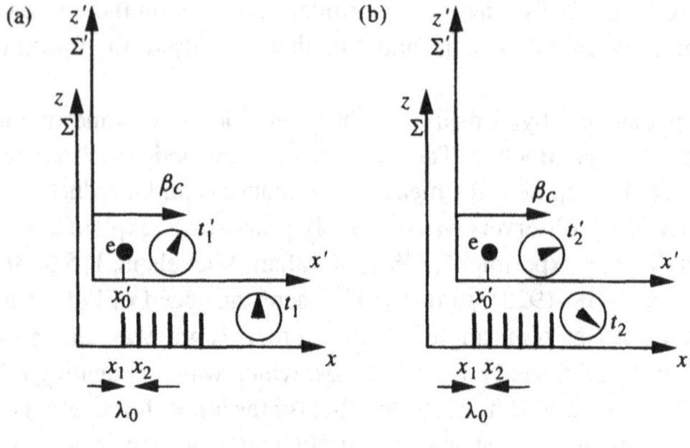

Fig. 3.2 Lorentz contraction.

result has been observed to be precisely correct in experiments where the lifetime of a sub-atomic particle, the μ meson, at rest has been compared with its lifetime when moving at high velocity.[2]

If we return to Fig. 3.2, in terms of distance, the observer in the laboratory sees that the electron is moving with velocity βc and the observer in the electron rest frame sees the magnetic poles moving with the same speed (but in the opposite direction), so that, from eqn (3.10),

$$\beta c(t_2 - t_1) = \beta c\gamma(t_2' - t_1').$$

Now $\lambda = \beta c(t_2' - t_1')$ is the distance between the poles as 'observed' by the electron at rest and $\lambda_0 = \beta c(t_2 - t_1)$ is the distance between the poles in the laboratory so that

$$\lambda = \frac{\lambda_0}{\gamma}. \tag{3.11}$$

Because γ is always >1, the distance between the poles is shorter (contracted) compared with the distance λ_0 measured in the frame in which the poles are at rest. The same result would be obtained by using the relation between space coordinates in eqns (3.6). In this approach, the coordinates in Σ are related to those in Σ' by

$$x_2 - x_1 = \beta c\gamma(t_2' - t_1'),$$

giving eqn (3.11) as before. We shall make use of these results in Chapter 14, where the production of X-rays from electrons moving through a magnetic undulator gives a precise confirmation of the correctness of this formalism.

The Doppler effect—in the forward direction

The frequency of the radiation emitted by the electron depends on the state of motion of the observer. This is the well-known Doppler effect (Christian Johann Doppler,

1803–53), familiar to all who have noticed that the pitch of the note emitted by an ambulance or a police siren depends on whether the vehicle carrying the siren is moving towards you or away from you. Qualitatively, what is happening is that as the siren moves towards you, the crests of the sound wave take less and less time to reach you as you stand on the pavement listening to the note. In other words, the number of waves per second arriving at your ear is higher than if the source had been at rest. As a consequence, the pitch of the note you hear is higher than that heard by the driver who is moving along with the vehicle carrying the siren. Conversely, if the siren is moving away from you, each successive crest takes a longer time to reach you because the source of the note is moving further away and so the pitch sounds lower to you. The numerical value of the frequency difference between the note you hear and the note the driver hears depends on the speed of the vehicle carrying the siren. The faster the vehicle is moving, the larger is the frequency difference.

Figure 3.3 shows how this works in detail, where we consider an electron which is emitting an electromagnetic wave.

The electron, at rest at the origin of coordinates in Σ', emits the crest of a wave at time t_1' and the next crest at time t_2' so that $t_2' - t_1' = 1/\nu_0$, where ν_0 is the frequency of the radiation emitted as measured by an observer who is also at rest in Σ'. Consider now the frame Σ in which the electron and Σ' are moving away from the observer with velocity βc, as shown in Fig. 3.3. In Σ the emission of the two successive wave crests occurs at times t_1 and t_2 and at positions x_1 and x_2 such that $x_2 - x_1 = \beta c(t_2 - t_1)$. An observer located at the origin of the frame Σ detects these two wave crests at the later times T_1 and T_2 which are related to the times of emission of the crests in Σ by

$$cT_1 = ct_1 + x_1, \qquad cT_2 = ct_2 + x_2$$

and

$$T_2 - T_1 = (t_2 - t_1) + \frac{x_2 - x_1}{c},$$

so that

$$T_2 - T_1 = (t_2 - t_1)(1 + \beta).$$

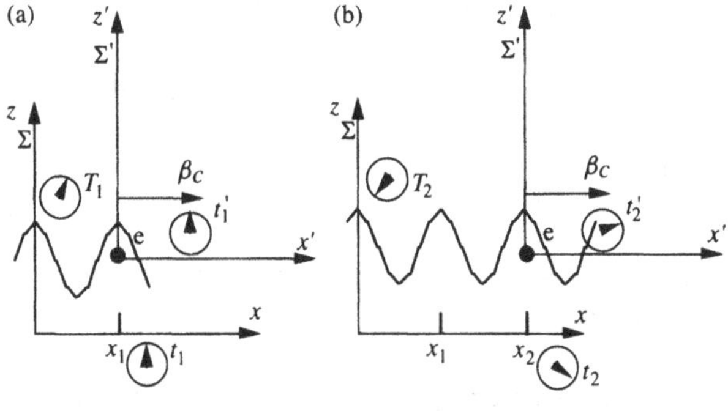

Fig. 3.3 The Doppler effect.

In the non-relativistic case, the time interval $t_2 - t_1$ is identical to the time interval $t_2' - t_1'$ so that, if ν is the frequency of the wave as measured by the observer in Σ, $T_2 - T_1 = 1/\nu$, and

$$\frac{1}{\nu} = \frac{1}{\nu_0}(1+\beta) \quad \text{or} \quad \lambda = \lambda_0(1+\beta). \tag{3.12a}$$

To first order in β

$$\nu \approx \nu_0(1-\beta). \tag{3.12b}$$

Equations (3.12) are the familiar equations for the Doppler frequency shift when the speed of the moving source is much less than the speed of light.

In the relativistic case, when β becomes close to 1, the time intervals $t_2 - t_1$ and $t_2' - t_1'$ are no longer equal but are related by the time dilation equation (3.10) so that

$$\frac{1}{\nu} = \frac{\gamma}{\nu_0}(1+\beta) \quad \text{or} \quad \lambda = \lambda_0\gamma(1+\beta). \tag{3.13}$$

In this case, since, by definition [eqn (3.7)],

$$\gamma = \frac{1}{\sqrt{1-\beta^2}},$$

we can express this as

$$\gamma(1+\beta) = \frac{1}{\gamma(1-\beta)} \tag{3.14}$$

and we can write, from eqn (3.13),

$$\nu = \nu_0\gamma(1-\beta), \tag{3.15}$$

which is an exact expression, in contrast to eqn (3.12).

Because β is almost equal to unity for an electron whose speed is almost equal to the speed of light, the quantity $(1-\beta)$ is difficult to calculate because it is the difference of two nearly equal quantities. However, from eqn (3.14) it is easy to see that $(1-\beta) = 1/2\gamma^2$ to a good approximation, when β is almost equal to 1, so that $\nu = \nu_0/2\gamma$. This expression will be useful later on when we shall show, by combining this result with that of the previous section, that the wavelength, λ, of the radiation from a magnetic undulator with magnetic wavelength λ_0 is given by $\lambda = \lambda_0/2\gamma^2$ in the forward direction.

The Doppler effect—when the source is viewed from an oblique angle

In the treatment of the Doppler effect in the previous section it has been assumed that the observer is looking straight at the source. But what happens when the source is

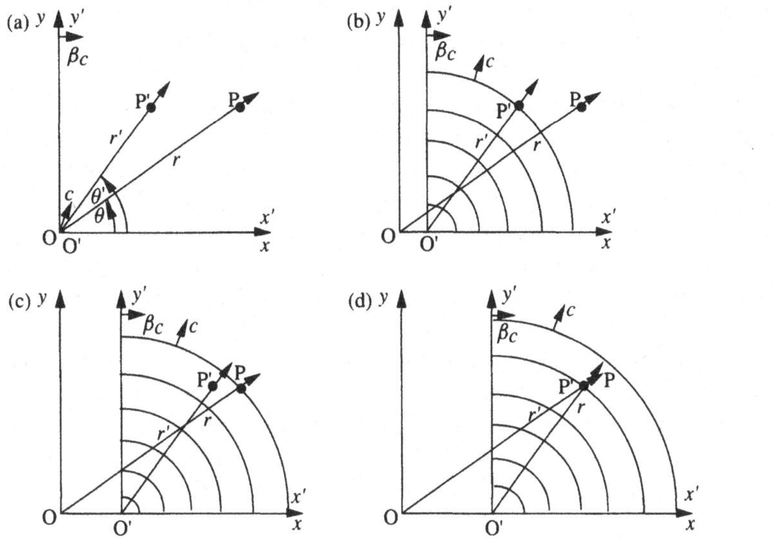

Fig. 3.4 Off-axis Doppler effect: (a) origins coincide at $t_0 = t_0' = 0$; (b) $t_1' = r'/c$ first crest reaches P′; (c) $t_1 = r/c$ first crest reaches P; (d) P′ and P coincide at times t_2' and t, respectively.

being viewed at an oblique angle? Examine Fig. 3.4(a). An observer at a point P′ who is moving along with the source (and for whom, therefore, the source is at rest), views the source at an angle θ' and detects radiation which has a frequency ν'. However, someone viewing the source at an angle θ from the point P, which is at rest in the laboratory detects radiation with frequency ν. From the point of view of the observer at P, the source is moving past P with velocity βc times $\cos\theta$. What is the relationship between ν and ν' in this case?

The answer can be found if each observer measures the time interval during which the same number of waves arrive at their respective observation points. This is the same thing as saying that each observer measures the time taken for the phase of the wave arriving at P or at P′ to change by the same amount. Let us be more precise about this. The amplitude of the wave as observed at P′ is given by $A = A_0 \cos 2\pi(\nu' t' - k' r')$ and the phase ϕ', of this wave is $2\pi(\nu' t' - k' r')$. Here r' is the distance from the source to the observation point which can be written as

$$r' = x' \cos\theta' + y' \sin\theta'. \tag{3.16}$$

It is useful to introduce the velocity of the wave, c, which, in the case of electromagnetic radiation, is the same for both observers, so that k', which is equal to $1/\lambda'$, becomes ν'/c and the phase of the wave is given by

$$\phi' = 2\pi\nu' \left(t' - \frac{r'}{c}\right).$$

The rate of change of phase with time, which is $\mathrm{d}\phi/\mathrm{d}t'$, is equal to $2\pi\nu'$, so that measuring the time taken for a known change of phase will tell us the frequency of the wave.

In Fig. 3.4(a), the origins of the two coordinate frames coincide at $t = t' = 0$ and at this moment the source begins to emit radiation. Neither of the two observers detect any radiation at this moment but some time $t_1' = r'/c$ later the observer at P′ notes that the first wave crest has arrived (shown in the figure as a quadrant of a circle spreading out from O′) and starts measuring the change of phase with time [Fig. 3.4(b)].

The observer at P has still not seen anything but at a time $t_1 = r/c$, that observer also notes the arrival of the first wave crest and starts the phase change measurement [Fig. 3.4(c)].

While all this is going on, P′ moves closer to P and sooner or later, at a time $t_2' = t' + r'/c$ according to P′ and $t_2 = t + r/c$ according to P, the two observers coincide [Fig. 3.4(d)]. At that moment, both observers again measure the phase of the wave. Although the time (t' for P′ and t for P) which has elapsed between the two phase measurements is different, both observers measure the same phase change and we can write

$$\nu\left(\nu - \frac{r}{c}\right) = \nu'\left(t' - \frac{r'}{c}\right)$$

which, from (3.16) can be expressed as

$$\nu(ct - x\cos\theta - y\sin\theta) = \nu'(ct' - x'\cos\theta' - y'\sin\theta'). \tag{3.17}$$

Because points P and P′ coincide in space, their coordinates in the two frames of reference are related through the Lorentz transformation. We can now apply this transformation to the right-hand side of eqn (3.17) so that, with some rearrangement,

$$\begin{aligned}
&\nu'(\gamma(ct - \beta x) - \gamma(x - \beta ct)\cos\theta' - y\sin\theta')\\
&\quad = \nu'(\gamma ct(1 + \beta\cos\theta') - \gamma x(\cos\theta' - \beta) - y\sin\theta')\\
&\quad = \nu(ct - x\cos\theta - y\sin\theta).
\end{aligned} \tag{3.18}$$

Now eqn (3.18) must be true for all pairs of observers looking at the same source so that the coefficients of x', y' and t' must be separately equal to each other. This gives the required relationships [eqns (3.19)] between the frequencies and angles of observation as seen by the two observers:

$$\begin{aligned}
\nu &= \nu_0\gamma(1 + \beta\cos\theta_0),\\
\nu\cos\theta &= \nu_0\gamma(\cos\theta_0 + \beta),\\
\nu\sin\theta &= \nu_0\sin\theta_0,
\end{aligned} \tag{3.19}$$

where, with an obvious change of notation, ν_0 and θ_0 have replaced ν' and θ' for the frequency and emission angle of the photon from the source at rest. We can rearrange these equations to give a relation [eqn (3.20)] between the angles of observation in the two reference frames:

$$\sin\theta = \frac{\sin\theta_0}{\gamma(1+\beta\cos\theta_0)} \quad \text{or} \quad \tan\theta = \frac{\sin\theta_0}{\gamma(\cos\theta_0+\beta)}. \tag{3.20}$$

These equations for the Doppler effect are particularly important because of the relationship $E = h\nu$, between frequency and energy. We can now express eqn (3.19) in terms of energy instead of frequency and obtain the energy of the radiation observed at an angle θ when the source is emitting radiation with energy E_0 at an angle θ_0 in its own rest frame. In particular [from eqn (3.15)], the energy of a photon in the forward direction is given by

$$E = E_0\gamma(1+\beta) = E_0\sqrt{\frac{1+\beta}{1-\beta}} \quad \text{when } \cos\theta = +1 \tag{3.21}$$

and, in the backward direction, by

$$E = E_0\gamma(1-\beta) = E_0\sqrt{\frac{1-\beta}{1+\beta}} \quad \text{when } \cos\theta = -1. \tag{3.22}$$

These results are an extension of our knowledge of the Doppler effect at low velocities. The observed energy of the radiation emitted by a source which is moving towards us at a speed approaching that of light ($\beta \approx 1$) is increased by a factor of approximately 2γ (compared with what would be observed when the source is at rest) and is reduced by the same factor if the source is moving away from us. A surprising result is that even when the source is viewed from the side, so that $\cos\theta = 0$, the observed energy of the emitted radiation is increased by a factor γ. These results are illustrated graphically in Fig. 3.5, in which a photon with energy E_0, frequency ν_0, emitted at an angle θ_0 from an object at rest is observed as a photon with energy E, frequency ν, at an angle θ when the source is moving towards the observer in the laboratory.

A moment's consideration of Fig. 3.5 may cause you to ask the question—where has the additional energy come from? After all, if an observer watching the source sees more energy emitted in the forward direction than it would emit if it were at rest, this extra energy must have come from somewhere. The answer must be that the total energy emitted, in a given time interval, summed over all angles, must be the same whether the source is at rest or whether it is moving. What the motion of the source does is to redistribute the energy while keeping the total amount the same.

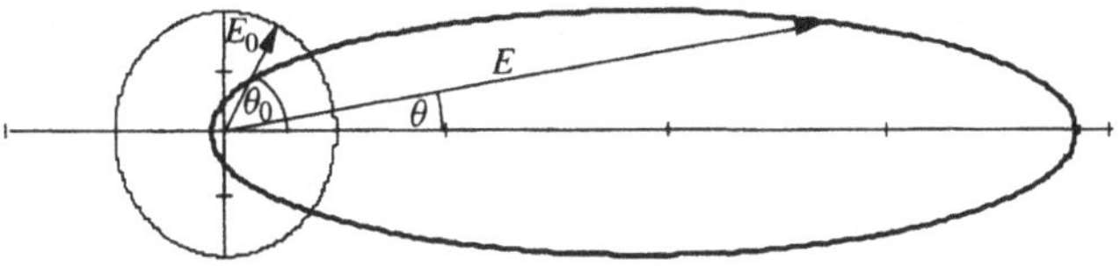

Fig. 3.5 Transformation of photon energy.

Lorentz transformation of four-vectors

The Lorentz transformation tells us how the space and time coordinates change as we go from one frame of reference to another. It also tells us how time intervals and the distances between points change when the transformation is applied. In the Galilean, or non-relativistic, transformation, distances between points did not change from one reference frame to another and a time noted by two observers was always the same regardless of their relative motion. In this new and different way of looking at the world, when the two reference frames are moving relative to each other at a speed approaching the speed of light, distances and time intervals are no longer invariant but are related by the eqns (3.10) and (3.11). However, the equations of the Lorentz transformation are such that the quantity $c^2t^2 - (x^2 + y^2 + z^2)$ is unchanged when the transformation is applied, as can be seen from applying eqns (3.4) and (3.5) to this expression and using the definition which relates γ and β. The term in brackets is just the distance from the point (x, y, z), obtained by applying the Pythagoras theorem to the distances to the origin of the coordinates along the three coordinate axes, x, y, and z. Apart from the minus sign, the expression $c^2t^2 - (x^2 + y^2 + z^2)$ is what we would obtain if, in an analogous way, we applied the Pythagoras theorem in a four-dimensional space in which the fourth dimension is the distance obtained by multiplying the time coordinate by the velocity of light. We can imagine a vector in this four-dimensional space, whose components are ct and x, y and z. We call this vector a four-vector and write it as

$$\boldsymbol{S} = (x_0, \boldsymbol{x}).$$

There are many such four-vectors in relativity theory (just as there are many ordinary, or three-vector quantities). Each four-vector is a combination of a scalar time component (in this case, $x_0 = ct$), and a vector (a three-vector), space component which, in this case, is $\boldsymbol{x} = (x, y, z)$ or (x_1, x_2, x_3). This four-vector, which we can call the four-distance, $\boldsymbol{S}$, has four components each transforming according to the Lorentz transformation. The length $|\boldsymbol{S}|$, of the four-vector is given by the scalar product $\boldsymbol{S} \cdot \boldsymbol{S}$. In particular, for each component,

$$\begin{aligned} x_0 &= \gamma\left(x'_0 - \beta x'_1\right), \\ x_1 &= \gamma\left(x'_1 - \beta x'_0\right), \\ x_2 &= x'_2, \\ x_3 &= x'_3. \end{aligned} \tag{3.23}$$

And, if we form the product $\boldsymbol{S} \cdot \boldsymbol{S}$ by analogy with the scalar product in the Appendix, then

$$\boldsymbol{S} \cdot \boldsymbol{S} = c^2t^2 - \boldsymbol{x} \cdot \boldsymbol{x} = c'^2t'^2 - \boldsymbol{x}' \cdot \boldsymbol{x}'. \tag{3.24}$$

Because the length stays constant from one frame to another even though the components of the four-vector change, the Lorentz transformation can be imagined as a rotation, which changes the direction of the four-vector but preserves its length. This is a general property of all four-vectors, that when they undergo the Lorentz transformation, their length remains unchanged but their four components change in accordance with the transformation equations (3.23).

Transformation of velocities

Imagine now an object which moves a distance $d\boldsymbol{x}$ in a time dt. This object will have a velocity $\boldsymbol{u} = d\boldsymbol{x}/dt$ in some reference frame. In this frame, the object can move under the influence of whatever forces are present so in addition to velocity it can also have acceleration. We deal with velocity first. If there exists a second reference frame moving, as usual, at a uniform velocity v along the x-axis relative to the first, how do components of $\boldsymbol{u}$ transform from one frame to the other? This situation is shown in Fig. 3.6.

In general the motion of the object is described differently in the two reference frames, though of course these two descriptions of position, velocity, and acceleration are related through the Lorentz transformation. There is however, one special reference frame, R, moving with the object, in which, by definition, the object is at rest. In this frame, we will call the time coordinate τ and the time interval $d\tau$, measured by an observer in the rest frame of the object, the proper time and the proper time interval. In this frame the square of the invariant four-distance, $|d\boldsymbol{S}|^2$, is given, as usual, by eqn (3.24) so that

$$d\boldsymbol{S}^2 = c^2 d\tau^2 - |d\boldsymbol{x}|^2. \tag{3.25}$$

In this frame the object is at rest so that $|d\boldsymbol{x}| = 0$ (the four-vector points along the time axis) and the length $|d\boldsymbol{S}|^2$ will be equal to $c^2 d\tau^2$. Since, through eqn (3.25), $|d\boldsymbol{S}|^2$ is the same for all reference frames, so is $c^2 d\tau^2$. Because, by definition and by experiment, the velocity of light, c, is itself a Lorentz invariant, the scalar quantity $d\tau$, the proper time interval measured in the object's rest frame, is also a Lorentz invariant quantity and we can make use of this to obtain the rate of change of the four-vector $\boldsymbol{S}$, $d\boldsymbol{S}/d\tau$, which we can call the four-velocity. How do we do this?

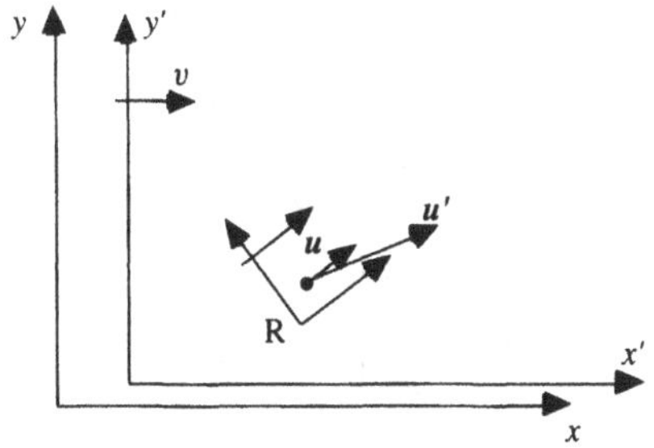

Fig. 3.6 Addition of velocities.

In the rest frame of the object we can write the components of the four-distance d$\boldsymbol{S}$ as $(c\mathrm{d}\tau, 0)$ and in the frame in which the object is moving the corresponding four-vector is $(c\mathrm{d}t, \mathrm{d}\boldsymbol{x})$. The individual components of any four-vector transform according to the Lorentz transformation so that, applying eqn (3.23) to the time component, we have that $c\mathrm{d}t = \gamma_u c\mathrm{d}\tau$, or $\mathrm{d}t/\mathrm{d}\tau = \gamma_u$, which is just the equation for time dilation obtained earlier with $\gamma_u = 1/\sqrt{(1-|\boldsymbol{u}|^2/c^2)}$.

Now if we write $\boldsymbol{S} = (ct, \boldsymbol{x})$ then, taking the first derivative,

$$\frac{\mathrm{d}\boldsymbol{S}}{\mathrm{d}\tau} = \left(\frac{c\mathrm{d}t}{\mathrm{d}\tau}, \frac{\mathrm{d}\boldsymbol{x}}{\mathrm{d}\tau}\right) = \left(\frac{c\mathrm{d}t}{\mathrm{d}\tau}, \frac{\mathrm{d}\boldsymbol{x}}{\mathrm{d}t}\frac{\mathrm{d}t}{\mathrm{d}\tau}\right).$$

Thus, the four-velocity $\boldsymbol{U}$ is given by

$$\boldsymbol{U} = (\gamma_u c, \gamma_u \boldsymbol{u}), \tag{3.26}$$

where we have used the definition of γ_u given above. The 'length' of the four-velocity, $\boldsymbol{U}\cdot\boldsymbol{U}$, or $|\boldsymbol{U}|^2$ is calculated in the same way as the length of the four-distance $|\boldsymbol{S}|^2$ so that

$$\boldsymbol{U}\cdot\boldsymbol{U} = (\gamma_u c, \gamma_u \boldsymbol{u})\cdot(\gamma_u c, \gamma_u \boldsymbol{u}) = \gamma_u^2 c^2 - \gamma_u^2 \boldsymbol{u}\cdot\boldsymbol{u} = c^2$$

from the definition of γ_u. This is what we would expect because U has the dimensions of velocity and c, the velocity of light, is the only invariant quantity with these dimensions.

Now we can find the velocity $\boldsymbol{u}'$ in a reference frame (Fig. 3.6) which is moving with velocity v and the transformation factor is $\gamma = 1/\sqrt{1-(v/c)^2}$ by applying the Lorentz transformation [eqns (3.23)] to the components of the four-velocity $\boldsymbol{U}$. The equations are

$$\gamma_{u'} c = \gamma(\gamma_u c + \beta u_x \gamma_u), \qquad \gamma_{u'} u_{x'} = \gamma(\gamma_u u_x + \beta c \gamma_u)$$

for the components of $\boldsymbol{U}'$ parallel to the x-axis, and

$$\gamma_{u'} u_{y'} = \gamma_u u_y, \qquad \gamma_{u'} u_{z'} = \gamma_u u_z$$

for the components of $\boldsymbol{U}'$ perpendicular to the x-axis. Division of the last three equations by the first one gives the required expressions for each velocity component, namely,

$$u'_x = \frac{u_x + v}{1 + u_x v/c^2}, \tag{3.27a}$$

$$u'_y = \frac{u_y}{\gamma(1 + u_x v/c^2)}, \tag{3.27b}$$

$$u'_z = \frac{u_z}{\gamma(1 + u_x v/c^2)}. \tag{3.27c}$$

When $v \ll c$, $\gamma \approx 1$ and the term $(u_x v/c^2)$ in the denominator of eqns (3.27) is small as well so that, in this case, $u'_x = u_x + v$, which is the simple formula for the addition of velocities. However, in the relativistic case, even if the object in question is moving in a direction perpendicular to the x-axis so that $u_x = 0$, the denominator in (3.27a) is equal to 1 and $u'_x = v$, the perpendicular components of u_x are still reduced by a factor γ which is an unexpected result.

We can use eqns (3.27) to obtain the magnitude and the direction of the velocity in the moving frame. For example, the size of the velocity $|u'|$ is obtained from $|u'|^2 = u_x'^2 + u_y'^2 + u_x'^2$. If we remember that $|u|^2 = u_x^2 + u_y^2 + u_x^2$, then

$$|u'|^2 = \left(\frac{1}{1 + u_x v/c^2}\right)^2 \left((u_x - v)^2 + \frac{1}{\gamma^2}\left(|u|^2 - u_x^2\right)\right). \tag{3.28}$$

Consider the particular case of a photon which is travelling with velocity c in some arbitrary direction as observed by someone who is moving along with the electron which has emitted the photon. That electron is travelling with velocity v in the laboratory. What will an experimenter in the laboratory expect to measure for the velocity of the photon? The electron velocity may itself be close to the speed of light and so, in the non-relativistic case, the photon in the laboratory could be moving with a velocity almost equal to twice the speed of light if the photon were emitted directly forward. Such a result is what we would expect intuitively but we already know from the fundamental rule of relativity that the velocity of light is the same for everybody whether they are moving along with the electron or sitting in the laboratory. What does the theory of relativity predict for $|u'|^2$ in this case? We can obtain the answer to this question by putting $|u|^2 = c^2$ in eqn (3.28). Because we do not know in what direction the photon is going when it is emitted by the electron, we do not know the value of the component of the photon velocity along the x-axis either, so we leave u_x in (3.28) just as it is. After a bit of elementary algebra, and putting $(1/\gamma^2) = 1 - (v/c)^2$, we find that u_x and v drop out completely and we are left with $|u'|^2 = c^2$. In other words, no matter in what direction the photon is going when it emerges from the electron (or from anything else) and no matter what the speed of the electron when it emits the photon, the photon is always observed to travel at the same velocity c, which is the speed of light.

Transformation of accelerations

What about the acceleration of this object? Newton's laws of motion tell us that the object continues for ever with no change to its velocity unless a force is applied to it. If a force is applied to the object, its velocity changes. If the object is being viewed in a reference frame in which it is at rest, the observer will see it starting to move in the direction of the applied force. We can ask the question—what will be the acceleration when the object is viewed from the standpoint of some other frame of reference? We can answer this question quite easily, because, now that we have introduced the idea of the four-velocity, we can say that the four-acceleration is the change of the four-velocity in unit time, so that $\boldsymbol{A} = \mathrm{d}\boldsymbol{U}/\mathrm{d}\tau$ or $\boldsymbol{A} = \mathrm{d}^2\boldsymbol{S}/\mathrm{d}\tau^2$. More explicitly, we can write

$$\frac{\mathrm{d}\boldsymbol{U}}{\mathrm{d}\tau} = \frac{\mathrm{d}}{\mathrm{d}\tau}(\gamma_u c, \gamma_u \boldsymbol{u}) = \left[c\frac{\mathrm{d}\gamma_u}{\mathrm{d}\tau}, \frac{\mathrm{d}(\gamma_u \boldsymbol{u})}{\mathrm{d}\tau}\right].$$

As before, $\mathrm{d}\tau$ is the invariant proper time interval and $\mathrm{d}t/\mathrm{d}\tau = \gamma_u$. If we insert this into the expression above and expand the second term as the derivative of a product, we find that

$$\frac{\mathrm{d}\boldsymbol{U}}{\mathrm{d}\tau} = \left[\gamma_u c\frac{\mathrm{d}\gamma_u}{\mathrm{d}t}, \gamma_u \frac{\mathrm{d}(\gamma_u \boldsymbol{u})}{\mathrm{d}t}\right]$$
$$= \left[\gamma_u c\frac{\mathrm{d}\gamma_u}{\mathrm{d}t}, \gamma_u \left(\gamma_u \frac{\mathrm{d}\boldsymbol{u}}{\mathrm{d}t} + \boldsymbol{u}\frac{\mathrm{d}\gamma_u}{\mathrm{d}t}\right)\right].$$

Because the velocity of the object under investigation is changing, the quantity $\mathrm{d}\gamma_u/\mathrm{d}t$, which occurs in all four components of the four-acceleration is not zero and must be calculated using the definition of γ_u:

$$\frac{\mathrm{d}\gamma_u}{\mathrm{d}t} = \frac{\mathrm{d}}{\mathrm{d}t}\left[\frac{1}{\left(1 - \boldsymbol{u}\cdot\boldsymbol{u}/c^2\right)^{1/2}}\right].$$

Using the rules for the differentiation of a function within a function and remembering that the order of the factors in a scalar product does not matter,

$$\frac{\mathrm{d}\gamma_u}{\mathrm{d}t} = -\frac{1}{2}\frac{1}{\left(1 - \boldsymbol{u}\cdot\boldsymbol{u}/c^2\right)^{3/2}}\frac{\mathrm{d}}{\mathrm{d}t}\left(1 - \frac{\boldsymbol{u}\cdot\boldsymbol{u}}{c^2}\right)$$
$$= -\frac{1}{2}\frac{1}{\left(1 - \boldsymbol{u}\cdot\boldsymbol{u}/c^2\right)^{3/2}}\left(-\frac{2}{c^2}\boldsymbol{u}\cdot\frac{\mathrm{d}\boldsymbol{u}}{\mathrm{d}t}\right)$$

so that

$$\frac{\mathrm{d}\gamma_u}{\mathrm{d}t} = \frac{\gamma_u^3}{c^2}\boldsymbol{u}\cdot\frac{\mathrm{d}\boldsymbol{u}}{\mathrm{d}t}.$$

Now we have all the ingredients to put together the expression for the four-acceleration. To make this appear simpler we substitute the three-vector $\boldsymbol{a}$ for the three-acceleration $\mathrm{d}\boldsymbol{u}/\mathrm{d}t$:

$$\boldsymbol{A} = \left[\frac{\gamma_u^4}{c}\boldsymbol{u}\cdot\boldsymbol{a}, \left(\gamma_u^2\boldsymbol{a} + \frac{\gamma_u^4}{c^2}\boldsymbol{u}(\boldsymbol{u}\cdot\boldsymbol{a})\right)\right]. \tag{3.29}$$

You can see that, as before, the acceleration four-vector combines a scalar quantity and a three-vector and these four components transform according to the Lorentz transformation. We can use the same arguments as we used for the components of the velocity, to determine how the components of the acceleration three-vector change when we switch from one reference frame to another. Equation (3.29) is rather complicated but we shall find that there are really only two particular cases which are important for us, $\boldsymbol{a}$ parallel to $\boldsymbol{u}$ and $\boldsymbol{a}$ perpendicular to $\boldsymbol{u}$. Let us look at these now.

Imagine a reference frame in which the object under consideration is momentarily (or instantaneously) at rest. In the laboratory, this frame is seen to be moving with velocity $\boldsymbol{v}$ parallel to the x-axis. In its own rest frame, the object has an acceleration $\boldsymbol{a}$ parallel to the same x-axis and therefore, in the same direction as $\boldsymbol{v}$. In the object's rest frame, $\gamma_u = 1$ and $\boldsymbol{u} = 0$ so that the components of the four-acceleration can be written from eqn (3.38) as

$$\boldsymbol{A}_{\text{rest}} = (0, a, 0, 0).$$

In the frame in which the object is moving, $\gamma_{u'} = \gamma$, and $\boldsymbol{u} = \beta c$. If we note that β has a component only along the x-axis, so that $\beta \cdot \boldsymbol{a}' = \beta a'_x$, we can write for the components of the four-acceleration in this frame:

$$\boldsymbol{A}_{\text{move}} = \left(\gamma^4 \beta a'_x, \gamma^2 a'_x + \gamma^4 \beta\beta a'_x, \gamma^2 a'_y, \gamma^2 a'_z\right),$$

where primes indicate the components of $\boldsymbol{a}$, viewed in the laboratory. By definition the components of $\boldsymbol{A}$ are related by the Lorentz transformation and $|\boldsymbol{A}_{\text{rest}}| = |\boldsymbol{A}_{\text{move}}|$. If we apply the transformation [eqn (3.23)] to the third and fourth components, we find that

$$\gamma^2 a'_y = 0, \qquad \gamma^2 a'_z = 0,$$

which shows that the transformation does not generate any acceleration perpendicular to the direction of motion, so that $a' = a'_x$, and

$$\gamma^4 \beta a' = \gamma \beta a, \qquad \gamma^2 a' + \gamma^4 \beta^2 a' = \gamma a.$$

Thus, in this case, when the acceleration is parallel to the direction of motion,

$$a' = \frac{a}{\gamma^3}.$$

In the second case, the acceleration is perpendicular to the direction of motion so that the components of the four-acceleration in the rest frame of the object are

$$\boldsymbol{A}_{\text{rest}} = \left(0, 0, a_y, a_z\right)$$

and the components of $\boldsymbol{A}_{\text{move}}$ are as before. Now apply the Lorentz transformation to the first two components so that

$$\gamma^4 \beta a'_x = 0, \qquad \gamma^2 a'_x + \gamma^4 \beta^2 a'_x = 0.$$

It is clear immediately that $a'_x = 0$, so that the transformation does not generate any acceleration component along the direction of motion. If we complete the exercise by writing down the transformation for the remaining two components we find

$$\gamma^2 a'_y = \gamma a_y, \qquad \gamma^2 a'_z = \gamma a_z;$$

so, when the acceleration is perpendicular to the direction of motion, a', which is just $\sqrt{a_x'^2 + a_y'^2 + a_z'^2}$, and similarly for the unprimed quantities, is given by

$$a' = \frac{a}{\gamma}.$$

The properties of the radiation emitted from an accelerating, electrically charged object (such as an electron) depend very strongly on the direction of the acceleration relative to the direction of motion of the object.

Energy and momentum

In classical, non-relativistic mechanics the two quantities energy and momentum are extremely important because the total energy and the total momentum of any closed system remain unchanged so long as the system is not acted upon by an external force. If such a force is applied to the system then the strength of that force is equal to the rate of change of momentum and the direction of the momentum change is the same as the direction of the force. For a single particle we define the momentum as the mass of the object times the speed with which it is moving, or $\boldsymbol{p} = m\boldsymbol{v}$, where, as usual, the use of bold italic type indicates vector quantities. In relativistic mechanics, we can use the definition of four-velocity [eqn (3.35)] to define the Lorentz invariant four-momentum as

$$\mathbf{P} = m_0\mathbf{U} = (\gamma_u m_0 c, \gamma_u m_0 \boldsymbol{u}).$$

The three-vector component of $\boldsymbol{P}$, $\gamma_u m_0 \boldsymbol{u}$ is equal to the three-momentum of the particle $m\boldsymbol{u}$, where the observed mass of the particle depends on the speed at which it is travelling. In general, when $\boldsymbol{v}$ is the velocity of the particle,

$$m = \gamma m_0 = \frac{m_0}{\sqrt{1 - v^2/c^2}}, \qquad \boldsymbol{p} = \gamma m_0 \boldsymbol{v}. \tag{3.30}$$

The scalar component of the four-momentum has the dimension of energy/velocity so that we can equate this to E/c and write

$$\mathbf{P} = \left(\frac{E}{c}, \boldsymbol{p}\right)$$

so that

$$E = \gamma m_0 c^2 = mc^2,$$

with m given by (3.42) as before. In non-relativistic mechanics, the energy of motion, or kinetic energy of a particle of mass m and moving with velocity u, is defined as $\frac{1}{2}mu^2$. Since, from eqn (3.9), when $u \ll c$, $E \approx m_0c^2 + \frac{1}{2}m_0u^2$, the relativistic definition of energy reduces to the classical kinetic energy with an additional term equal to m_0c^2 which acts as an energy baseline. In classical problems, the term m_0c^2 drops out of the equation of energy conservation because the total mass of any closed system is assumed to remain constant. However, it is clear that even when $u = 0$, the relativistic definition of energy demands that any object has a rest energy, E_0, equal to m_0c^2. When an object is moving, its total energy E is the sum of its rest energy and its kinetic energy T, so that $E = E_0 + T$.

Because m_0 is a Lorentz invariant so also is E_0 and the length of the four-momentum vector, which is defined, as usual, by the quantity $|\boldsymbol{P}|^2$, is equal to E_0^2/c^2, so that

$$E_0^2 = m_0^2c^4 = E^2 - p^2c^2. \tag{3.31}$$

It will be useful to notice, from eqn (3.31), that the Lorentz transformation factor and the particle velocity are given, respectively by

$$\gamma = \frac{E}{m_0c^2} \quad \text{and} \quad \frac{v}{c} = \beta = \frac{pc}{E}.$$

In the particular case of a photon, which has zero rest mass, $\beta = 1$, γ is undefined and $E = pc$.

The four-force

We can continue further with the analogy with classical mechanics and define the relativistic four-force, often called the Minkowski four-Force ($\boldsymbol{F}_\mathrm{M}$), after Hermann Minkowski (1864–1909). In Newtonian mechanics force is often defined as mass times acceleration but it is more precise to take the definition as rate of change of momentum with time, which amounts to the same thing, so long as the mass is constant. In the mechanics of Special Relativity, we can define $\boldsymbol{F}_\mathrm{M}$ as $\mathrm{d}\boldsymbol{P}/\mathrm{d}\tau$, so that $\boldsymbol{F}_\mathrm{M}$ is also a four-vector whose length does not change during a Lorentz transformation. We may write

$$\boldsymbol{F}_\mathrm{M} = \left(\frac{1}{c}\frac{\mathrm{d}E}{\mathrm{d}\tau}, \frac{\mathrm{d}\boldsymbol{p}}{\mathrm{d}\tau}\right);$$

as before, $\mathrm{d}t = \gamma\mathrm{d}\tau$, so that

$$\boldsymbol{F}_\mathrm{M} = \left(\frac{1}{c}\frac{\gamma\mathrm{d}E}{\mathrm{d}t}, \frac{\gamma\mathrm{d}\boldsymbol{p}}{\mathrm{d}t}\right). \tag{3.32}$$

However, if $\boldsymbol{F}$ is the three-vector force acting on the object,

$$\boldsymbol{F} = \frac{\mathrm{d}\boldsymbol{p}}{\mathrm{d}t}$$

and, because, the force, applied over a distance ds, changes the energy of motion by an amount dT given by

$$\mathrm{d}T = \boldsymbol{F}\cdot\mathrm{d}s,$$

so that

$$\frac{\mathrm{d}E}{\mathrm{d}t} = \frac{\mathrm{d}}{\mathrm{d}t}(E_0 + T) = \frac{\mathrm{d}T}{\mathrm{d}t} = \boldsymbol{F} \cdot \frac{\mathrm{d}\boldsymbol{s}}{\mathrm{d}t} = \boldsymbol{F} \cdot \boldsymbol{u},$$

it follows that

$$F_{\mathrm{M}} = \left(\frac{\gamma}{c}\boldsymbol{F} \cdot \boldsymbol{u}, \gamma \boldsymbol{F}\right). \tag{3.33}$$

What is the relationship between eqn (3.33) for the four-force and (3.29) for the four-acceleration? We would expect that, for the analogy to be complete, the four-force would be equal to the rest mass times the four-acceleration, i.e.

$$\boldsymbol{F}_{\mathrm{M}} = m_0 \boldsymbol{A}.$$

That this is the case can be seen by calculating the four-acceleration from eqn (3.32):

$$\frac{\mathrm{d}E}{\mathrm{d}t} = \frac{\mathrm{d}}{\mathrm{d}t}(\gamma m_0 c^2) = m_0 \gamma^3 \boldsymbol{u} \cdot \boldsymbol{a}, \tag{3.34}$$

where we have used the expression for $\mathrm{d}\gamma/\mathrm{d}t$ calculated earlier in this chapter. Similarly,

$$\frac{\mathrm{d}\boldsymbol{p}}{\mathrm{d}t} = \frac{\mathrm{d}}{\mathrm{d}t}(\gamma m_0 \boldsymbol{u}) = m_0 \left(\gamma \boldsymbol{a} + \frac{\gamma^3}{c^2}\boldsymbol{u}(\boldsymbol{u} \cdot \boldsymbol{a})\right),$$

so that, by insertion into (3.32):

$$\boldsymbol{F}_{\mathrm{M}} = m_0 \left(\frac{\gamma^4}{c}\boldsymbol{u} \cdot \boldsymbol{a}, \left(\gamma^2 \boldsymbol{a} + \frac{\gamma^4}{c^2}\boldsymbol{u}(\boldsymbol{u} \cdot \boldsymbol{a})\right)\right).$$

Comparison with eqn (3.29) shows instantly that this result is exactly what we would expect to find.

References

1. J. D. Jackson, *Classical electrodynamics* (2nd edn), Chapter 11. John Wiley & Sons, New York, USA (1975).
2. R. G. V. Rosser, *Introductory relativity*, Chapter 3, p. 70. Butterworths, London (1967); D. H. Perkins, *Introduction to high energy physics*, Chapter 5, p. 192. Addison-Wesley, Reading, MA, USA (1972).

4

Radiation from moving electrons

Electromagnetic waves in free space—no electric charges or currents

In Chapter 2 we showed that Maxwell's equations for the electromagnetic field could be rearranged as two equations which described the progress of an electric and a magnetic field throughout the whole of space and time at a speed which experiment showed was equal to the speed of light. We call these two waves electromagnetic radiation. Let us consider their properties.

In free space, where there are no electric charges or currents, we can put ρ and $\boldsymbol{J}$ equal to zero and Maxwell's equations take a very simple form:

$$\begin{aligned}
\nabla \cdot \boldsymbol{E} &= 0, \\
\nabla \cdot \boldsymbol{B} &= 0, \\
\nabla \times \boldsymbol{E} &= -\frac{\partial \boldsymbol{B}}{\partial t}, \\
\nabla \times \boldsymbol{B} &= \mu_0 \varepsilon_0 \frac{\partial \boldsymbol{B}}{\partial t},
\end{aligned} \tag{4.1}$$

and the wave equations are

$$\begin{aligned}
\nabla^2 \boldsymbol{E} - \frac{1}{c^2}\frac{\partial^2 \boldsymbol{E}}{\partial t^2} &= 0, \\
\nabla^2 \boldsymbol{B} - \frac{1}{c^2}\frac{\partial^2 \boldsymbol{B}}{\partial t^2} &= 0,
\end{aligned} \tag{4.2}$$

where $\mu_0 \varepsilon_0$ has been replaced by $1/c^2$.

These two wave equations can be written out in full as

$$\frac{\partial^2 \boldsymbol{E}}{\partial x^2} + \frac{\partial^2 \boldsymbol{E}}{\partial y^2} + \frac{\partial^2 \boldsymbol{E}}{\partial z^2} - \frac{1}{c^2}\frac{\partial^2 \boldsymbol{E}}{\partial t^2} = 0 \tag{4.3}$$

and

$$\frac{\partial^2 \boldsymbol{B}}{\partial x^2} + \frac{\partial^2 \boldsymbol{B}}{\partial y^2} + \frac{\partial^2 \boldsymbol{B}}{\partial z^2} - \frac{1}{c^2}\frac{\partial^2 \boldsymbol{B}}{\partial t^2} = 0. \tag{4.4}$$

There are many possible solutions to these equations which give the values of $\boldsymbol{E}$ and $\boldsymbol{B}$ at a particular point in space and time. The actual solution in any particular problem depends on the conditions which started the fields $\boldsymbol{E}$ and $\boldsymbol{B}$ in the first place. The most general solutions have the form $\boldsymbol{E}(\boldsymbol{x}, t) = f(\boldsymbol{k}\cdot\boldsymbol{x} - \omega t) + g(\boldsymbol{k}\cdot\boldsymbol{x} + \omega t)$ and a similar equation for $\boldsymbol{B}$, in which f and g are any function of $\boldsymbol{x}$ and t, the vector $\boldsymbol{x}$ stands for the space coordinates (x, y, z), and $\boldsymbol{k}$ and ω are constants such that $\omega/k = c$, the velocity of the disturbance through space. The function f represents a disturbance which has some value $f(0)$ at the origin of coordinates and for all points at which $\boldsymbol{k}\cdot\boldsymbol{x} = ct$ so that the wave travels outwards, away from the origin along the positive direction of the coordinate axes so that it gets further from its starting point at later times. The function g represents a disturbance which moves closer to the origin at later times. Because we can use Fourier's theorem to decompose any reasonable function into a sum of sine and cosine functions, we can make use of a solution which is easy to handle mathematically such as

$$\boldsymbol{E}(\boldsymbol{x}, t) = \boldsymbol{E}_0 \exp(\mathrm{i}\boldsymbol{k}\cdot\boldsymbol{x} - \mathrm{i}\omega t) \tag{4.5}$$

and

$$\boldsymbol{B}(\boldsymbol{x}, t) = \boldsymbol{B}_0 \exp(\mathrm{i}\boldsymbol{k}\cdot\boldsymbol{x} - \mathrm{i}\omega t). \tag{4.6}$$

In eqns (4.5) and (4.6), $\boldsymbol{E}_0$ and $\boldsymbol{B}_0$ are constant vectors, independent of $\boldsymbol{x}$ and t, which describe the amplitude of the electric and magnetic waves. When these vectors are multiplied by the scalar exponential function, they give the magnitudes and directions of the vectors $\boldsymbol{E}$ and $\boldsymbol{B}$ at the point in space and time whose coordinates are $\boldsymbol{x}$ and t.

As was explained in Chapter 1, the physical meaning of this solution is obtained by taking the real part of the expression for $\boldsymbol{E}$ or $\boldsymbol{B}$. Equations (4.5) and (4.6) are solutions of (4.3) and (4.4), i.e. the three components of $\boldsymbol{E}$ and the three components of $\boldsymbol{B}$, respectively, satisfy the wave equations provided that $\boldsymbol{k}\cdot\boldsymbol{k} = k^2 = \omega^2/c^2$ so that $k = 2\pi/\lambda$ is the wavenumber and $\omega = 2\pi\nu$ is the frequency of the periodic disturbance of the electric field. The direction of the vector $\boldsymbol{k}$ is the direction of propagation of the wave and the minus sign in the phase factor $(\boldsymbol{k}\cdot\boldsymbol{x} - \omega t)$ indicates that as time increases (gets later), the wave travels towards larger values of the space coordinates.

The solutions of eqns (4.3) and (4.4) for the electric and magnetic field vectors must satisfy the Maxwell equations (4.1), so that, for waves travelling in a region free of charges and currents, both div $\boldsymbol{E}$ and div $\boldsymbol{B}$ must be equal to zero everywhere. Let us calculate div $\boldsymbol{E}$ from eqn (4.3). We use the rule for the differentiation of a product, generalized to the vector calculus:

$$\mathrm{div}(\phi\boldsymbol{A}) = \nabla\cdot(\phi\boldsymbol{A}) = \boldsymbol{A}\cdot\nabla\phi + \phi\nabla\cdot\boldsymbol{A}, \tag{4.7}$$

where ϕ is any scalar function and $\boldsymbol{A}$ is any vector field. In the case of the propagation of the electric field:

$$\nabla\cdot\boldsymbol{E}(\boldsymbol{x}, t) = \nabla\cdot(\exp(\mathrm{i}\boldsymbol{k}\cdot\boldsymbol{x} - \mathrm{i}\omega t)\boldsymbol{E}_0) = 0,$$

which, since $\nabla\cdot\boldsymbol{E}_0 = 0$, reduces to

$$\mathrm{i}\boldsymbol{k}\cdot\boldsymbol{E}(\boldsymbol{x}, t) = 0$$

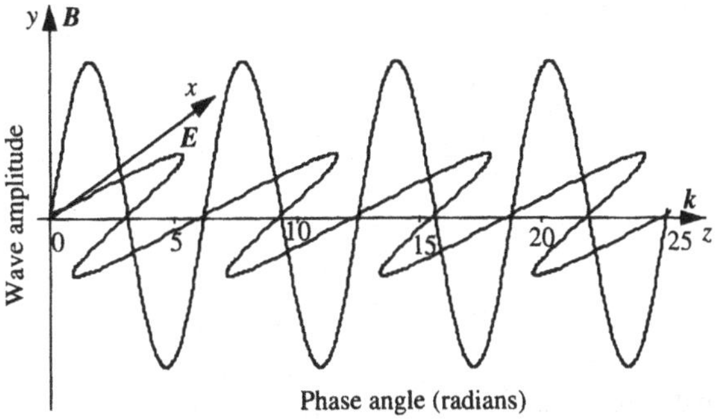

Fig. 4.1 Electromagnetic wave.

so that $\boldsymbol{E}$ must be perpendicular to $\boldsymbol{k}$. In other words, the component of $\boldsymbol{E}$ along the direction of wave propagation defined by $\boldsymbol{k}$ must be zero. A similar argument shows that the magnetic field vector $\boldsymbol{B}$ must be perpendicular to $\boldsymbol{k}$ as well.

The equations for the curl of $\boldsymbol{E}$ and $\boldsymbol{B}$ place further restrictions on these vectors. There is a rule similar to (4.7) for the curl of the product of a scalar function ϕ with a vector function $\boldsymbol{A}$:

$$\text{curl}(\phi\boldsymbol{A}) = \nabla \times (\phi\boldsymbol{A}) = \nabla\phi \times \boldsymbol{A} + \phi\nabla \times \boldsymbol{A}. \tag{4.8}$$

The application of this rule gives

$$\nabla \times \boldsymbol{E} = \text{i}\boldsymbol{k} \times \boldsymbol{E},$$

which, from (4.1), is equal to $-\partial\boldsymbol{B}/\partial t = \text{i}\omega\boldsymbol{B}$, so that

$$\omega\boldsymbol{B} = \boldsymbol{k} \times \boldsymbol{E}, \tag{4.9}$$

which means that the magnetic field at any point along the wave is always pointing at right angles to both the electric field and the direction of propagation. Since $\omega k = 1/c$, eqn (4.9) implies that $B = E/c$. The plane containing the vectors $\boldsymbol{k}$ and $\boldsymbol{E}$ is called the plane of polarization. This is illustrated in Fig. 4.1.

The $\boldsymbol{E}$ and $\boldsymbol{k}$ vectors define the plane of polarization, which is the x–z plane in this case. The $\boldsymbol{B}$ vector is at right angles to the x–z plane, parallel to the y-axis. Equation (4.9) also implies that $\boldsymbol{E}$ and $\boldsymbol{B}$ have the same phase.

Electromagnetic waves produced by currents and charges

In order to proceed further, we need to consider the sources of the electromagnetic waves so that we must return to the full set of Maxwell's equations including charges and currents which we rewrite as

$$\begin{aligned}\nabla \cdot \boldsymbol{E} &= \frac{\rho}{\varepsilon_0},\\ \nabla \cdot \boldsymbol{B} &= 0,\\ \nabla \times \boldsymbol{E} &= -\frac{\partial \boldsymbol{B}}{\partial t},\\ \nabla \times \boldsymbol{B} &= \mu_0\left(\boldsymbol{J} + \varepsilon_0 \frac{\partial \boldsymbol{E}}{\partial t}\right).\end{aligned} \tag{4.10}$$

Consider first what happens if charge is flowing at a uniform rate, so that no radiation is generated. From the first equation, this rate of flow of charge produces a rate of change of the divergence of the electric field given by

$$\frac{\partial \rho}{\partial t} = \varepsilon_0 \frac{\partial}{\partial t}(\nabla \cdot \boldsymbol{E}).$$

We can obtain the rate of change of the divergence of the electric field from the fourth equation as well:

$$\nabla \cdot \nabla \times \boldsymbol{B} = \mu_0 \nabla \cdot \left(\boldsymbol{J} + \varepsilon_0 + \frac{\partial \boldsymbol{E}}{\partial t}\right).$$

The left-hand side of this equation is always zero because it is a scalar triple product containing two identical vectors. Since the vector operation ∇ is time independent, we can combine these two equations to give

$$\frac{\partial \rho}{\partial t} + \nabla \cdot \boldsymbol{J} = 0. \tag{4.11}$$

Equation (4.11) tells us that the rate of change of the charge density at any point in space is equal to the divergence of the electric current density at that point. Suppose we imagine a volume τ surrounding the point in question. We can integrate over this volume to obtain the rate at which the total charge is changing within the volume, so that

$$\frac{\partial}{\partial t}\int \rho \, \mathrm{d}\tau + \int \nabla \cdot \boldsymbol{J} \, \mathrm{d}\tau = 0$$

but, by the divergence theorem, the second integral can be written as the integral of the current density over the surface $\boldsymbol{S}$ which encloses the volume τ:

$$\frac{\partial}{\partial t}\int \rho \, \mathrm{d}\tau + \int \boldsymbol{J} \cdot \mathrm{d}\boldsymbol{S} = 0. \tag{4.12}$$

Equation (4.12) is another statement of the continuity equation and expresses, in a formal way, the experimental observation that electric charge is conserved. It tells us that if the charge inside the volume is increasing, the rate of increase is equal to the inflow of electric current through the surface enclosing the volume.

The process which has just been described implies an energy equation. Suppose the charge is being brought into the volume by the application of an electric field. This field does mechanical work to bring the charge from outside the volume, through the surface to the inside of the volume. Some of this energy is used to generate a magnetic field produced by the moving current. If the charge is being accelerated (as is inevitable when the process starts or finishes), some electromagnetic radiation is generated which could transfer energy (via the changing electric and magnetic fields) to some distant charges. We believe, as a result of experiment, that energy, as well as charge, is conserved during this process so there should exist a continuity equation for energy flow separate from that for charge.

Suppose that U denotes the energy density located at some point in the volume τ. Left to themselves, the charges adjust their position under the influence of the electric and magnetic fields in the volume so that the work done by the fields on the charges (conversion of field energy to energy stored in charge movement) is equal to the reduction in the energy density, U, stored in the fields, plus the field energy radiated through the surface. Let us express this statement as a continuity equation. We remember that the force acting on a charge q, produced by an electric field $\boldsymbol{E}$, is equal to $q\boldsymbol{E}$ and the energy which must be supplied by the field to move the charge through a small distance $\mathrm{d}\boldsymbol{x}$ will be $q\boldsymbol{E}\cdot\mathrm{d}\boldsymbol{x}$. If the field acts over a small time $\mathrm{d}t$ then the rate at which energy is being supplied by the field is $q\boldsymbol{E}\cdot\mathrm{d}\boldsymbol{x}/\mathrm{d}t$ which is $q\boldsymbol{E}\cdot\boldsymbol{v}$. If there are N charges in unit volume, then this expression is equal to $\boldsymbol{E}\cdot\boldsymbol{J}$ because $\boldsymbol{J}=Nq\boldsymbol{v}$. If U is the energy density residing in the field and S is the rate of outward flow of energy across unit surface area, then we can express the energy balance by

$$-\frac{\partial}{\partial t}\left(\int U\cdot\mathrm{d}\tau\right)=\int S\cdot\mathrm{d}\sigma+\int \boldsymbol{E}\cdot\boldsymbol{J}\,\mathrm{d}\tau,$$

where the minus sign on the right-hand side indicates that this term represents the rate of loss of field energy. Application of the divergence theorem to the first term on the right-hand side, to transform it from an integral over a surface to an integral throughout a volume, enables us to write

$$-\frac{\partial U}{\partial t}=\nabla\cdot\boldsymbol{S}+\boldsymbol{E}\cdot\boldsymbol{J}. \tag{4.13}$$

It is perhaps surprising that this equation contains no magnetic field term corresponding to the electric field term $\boldsymbol{E}\cdot\boldsymbol{J}$. The force acting on a charge q, moving with velocity $\boldsymbol{v}$ in a magnetic field $\boldsymbol{B}$ is $q\boldsymbol{v}\times\boldsymbol{B}$ and, by the same argument given above for the electric field, the rate at which energy is supplied to the moving charge by the magnetic field must be $q\boldsymbol{v}\times\boldsymbol{B}\cdot\boldsymbol{v}$ which is zero.

Electromagnetic waves transfer energy—the Poynting vector

The next stage in the argument is to obtain expressions for U and $\boldsymbol{S}$ in terms of $\boldsymbol{E}$ and $\boldsymbol{B}$. We can do this by turning once again to the Maxwell equations and, in particular, to the second pair of equations which relate the spatial variation of $\boldsymbol{B}$ to the rate of change of $\boldsymbol{E}$ with time and the spatial variation of $\boldsymbol{E}$ to the rate of change of $\boldsymbol{B}$ with time. We rewrite these eqns (4.10) below:

$$\nabla \times \boldsymbol{B} - \frac{1}{c^2}\frac{\partial \boldsymbol{E}}{\partial t} = \mu_0 \boldsymbol{J},$$
$$\nabla \times \boldsymbol{E} + \frac{\partial \boldsymbol{B}}{\partial t} = 0.$$

If we take the scalar product of the first of the above equations with $\boldsymbol{E}$ (to obtain an equation for $\boldsymbol{E}\cdot\boldsymbol{J}$) and the second with $\boldsymbol{B}$, take the difference between the two, and group similar terms together, we obtain

$$\boldsymbol{E}\cdot\nabla\times\boldsymbol{B} - \boldsymbol{B}\cdot\nabla\times\boldsymbol{E} - \left(\boldsymbol{E}\cdot\frac{\partial \boldsymbol{B}}{\partial t} + \frac{1}{c^2}\boldsymbol{B}\cdot\frac{\partial \boldsymbol{E}}{\partial t}\right) = \mu_0 \boldsymbol{E}\cdot\boldsymbol{J}.$$

The first term in this equation is equal to $-\nabla\cdot\boldsymbol{E}\times\boldsymbol{B}$ and the second to $+\nabla\cdot\boldsymbol{E}\times\boldsymbol{B}$; we can rearrange the expression further to give

$$-\frac{\partial}{\partial t}\left(\frac{1}{2}\varepsilon_0 E^2 + \frac{1}{2\mu_0}B^2\right) = \nabla\cdot\frac{\boldsymbol{E}\times\boldsymbol{B}}{\mu_0} + \boldsymbol{E}\cdot\boldsymbol{J} \qquad (4.14)$$

to be compared with eqn (4.13) to give the required solution for U and $\boldsymbol{S}$. We can write eqn (4.14) in the form of an integral over the surface area σ, enclosing a volume τ:

$$-\frac{\partial}{\partial t}\int\left(\frac{1}{2}\varepsilon_0 E^2 + \frac{1}{2\mu_0}B^2\right)\mathrm{d}\tau = \int \nabla\cdot\frac{\boldsymbol{E}\times\boldsymbol{B}}{\mu_0}\mathrm{d}\tau + \int \boldsymbol{E}\cdot\boldsymbol{J}\,\mathrm{d}\tau,$$

which, with the application of the divergence theorem to the first term on the right-hand side of the equation, yields

$$-\frac{\partial}{\partial t}\int\left(\frac{1}{2}\varepsilon_0 E^2 + \frac{1}{2\mu_0}B^2\right)\mathrm{d}\tau = \int \frac{\boldsymbol{E}\times\boldsymbol{B}}{\mu_0}\cdot\mathrm{d}\sigma + \int \boldsymbol{E}\cdot\boldsymbol{J}\,\mathrm{d}\tau. \qquad (4.15)$$

Equation (4.15) makes it clear that the vector quantity $(\boldsymbol{E}\times\boldsymbol{B})/\mu_0$ can be equated to the rate of flow across unit area of the surface, and the scalar quantity under the integral on the left-hand side of eqn (4.15) can be regarded as the energy density within the volume enclosed by the surface. In other words,

$$U = \frac{1}{2}\varepsilon_0 E^2 + \frac{1}{2\mu_0}B^2,$$
$$\boldsymbol{S} = \frac{\boldsymbol{E}\times\boldsymbol{B}}{\mu_0} = \varepsilon_0 c^2 \boldsymbol{E}\times\boldsymbol{B}. \qquad (4.16)$$

$\boldsymbol{S}$ is often called the Poynting vector after its discoverer, John Henry Poynting (1852–1914, Professor of Physics at Birmingham University, England).

Let us apply this to the plane electromagnetic waves described earlier in this chapter. We saw [eqn (4.9)] that the electric and magnetic fields at any point along the wave are related by $\omega \boldsymbol{B} = \boldsymbol{k} \times \boldsymbol{E}$. It is convenient to rewrite this expression in the form $\omega \boldsymbol{B} = k\boldsymbol{n} \times \boldsymbol{E}$, where $\boldsymbol{n}$ is a unit vector pointing along the direction of wave propagation. Then, since $k = \omega/c$,

$$c\boldsymbol{B} = \boldsymbol{n} \times \boldsymbol{E}$$

and, because, in turn, $c = 1/\sqrt{\varepsilon_0 \mu_0}$, we can write eqns (4.16) in the form:

$$U = \varepsilon_0 E^2, \qquad \boldsymbol{S} = \varepsilon_0 c E^2 \boldsymbol{n}. \tag{4.17}$$

The electromagnetic wave is transporting an energy density U with velocity c in the direction $\boldsymbol{n}$ along which the wave is being propagated. Reference to Fig. 4.1 shows that at any particular point, the values of U and S fluctuate in time so that an observer at any particular place (e.g. $\boldsymbol{x} = 0$), observes an electric field strength equal to the average of the real part of E, which is $|\mathrm{Re}(E)|$ and a wave energy proportional to the average of $|\mathrm{Re}(E)|^2$. We can calculate this in the usual way, for example,

$$\begin{aligned} U_{\mathrm{rms}} &= \varepsilon_0 \frac{\int_0^{2\pi/\omega} |\mathrm{Re}\, E|^2 \mathrm{d}t}{\int_0^{2\pi/\omega} \mathrm{d}t} = \varepsilon_0 E_0^2 \frac{\int_0^{2\pi/\omega} \cos^2 \omega t \mathrm{d}t}{\int_0^{2\pi/\omega} \mathrm{d}t} \\ &= \tfrac{1}{2} \varepsilon_0 E_0^2 \end{aligned} \tag{4.18}$$

and

$$S_{\mathrm{rms}} = \tfrac{1}{2} \varepsilon_0 c E_0^2. \tag{4.19}$$

In Fig. 4.2, U_{rms} is the energy density in a volume V of length L and cross section A . The total energy in the volume is LU_{rms}. This amount of energy must flow out through the end face of the volume in a time L/c so that the rate of energy flow per unit area through the end face is equal to

$$\frac{VU_{\mathrm{rms}}c}{LA} = cU_{\mathrm{rms}},$$

which is equal to S_{rms} by definition, in agreement with eqns (4.18) and (4.19).

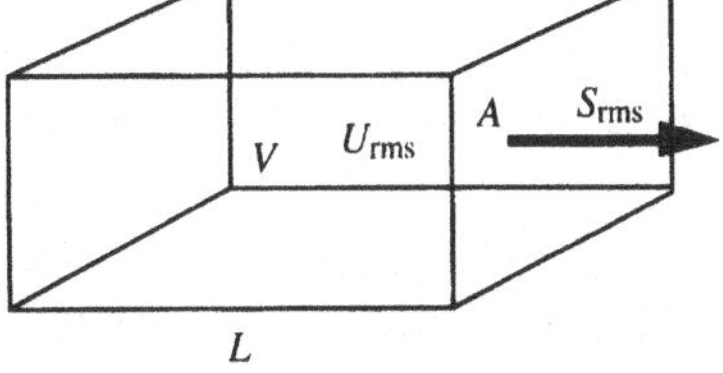

Fig. 4.2 Energy density and flux.

Electromagnetic waves transfer momentum

Besides carrying energy, the EM wave must also carry momentum. If we imagine a charge q within the volume τ, the changing electric and magnetic fields generate a force $\boldsymbol{F}$, on the charge given by the Lorentz equation:

$$\boldsymbol{F} = q(\boldsymbol{E} + \boldsymbol{v} \times \boldsymbol{B}). \tag{4.20}$$

If ρ is the charge density in the volume, $\rho\boldsymbol{v}$ is the current density and the right-hand side of eqn (4.20) can be written as an integral over the volume and set equal to $\mathrm{d}\boldsymbol{G}/\mathrm{d}t$, the rate of change of momentum of the charges in the volume:

$$\frac{\mathrm{d}\boldsymbol{G}}{\mathrm{d}t} = \int (\rho\boldsymbol{E} + \boldsymbol{J} \times \boldsymbol{B})\,\mathrm{d}\tau.$$

We now use the Maxwell equations (4.10)

$$\nabla \cdot \boldsymbol{E} = \frac{\rho}{\varepsilon_0},$$

$$\nabla \times \boldsymbol{B} - \mu_0\varepsilon_0\frac{\partial \boldsymbol{E}}{\partial t} = \mu_0\boldsymbol{J}$$

to eliminate ρ and $\boldsymbol{J}$ so that

$$\frac{\mathrm{d}\boldsymbol{G}}{\mathrm{d}t} = \int \left(\varepsilon_0(\nabla \cdot \boldsymbol{E})\boldsymbol{E} + \frac{(\nabla \times \boldsymbol{B})\boldsymbol{B}}{\mu_0} - \varepsilon_0\left(\frac{\partial \boldsymbol{E}}{\partial t} \times \boldsymbol{B}\right)\right)\mathrm{d}\tau.$$

We would like to eliminate the time derivatives from the right-hand side of this equation so that we can interpret these terms as the rate of change of momentum quantities. To do this, we note that

$$\frac{\partial}{\partial t}(\boldsymbol{E} \times \boldsymbol{B}) = \frac{\partial \boldsymbol{E}}{\partial t} \times \boldsymbol{B} + \boldsymbol{E} \times \frac{\partial \boldsymbol{B}}{\partial t}$$

and, again from Maxwell equation

$$\nabla \times \boldsymbol{E} = -\frac{\partial \boldsymbol{B}}{\partial t}$$

so that, by substitution,

$$\frac{\mathrm{d}\boldsymbol{G}}{\mathrm{d}t} + \frac{\partial}{\partial t}\left(\int \varepsilon_0(\boldsymbol{E} \times \boldsymbol{B})\,\mathrm{d}\tau\right)$$
$$= \int \left(\varepsilon_0(\nabla \cdot \boldsymbol{E})\boldsymbol{E} + \frac{(\nabla \times \boldsymbol{B})\boldsymbol{B}}{\mu_0} + \varepsilon_0(\nabla \times \boldsymbol{E}) \times \boldsymbol{E}\right)\mathrm{d}\tau. \tag{4.21}$$

This equation expresses the conservation of linear momentum carried by the electromagnetic field. The first term on the left-hand side is the rate of change of momentum which the charges within the volume V possess by virtue of their motion. The second term on the left-hand side is the rate of change of the momentum carried by the electromagnetic field. Even if there are no charges within the volume, this quantity need not be zero. The momentum density $\boldsymbol{g}$ is given by

$$\boldsymbol{g} = \varepsilon_0 \boldsymbol{E} \times \boldsymbol{B}. \tag{4.22}$$

The expression under the integral on the right-hand side of eqn (4.21) can be expressed as the divergence of a quantity called the Maxwell stress tensor. The latter can be written as an array of nine quantities corresponding to the flow of momentum, per unit area, transmitted across the surface enclosing the volume V.

The expression [eqn (4.22)] which we have obtained for the momentum density carried by an electromagnetic field is of more than just academic interest. It is obvious, from eqn (4.16), that the definition of $\boldsymbol{g}$ is equivalent to

$$\boldsymbol{g} = \frac{\boldsymbol{S}}{c^2} \quad \text{or} \quad g = \frac{U}{c}. \tag{4.23}$$

This result enables us to make the link between the energy and momentum of a photon, the quantum of electromagnetic radiation. Each photon carries an energy $E = h\nu$, so for the quantum theory to be consistent with the classical theory of electromagnetism, the photon must carry a momentum $p = h\nu/c$, from eqn (4.23), which is equal to h/λ. Furthermore, we know from the theory of relativity that for an object of mass m, its energy and momentum are related through the equation

$$E^2 = p^2c^2 + m^2c^4.$$

In the case of the photon, the mass m is zero so that $E = pc$, which is again consistent with eqn (4.23), just as it should be.

These results also give us a prescription for calculating the number of photons emitted by an accelerating electron (or by any accelerating charge). Equation (4.16) relates the energy flow carried by an electromagnetic wave across unit area per second to the strength of the electric field at the point in question. It follows that if we can calculate this field strength, then we can obtain the energy flow and the photon flux. The calculation of the electric field at a point distant from an accelerating electric charge is not trivial. It is to that calculation we must now turn.

Electromagnetic waves generated by a distant source

The calculation of the electric and magnetic field strengths is easier if we make use of the potential function. We showed in Chapter 2 that in the static case the electric field $\boldsymbol{E}$ could be described by a scalar potential function V with the relationship

$$\boldsymbol{E} = -\nabla V. \tag{4.24}$$

This is possible because, from Maxwell equations,

$$\nabla \times \boldsymbol{E} = -\frac{\partial \boldsymbol{B}}{\partial t} \tag{4.25}$$

which is zero in the static case, when all time derivatives are zero. To see this, replace $\boldsymbol{E}$ with $-\nabla V$ from eqn (4.24) in eqn (4.25). We are left with $-\nabla \times \nabla V$, which is always zero. In the dynamic situation, when the charges and fields are changing with time as well as with position, we can replace the magnetic field $\boldsymbol{B}$ in eqn (4.25) by the curl of a vector potential function $\boldsymbol{A}$,

$$\boldsymbol{B} = \nabla \times \boldsymbol{A} \tag{4.26}$$

and eqn (4.25) becomes

$$\nabla \times \left(\boldsymbol{E} + \frac{\partial \boldsymbol{A}}{\partial t}\right) = 0 \tag{4.27}$$

so that grad V is now equal to the term in brackets in eqn (4.27) or

$$\boldsymbol{E} = -\nabla V - \frac{\partial \boldsymbol{A}}{\partial t}. \tag{4.28}$$

That eqn (4.26) is justified as a prescription for calculating $\boldsymbol{B}$ is clear from the Maxwell equation which states that $\nabla \cdot \boldsymbol{B} = 0$, which is always the case when $\boldsymbol{B} = \nabla \times \boldsymbol{A}$ because $\nabla \cdot \boldsymbol{B} = \nabla \cdot \nabla \times \boldsymbol{A}$ is always zero.

The potential functions which we have chosen satisfy the two homogeneous Maxwell equations from eqns (4.1) (so called because they have zero on the right-hand side and contain only first order derivatives), these are

$$\nabla \cdot \boldsymbol{B} = 0,$$
$$\nabla \times \boldsymbol{E} - \frac{\partial \boldsymbol{B}}{\partial t} = 0,$$

but what effect do they have on the two inhomogeneous eqns (4.10)? We can write these in the form

$$\begin{aligned} \nabla \cdot \boldsymbol{E} &= \frac{\rho}{\varepsilon_0}, \\ \nabla \times \boldsymbol{B} - \frac{1}{c^2}\frac{\partial \boldsymbol{E}}{\partial t} &= \mu_0 \boldsymbol{J}. \end{aligned} \tag{4.29}$$

Let us replace $\boldsymbol{E}$ and $\boldsymbol{B}$ by the potential functions according to the definitions above in eqns (4.26) and (4.27). This gives

$$\nabla^2 V + \frac{\partial}{\partial t}(\nabla \cdot \boldsymbol{A}) = -\frac{\rho}{\varepsilon_0},$$
$$\nabla \times (\nabla \times \boldsymbol{A}) + \frac{1}{c^2}\left(\frac{\partial^2 \boldsymbol{A}}{\partial t^2}\right) + \frac{1}{c^2}\frac{\partial}{\partial t}(\nabla V) = \mu_0 \boldsymbol{J}.$$

These two equations can now be simplified as follows. We know that $\nabla \times \nabla \times \boldsymbol{A} = \nabla\nabla \cdot \boldsymbol{A} - \nabla^2 \boldsymbol{A}$ by the usual rule for a triple vector product so we can insert this in the second equation, and, at the same time, rewrite the first of the above equations to obtain

$$\nabla^2 V - \frac{1}{c^2}\frac{\partial^2 V}{\partial t^2} + \frac{\partial}{\partial t}\left(\nabla \cdot \boldsymbol{A} + \frac{1}{c^2}\frac{\partial V}{\partial t}\right) = -\frac{\rho}{\varepsilon_0},$$

$$\nabla^2 \boldsymbol{A} - \frac{1}{c^2}\frac{\partial^2 \boldsymbol{A}}{\partial t^2} - \nabla\left(\nabla \cdot \boldsymbol{A} + \frac{1}{c^2}\frac{\partial V}{\partial t}\right) = -\mu_0 \boldsymbol{J}.$$

If we now apply to the potential functions the condition that

$$\nabla \cdot \boldsymbol{A} + \frac{1}{c^2}\frac{\partial V}{\partial t} = 0 \tag{4.30}$$

everywhere [this restriction, eqn (4.30), is called the Lorentz condition], we have two separate inhomogeneous equations which relate the potential functions to their charge and current sources:

$$\nabla^2 V - \frac{1}{c^2}\frac{\partial^2 V}{\partial t^2} = -\frac{\rho}{\varepsilon_0}, \tag{4.31a}$$

$$\nabla^2 \boldsymbol{A} - \frac{1}{c^2}\frac{\partial^2 \boldsymbol{A}}{\partial t^2} = -\mu_0 \boldsymbol{J}. \tag{4.31b}$$

It is very convenient to invoke eqn (4.30) because its use decouples the equations and so simplifies the problem of determining V and $\boldsymbol{A}$, but why are we allowed to do this? To answer this question we note first of all that in order to specify the vector potential function $\boldsymbol{A}$ we must specify both the curl and the divergence of $\boldsymbol{A}$. The curl of $\boldsymbol{A}$ is given by eqn (4.26) and the divergence of $\boldsymbol{A}$ by (4.30) so that the use of the Lorentz condition completes the definition of $\boldsymbol{A}$. However, this definition is not unique because, if $\boldsymbol{A}_0$ and V_0 are solutions of eqns (4.31) then so are $\boldsymbol{A}_0 + \nabla\psi$ and $V_0 - \partial\psi/\partial t$, where ψ is any scalar function. This can be seen by inserting these transformed values of the potential functions into eqns (4.26) and (4.28) so that

$$\boldsymbol{B} = \nabla \times \boldsymbol{A} = \nabla \times \boldsymbol{A}_0 + \nabla \times \nabla\psi = \nabla \times \boldsymbol{A}_0$$

because $\nabla \times \nabla = 0$, and

$$\begin{aligned}\boldsymbol{E} = -\nabla V - \frac{\partial \boldsymbol{A}}{\partial t} &= -\nabla V_0 - \nabla\frac{\partial V}{\partial t} - \frac{\partial \boldsymbol{A}_0}{\partial t} + \frac{\partial(\nabla V)}{\partial t}\\ &= -\nabla V_0 - \frac{\partial \boldsymbol{A}_0}{\partial t}\end{aligned}$$

because the gradient operator contains no time dependence. This means that there are a large number of solutions for the vector potential, all of them related by what is called a gauge transformation. We are at liberty to choose a gauge which is convenient for our purpose, in this case, the solution which yields eqns (4.31).

The particular choice we make implies a choice of the scalar function ψ. In this case, when we impose the Lorentz condition on A_0 and V_0, the function ψ must satisfy the homogeneous wave equation

$$\nabla^2\psi - \frac{1}{c^2}\frac{\partial^2\psi}{\partial t^2} = 0$$

because, with the definitions just given,

$$\nabla \cdot \boldsymbol{A} = \nabla \cdot \boldsymbol{A}_0 + \nabla^2\psi,$$
$$\frac{1}{c^2}\frac{\partial\phi}{\partial t} = \frac{1}{c^2}\frac{\partial\phi_0}{\partial t} - \frac{1}{c^2}\frac{\partial^2\psi}{\partial t^2};$$

so, using the Lorentz condition we have

$$\left(\nabla \cdot \boldsymbol{A} + \frac{1}{c^2}\frac{\partial\phi}{\partial t}\right) = \left(\nabla \cdot \boldsymbol{A}_0 + \frac{1}{c^2}\frac{\partial\phi_0}{\partial t}\right) + \left(\nabla^2\psi - \frac{1}{c^2}\frac{\partial^2\psi}{\partial t^2}\right).$$

The transformed potentials satisfy the Lorentz condition provided that in the above equation the function ψ makes the second term on the right-hand side equal to zero so that the gauge transformation does not depend on the charges and currents.

The next stage in the argument is to solve eqns (4.31) to obtain the scalar and vector potential functions which are related to the fields through eqns (4.26) and (4.28), namely,

$$\boldsymbol{B} = \nabla \times \boldsymbol{A}, \qquad \boldsymbol{E} = -\nabla V - \frac{\partial \boldsymbol{A}}{\partial t}.$$

Let us examine eqns (4.31) which we rewrite below:

$$\nabla^2 V - \frac{1}{c^2}\frac{\partial^2 V}{\partial t^2} = -\frac{\rho}{\varepsilon_0},$$
$$\nabla^2 \boldsymbol{A} - \frac{1}{c^2}\frac{\partial^2 \boldsymbol{A}}{\partial t^2} = -\mu_0 \boldsymbol{J}.$$

Both of these equations have the same general form. The left-hand side is the familiar wave equation and, when there is no time dependence, they reduce to those for static electricity and magnetism:

$$\nabla^2 V = -\frac{\rho}{\varepsilon_0}, \qquad \nabla^2 \boldsymbol{A} = -\mu_0 \boldsymbol{J}. \tag{4.32}$$

The first of these is the Poisson equation and relates the electrostatic potential function to the source of the electric field. The solution of this static equation is

$$V(\boldsymbol{r}) = \frac{1}{4\pi\varepsilon_0}\int \frac{\rho(\boldsymbol{r}')}{|\boldsymbol{r} - \boldsymbol{r}'|}\,\mathrm{d}\tau, \tag{4.33}$$

which gives the value of the scalar potential at a point P defined by the vector $\boldsymbol{r}$ which is generated by a charge density $\rho(\boldsymbol{r}')$ at the point $\boldsymbol{r}'$. The integral is carried out over the volume of space, τ, occupied by electric charges as shown in Fig. 4.3.

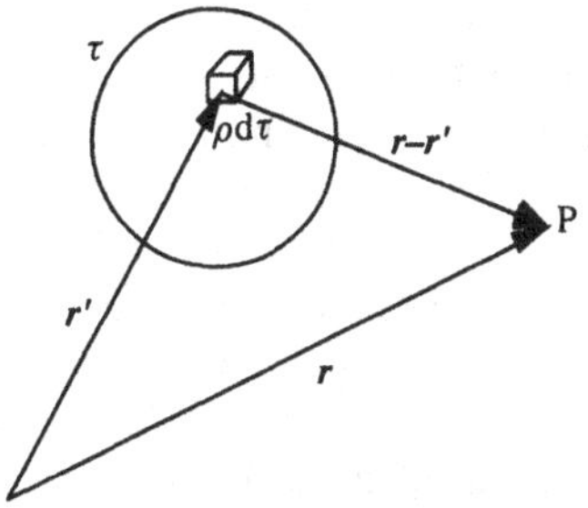

Fig. 4.3 Potential at P.

In free space, where there are neither charges nor currents, eqns (4.31) have the same form as the wave equations (4.2) for the electric and magnetic fields. A disturbance in the electromagnetic field, travelling through space with the velocity of light, can be represented by potential functions V and $\boldsymbol{A}$ moving with the same velocity. When charges and currents are present, these act as generators of the potentials which are now functions of time as well as position; so, by analogy with the solution of Poisson's equation (4.32) we must write

$$V(\boldsymbol{r}, t) = \frac{1}{4\pi\varepsilon_0} \int \frac{\rho(\boldsymbol{r}', t')}{|\boldsymbol{r} - \boldsymbol{r}'|} \, d\tau \tag{4.34}$$

where $t' = t - |\boldsymbol{r} - \boldsymbol{r}'|/c$ is called the retarded time, or the emission time. In other words, in order to calculate the potential at a point P, distance $|\boldsymbol{r} - \boldsymbol{r}'|$ from a charge $\rho\, d\tau$ we must insert into the integral the value of the charge at the retarded time t'. The value of the potential function at P at time t is produced by the charge $\rho\, d\tau$ as it was at the earlier time t'. The time difference $t - t'$ is the time taken for the electromagnetic disturbance to travel to P through the distance $|\boldsymbol{r} - \boldsymbol{r}'|$. In the static case $\rho(\boldsymbol{r}', t') = \rho(\boldsymbol{r}', t)$ so that we get back the solution of Poisson's equation which is eqn (4.33). We also arrive back at eqn (4.33) in the limit when $t' \to t$ which is what we would expect. All this seems very satisfactory. It appears that eqn (4.34) is what we would expect intuitively. An observer at P sees the distribution of charge, not as it is at the time of observation but as it was at an earlier time, defined by the time taken for the signal to reach the point P. For example, we see the stars not as they are now but as they were many thousands or even millions of years ago, at the moment when the light was emitted. The precise time depends on their distance from us.

However, as it sometimes happens, things are not quite what they seem. Why is that so in this case? We know that the solution (4.33) to Poisson's equation generates the inverse square law for the electric field from a spherical charge distribution occupying the volume τ; in particular, for a point charge $\rho\, d\tau = q$ at $\boldsymbol{r}$, the strength of the electric field $\boldsymbol{E}$ at $\boldsymbol{r}'$ is equal to $q/(4\pi\varepsilon_0 r^2)$, where $r = |\boldsymbol{r} - \boldsymbol{r}'|$. Suppose that q is suddenly moved through a distance small compared with r and returned to its former position at rest. The movement of the charge would be observed as a change in the electromagnetic field at P at a time r/c later given by eqns (4.32), which, by analogy with the solution to Poisson's equation, we would expect to be observed as a momentary change in field strength with an amplitude decreasing as $1/r^2$. However, the intensity of the electromagnetic wave generated by the charge movement would be expected to decrease as the inverse fourth power of the distance from the charge, because the intensity (the energy carried by the wave) is proportional to $\boldsymbol{E}^2$. This contradicts observation because we know from experiment that the intensity of an electromagnetic wave from a point source decreases with distance according to the inverse square law. The intuitive argument contradicts simple observation!

This contradiction is resolved later but it is important at this stage to be sure that the intuitive expression [eqn (4.33)] for the potential generated by a charge distribution varying with time is the correct one. Much effort has gone into a rigorous proof of this result using a method of solution first developed by George Green (miller and mathematician, 1793–1841). In outline, the proof proceeds as follows.

We are required to find the potential function $V(\boldsymbol{r}, t)$ for a known charge density distribution $\rho(\boldsymbol{r}, t)$, satisfying the eqn (4.31a). We begin by using the Fourier transformation to express the time-dependent components of eqn (4.31a) as an integral over frequency ω so that, for example,

$$V(\boldsymbol{r}, t) = \sqrt{\frac{1}{2\pi}} \int_{-\infty}^{+\infty} \bar{V}(\boldsymbol{r}, \omega) \exp(-\mathrm{i}\omega t)\, \mathrm{d}t,$$
$$\rho(\boldsymbol{r}, t) = \sqrt{\frac{1}{2\pi}} \int_{-\infty}^{+\infty} \bar{\rho}(\boldsymbol{r}, \omega) \exp(-\mathrm{i}\omega t)\, \mathrm{d}t, \tag{4.35}$$

and the inverse transformations

$$\bar{V}(\boldsymbol{r}, \omega) = \sqrt{\frac{1}{2\pi}} \int_{-\infty}^{+\infty} V(\boldsymbol{r}, t) \exp(\mathrm{i}\omega t)\, \mathrm{d}t,$$
$$\bar{\rho}(\boldsymbol{r}, \omega) = \sqrt{\frac{1}{2\pi}} \int_{-\infty}^{+\infty} \rho(\boldsymbol{r}, t) \exp(\mathrm{i}\omega t)\, \mathrm{d}t,$$

in which $\bar{V}$ and $\bar{\rho}$ denote the Fourier transforms of V and ρ, respectively. When we insert the Fourier transforms for V from eqn (4.35) into eqn (4.31), we obtain, for the terms on the left-hand side of the first equation,

$$\nabla^2 V(\boldsymbol{r}, t) = \sqrt{\frac{1}{2\pi}} \int_{-\infty}^{+\infty} \nabla^2 \bar{V}(\boldsymbol{r}, \omega) \exp(-\mathrm{i}\omega t)\, \mathrm{d}t,$$
$$\frac{\partial^2 V(\boldsymbol{r}, t)}{\partial t^2} = -\sqrt{\frac{1}{2\pi}} \int_{-\infty}^{+\infty} \omega^2 \bar{V}(\boldsymbol{r}, \omega) \exp(-\mathrm{i}\omega t)\, \mathrm{d}t,$$

so that, when we insert the Fourier transform $\bar{\rho}$ we can eliminate the integrals and write, for any value of ω,

$$(\nabla^2 + k^2)\bar{V} = -\frac{\bar{\rho}}{\varepsilon_0}, \tag{4.36}$$

where $k = \omega/c = 2\pi/\lambda$.

If the charge density at $\boldsymbol{r}'$ (see Fig. 4.3) is contained within a small sphere, radius r_0 and $|\boldsymbol{r}-\boldsymbol{r}'| \gg r_0$, then outside the sphere, $\bar{\rho}=0$ and eqn (4.36) reduces to

$$(\nabla^2 + k^2)\bar{V} = 0. \tag{4.37}$$

Equation (4.37), known as the Helmholtz equation, is similar to the Laplace equation but with ∇^2 replaced by $\nabla^2 + k^2$, and its solution must be spherically symmetrical, depending only on $r = |\boldsymbol{r}|$. In spherical coordinates, the r dependence of the Laplacian operator, acting on any function ψ, has the form

$$\nabla^2\psi = \frac{1}{r^2}\frac{\partial}{\partial r}\left(r^2\frac{\partial\psi}{\partial r}\right) = \frac{1}{r}\frac{\partial^2}{\partial r^2}(r\psi) \tag{4.38}$$

so that eqn (4.37) reduces to

$$\frac{1}{r}\frac{\partial^2}{\partial r^2}(r\bar{V}) + k^2\bar{V} = 0.$$

The general solution to this equation is

$$r\bar{V} = A\exp(ikr) + B\exp(-ikr),$$

where A and B are constants whose values depend on the boundary conditions and $r = |\boldsymbol{r}-\boldsymbol{r}'|$. As we shall show very soon, the second term would correspond to an electromagnetic disturbance whose arrival at P would precede the onset of the time variation which caused it, violating causality. It is reasonable to assume therefore that $B = 0$. To obtain the value of A, we examine the first term and see that for $r_0 \to 0$ and $r \gtrsim r_0$, the solution for $\bar{V}$ must lead to eqn (4.33), so $A = \rho/4\pi\varepsilon_0$ and

$$\bar{V}(\boldsymbol{r},\omega) = \frac{\bar{\rho}(\boldsymbol{r}',\omega)\exp(\mathrm{i}k|\boldsymbol{r}-\boldsymbol{r}'|)}{4\pi\varepsilon_0|\boldsymbol{r}-\boldsymbol{r}'|}. \tag{4.39}$$

Now we must take the inverse Fourier transform of eqn (4.39) in order to obtain the function V which is the solution of eqn (4.31) for the scalar potential. We substitute eqn (4.39) into eqn (4.35) and write

$$\begin{aligned}
V(\boldsymbol{r},t) &= \sqrt{\frac{1}{2\pi}}\int_{-\infty}^{+\infty}\frac{\bar{\rho}(\boldsymbol{r}',\omega)}{4\pi\varepsilon_0}\frac{\exp(\mathrm{i}k|\boldsymbol{r}'-\boldsymbol{r}|)}{|\boldsymbol{r}'-\boldsymbol{r}|}\exp(-\mathrm{i}\omega t)\,\mathrm{d}t \\
&= \sqrt{\frac{1}{2\pi}}\int_{-\infty}^{+\infty}\frac{\bar{\rho}(\boldsymbol{r}',\omega)}{4\pi\varepsilon_0}\frac{\exp(-\mathrm{i}\omega(t-|\boldsymbol{r}'-\boldsymbol{r}|/c))}{|\boldsymbol{r}'-\boldsymbol{r}|}\,\mathrm{d}t \\
&= \sqrt{\frac{1}{2\pi}}\int_{-\infty}^{+\infty}\frac{\bar{\rho}(\boldsymbol{r}',\omega)}{4\pi\varepsilon_0}\frac{\exp(-\mathrm{i}\omega t')}{|\boldsymbol{r}'-\boldsymbol{r}|}\,\mathrm{d}t',
\end{aligned}$$

where $t' = t - |\boldsymbol{r}-\boldsymbol{r}'|/c$ is the retarded time, as before. But the expression under the integral sign, multiplied by $\sqrt{1/2\pi}$, is just the Fourier transform of $\rho(\boldsymbol{r}',t')$, identical to eqn (4.34) so that

$$V(\boldsymbol{r},t) = \frac{1}{4\pi\varepsilon_0}\int\frac{\rho(\boldsymbol{r}',t')}{|\boldsymbol{r}-\boldsymbol{r}'|}\,\mathrm{d}\tau,$$

where the integral over the volume τ adds up the contributions to the scalar potential function from all the charges at each separate point.

A similar procedure can be applied to each component of the vector potential function $\boldsymbol{A}$, which is a solution of eqn (4.31b) to show that $\boldsymbol{A}(\boldsymbol{r}, t)$ is given by

$$\boldsymbol{A}(\boldsymbol{r}, t) = \frac{\mu_0}{4\pi} \int \frac{\boldsymbol{J}(\boldsymbol{r}', t')}{|\boldsymbol{r} - \boldsymbol{r}'|} \, \mathrm{d}\tau.$$

These two expressions for V and $\boldsymbol{A}$ are called the retarded potentials and are often written in the form

$$\begin{aligned} V(\boldsymbol{r}, t) &= \frac{1}{4\pi\varepsilon_0} \int \frac{[\rho]}{|\boldsymbol{r} - \boldsymbol{r}'|} \, \mathrm{d}\tau, \\ \boldsymbol{A}(\boldsymbol{r}, t) &= \frac{\mu_0}{4\pi} \int \frac{[\boldsymbol{J}]}{|\boldsymbol{r} - \boldsymbol{r}'|} \, \mathrm{d}\tau, \end{aligned} \tag{4.40}$$

where the square brackets are used to denote quantities which must be calculated not at time t but at the earlier time $t' = t - |\boldsymbol{r} - \boldsymbol{r}'|/c$.

Let us now use these expressions to calculate the potentials at the point P generated by the distribution of moving charges in Fig. (4.3). In order to do this, we must obtain the state of motion of the charge density at the time t' and integrate this over the volume whose charges contribute to the potential functions at P. This must be done carefully because although the observer is at a fixed point in space and time, the charge distribution fills a region of space τ so that each small volume of charge $\rho \, \mathrm{d}\tau$ in the distribution makes its contribution to the potential at P at a different time t'.

The process of adding these contributions can be visualized by allowing an imaginary sphere, centred on P, to contract onto the point P at a speed equal to that of light as shown in Fig. 4.4.

The sphere has the fictitious property that points on its surface can remember all the charges traversed by the surface and retain that memory until the sphere reaches the observer at P. At the moment when the radius of the sphere is equal to $|\boldsymbol{r} - \boldsymbol{r}'|$, those portions of the charge distribution intersecting the surface of the sphere at time t' reach P at a time $t = t' + |\boldsymbol{r} - \boldsymbol{r}'|/c$. The sphere has become a detector of the retarded charge distribution $[\rho]$ which features in the integral of eqns (4.40). The process of integration is the same as allowing the radius of the sphere to start at

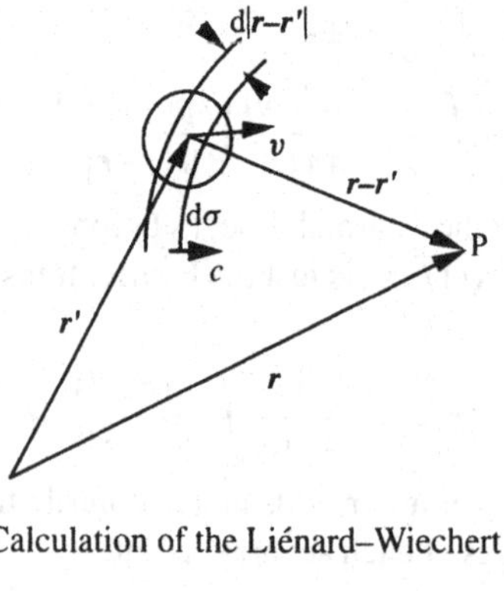

Fig. 4.4 Calculation of the Liénard–Wiechert potentials.

infinity and end at zero where the observer is situated. The elementary volume $\mathrm{d}\tau$ in the integral is equal to $\mathrm{d}\sigma\,\mathrm{d}|\boldsymbol{r}-\boldsymbol{r}'|$ and, if the charge distribution is static, the amount of charge traversed by the sphere $[\rho]\,\mathrm{d}\tau$ will be equal to $[\rho]\,\mathrm{d}\sigma\,\mathrm{d}|\boldsymbol{r}-\boldsymbol{r}'|$. However, if the charge distribution is not static, some of the charge is lost (or gained) by the elementary volume during the time $\mathrm{d}t'$ taken by the surface of the sphere to contract through a distance $\mathrm{d}|\boldsymbol{r}-\boldsymbol{r}'|$. This time is $\mathrm{d}|\boldsymbol{r}-\boldsymbol{r}'|/c$. If $[\boldsymbol{v}]$ is the velocity of the charge in this volume at t' then the amount of charge lost from the volume through the surface $\mathrm{d}\sigma$ during the time $\mathrm{d}t'$ is the amount of charge in the volume times the rate at which charge is flowing across the surface. This last quantity is the component of $[\boldsymbol{v}]$ along the radius of the sphere and pointing towards the centre of the sphere. If $\boldsymbol{n}$ is a unit vector in this direction, then

$$[\boldsymbol{n}] = \frac{[\boldsymbol{r}-\boldsymbol{r}']}{[|\boldsymbol{r}-\boldsymbol{r}'|]},$$

and the amount of charge lost through the surface area $\mathrm{d}\sigma$ is

$$[\rho][\boldsymbol{n}\cdot\boldsymbol{v}]\,\mathrm{d}\sigma\,\mathrm{d}t',$$

which is equal to

$$[\rho][\boldsymbol{n}\cdot\boldsymbol{v}]\,\mathrm{d}\sigma\frac{\mathrm{d}|\boldsymbol{r}-\boldsymbol{r}'|}{c};$$

it follows that the amount of charge collected by the sphere as its radius contracts by an amount $\mathrm{d}|\boldsymbol{r}-\boldsymbol{r}'|$ is equal to the amount present at radius $|\boldsymbol{r}-\boldsymbol{r}'|$ reduced by the amount lost during the time $\mathrm{d}t'$, which is

$$\begin{aligned}\mathrm{d}q &= [\rho]\,\mathrm{d}\sigma\,\mathrm{d}|\boldsymbol{r}-\boldsymbol{r}'| - [\rho][\boldsymbol{n}\cdot\boldsymbol{v}]\,\mathrm{d}\sigma\frac{\mathrm{d}|\boldsymbol{r}-\boldsymbol{r}'|}{c}\\ &= [\rho]\,\mathrm{d}\tau - [\rho]\frac{[\boldsymbol{n}\cdot\boldsymbol{v}]}{c}\,\mathrm{d}\tau.\end{aligned} \tag{4.41}$$

The integral [eqn (4.40)] which we have to determine is

$$\int\frac{[\rho]}{|\boldsymbol{r}-\boldsymbol{r}'|}\,\mathrm{d}\tau$$

and, from eqn (4.41),

$$\int\frac{[\rho]}{|\boldsymbol{r}-\boldsymbol{r}'|}\,\mathrm{d}\tau = \int\frac{\mathrm{d}q}{[|\boldsymbol{r}-\boldsymbol{r}'|]\,[1-[\boldsymbol{n}\cdot\boldsymbol{v}]/c]}. \tag{4.42}$$

In the case when the charge distribution reduces to a point charge at $\boldsymbol{r}'$ then the denominator of the integral on the right-hand side of eqn (4.42) is constant and $\int\mathrm{d}q$

is equal to the value of the charge at $\boldsymbol{r}'$ which is q, so the expressions for the potential functions $V(\boldsymbol{r}, t)$ and $\boldsymbol{A}(\boldsymbol{r}, t)$ from eqns (4.40) are (remembering that $[\boldsymbol{j}] = q[\boldsymbol{v}]$)

$$V(\boldsymbol{r}, t) = \frac{1}{4\pi\varepsilon_0}\left\{\frac{q}{[|\boldsymbol{r} - \boldsymbol{r}'|]\,[1 - [\boldsymbol{n}\cdot\boldsymbol{v}]/c]}\right\}, \tag{4.43a}$$

$$\boldsymbol{A}(\boldsymbol{r}, t) = \frac{\mu_0}{4\pi}\left\{\frac{q[\boldsymbol{v}]}{[|\boldsymbol{r} - \boldsymbol{r}'|]\,[1 - [\boldsymbol{n}\cdot\boldsymbol{v}]/c]}\right\}. \tag{4.43b}$$

Equations (4.43) are expressions for the potential functions of a moving electric charge and are known as the Liénard–Wiechert potentials (A. Liénard, 1869–?; Emil Wiechert, 1861–1928). As before, the square brackets denote retarded quantities.

The next stage in the argument is to use these expressions to obtain the electric and magnetic fields generated by the moving charge by making use of the relationship between the potential functions and the fields which are described by these functions through eqns (4.25) and (4.27) which are

$$\boldsymbol{B} = \nabla \times \boldsymbol{A}, \qquad \boldsymbol{E} = -\nabla V - \frac{\partial \boldsymbol{A}}{\partial t}.$$

To determine the electric field $\boldsymbol{E}$ we must obtain the gradient of V and the time derivative of $\boldsymbol{A}$ at the point P, whose coordinates are $(\boldsymbol{r}, t)$. However, because the potentials of eqns (4.43) contain retarded quantities measured at the point $(\boldsymbol{r}', t')$, we must first obtain expressions which relate changes to $\boldsymbol{r}$ and t to changes in $\boldsymbol{r}'$ and t', where these coordinates are related by

$$t' = t - \frac{|\boldsymbol{r} - \boldsymbol{r}'|}{c}. \tag{4.44}$$

We start by obtaining $\partial \boldsymbol{A}/\partial t$. The use of the partial derivative implies that we are asking what change would be observed in the vector potential function at P, with a fixed position vector $\boldsymbol{r}$, during the infinitesimal time interval ∂t. From the definition of t' [eqn (4.44)], we have

$$\frac{\partial t'}{\partial t} = 1 - \frac{1}{c}\frac{\partial}{\partial t}|\boldsymbol{r} - \boldsymbol{r}'| = 1 - \frac{1}{c}\frac{\partial}{\partial t'}|\boldsymbol{r} - \boldsymbol{r}'|\frac{\partial t'}{\partial t}$$

but

$$\frac{\partial}{\partial t'}|\boldsymbol{r} - \boldsymbol{r}'| = [\boldsymbol{n}\cdot\boldsymbol{v}]$$

[remember, $\boldsymbol{r}$ is fixed so $\partial \boldsymbol{r}/\partial t' = 0$ and the rate of change of the length of the vector linking the retarded position of the charge to the point P (Fig. 4.4) must be equal to the component of the velocity of the charge in the direction of the vector $\boldsymbol{r} - \boldsymbol{r}'$] so that

$$\frac{\partial t'}{\partial t} = \frac{1}{1 - [\boldsymbol{n}\cdot\boldsymbol{v}]/c} \tag{4.45}$$

and the time derivative $\partial \boldsymbol{A}/\partial t$ is given by

$$\frac{\partial \boldsymbol{A}}{\partial t} = \frac{\partial \boldsymbol{A}}{\partial t'}\frac{\partial t'}{\partial t} = \frac{1}{1 - [\boldsymbol{n}\cdot\boldsymbol{v}]/c}\frac{\partial \boldsymbol{A}}{\partial t'}. \tag{4.46}$$

Now, from eqn (4.43b)

$$\frac{\partial A}{\partial t'} = \frac{\mu_0 q}{4\pi}\frac{\partial}{\partial t'}\left[\frac{\boldsymbol{v}}{|\boldsymbol{r}-\boldsymbol{r}'|\,(1-\boldsymbol{n}\cdot\boldsymbol{v}/c)}\right]; \tag{4.47}$$

so, it is useful to obtain separately

$$\begin{aligned}\frac{\partial}{\partial t'}\left[|\boldsymbol{r}-\boldsymbol{r}'|\left(1-\frac{\boldsymbol{n}\cdot\boldsymbol{v}}{c}\right)\right] &= \frac{\partial}{\partial t'}\left[|\boldsymbol{r}-\boldsymbol{r}'| - \frac{\boldsymbol{v}\cdot(\boldsymbol{r}-\boldsymbol{r}')}{c}\right]\\ &= \left[-\boldsymbol{n}\cdot\boldsymbol{v} + \frac{v^2}{c} - \frac{\dot{\boldsymbol{v}}\cdot(\boldsymbol{r}-\boldsymbol{r}')}{c}\right].\end{aligned} \tag{4.48}$$

We can now make use of eqn (4.48) to evaluate eqn (4.47) and obtain $\partial \boldsymbol{A}/\partial t$ with the help of eqn (4.46). The result, after some algebra, is

$$\begin{aligned}\frac{\partial A}{\partial t} &= \frac{\mu_0 q}{4\pi}\left[\frac{1}{(1-\boldsymbol{n}\cdot\boldsymbol{v}/c)^3}\right]\\ &\times\left[\frac{c\boldsymbol{v}\left((\boldsymbol{n}\cdot\boldsymbol{v}/c) - v^2/c^2\right)}{|\boldsymbol{r}-\boldsymbol{r}'|^2} + \frac{\dot{\boldsymbol{v}}\,(1-\boldsymbol{n}\cdot\boldsymbol{v}/c) + \boldsymbol{v}\,(\boldsymbol{n}\cdot\dot{\boldsymbol{v}}/c)}{|\boldsymbol{r}-\boldsymbol{r}'|}\right].\end{aligned} \tag{4.49}$$

In eqn (4.49), and elsewhere, $\dot{\boldsymbol{v}} = \partial\boldsymbol{v}/\partial t' = [\partial\boldsymbol{v}/\partial t]$.

We must now obtain the gradient of the potential function, ∇V or grad$V$, at the retarded time. If we imagine the detection point P to be moved through an infinitesimal distance dr, in some arbitrary direction, so that this movement can be denoted by the vector d$\boldsymbol{r}$, then the gradient of any scalar function ψ in this direction is given, according to the definition of $\nabla\psi$ or grad ψ, by

$$\frac{\mathrm{d}\boldsymbol{r}}{\mathrm{d}|\boldsymbol{r}|}\cdot \mathrm{grad}\,\psi = \frac{\mathrm{d}\boldsymbol{r}}{\mathrm{d}|\boldsymbol{r}|}\cdot\nabla\psi = \frac{\mathrm{d}\psi}{\mathrm{d}|\boldsymbol{r}|}. \tag{4.50}$$

In eqn (4.50), $\mathrm{d}\boldsymbol{r}/\mathrm{d}|\boldsymbol{r}|$ is a unit vector in the direction d$\boldsymbol{r}$ so that $\mathrm{d}\psi/\mathrm{d}|\boldsymbol{r}|$ is the component of grad ψ in this same direction. In our case, the function we must insert for ψ in eqn (4.50) is the same one whose time derivative we obtained previously [eqn (4.48)], only now we need the spatial derivative at fixed time. We are required therefore to find

$$\begin{aligned}\frac{\partial}{\partial|\boldsymbol{r}|}\left[|\boldsymbol{r}-\boldsymbol{r}'|\left(1-\frac{\boldsymbol{n}\cdot\boldsymbol{v}}{c}\right)\right] &= \frac{\partial}{\partial|\boldsymbol{r}|}\left[|\boldsymbol{r}-\boldsymbol{r}'| - \frac{\boldsymbol{v}\cdot(\boldsymbol{r}-\boldsymbol{r}')}{c}\right]\\ &= \frac{\partial}{\partial|\boldsymbol{r}|}\left[|\boldsymbol{R}| - \frac{\boldsymbol{v}\cdot\boldsymbol{R}}{c}\right],\end{aligned} \tag{4.51}$$

where the vector $\boldsymbol{r}-\boldsymbol{r}'$ has been replaced by $\boldsymbol{R}$. The physical basis of this piece of algebra is shown in Fig. 4.5.

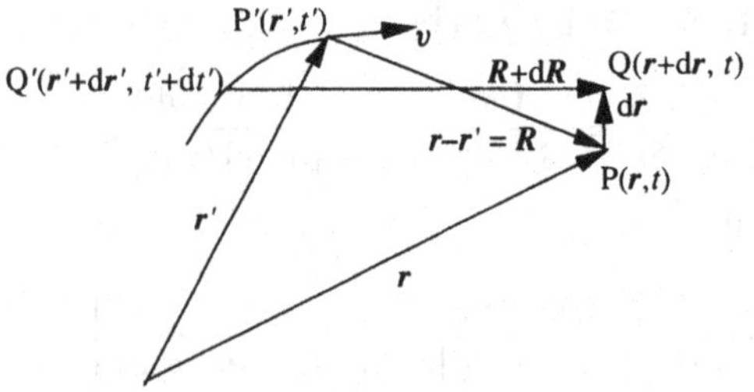

Fig. 4.5 Determination of grad V.

The determination of the gradient implies the comparison of the potential function at points P and Q separated by d$\boldsymbol{r}$, but at the same time t. An instrument at P measures the state of the field which started out from the point P$'$ at position $\boldsymbol{r}'$ and time t', which are the retarded coordinates. An instrument at Q is detecting the field which starts out from Q$'$. Although P and Q have different space coordinates, they have the same time coordinate so that the time difference, dt', between P$'$ and Q$'$ is given by the difference in the path length between PP$'$ and QQ$'$, divided by the velocity of light, so that $\mathrm{d}|\boldsymbol{R}| = -c\,\mathrm{d}t'$. The minus sign comes about because Q$'$ is earlier in time than P$'$. During the time dt', the electric charge, which is generating the fields observed at P and Q, moves from Q$'$ to P$'$, and its position vector changes from $\boldsymbol{r}'$ to $\boldsymbol{r}' + \mathrm{d}\boldsymbol{r}'$ so that the velocity of the charge is just given by $\boldsymbol{v} = \mathrm{d}\boldsymbol{r}'/\mathrm{d}t'$.

We can use these relationships to evaluate the expression in eqn (4.51). Let us look at each component of eqn (4.51) in turn:

$$
\begin{aligned}
\frac{\mathrm{d}|\boldsymbol{R}|}{\mathrm{d}|\boldsymbol{r}|} &= -c\frac{\mathrm{d}t'}{\mathrm{d}|\boldsymbol{r}|},\\
\frac{\mathrm{d}}{\mathrm{d}|\boldsymbol{r}|}\left(\boldsymbol{R}\cdot\frac{\boldsymbol{v}}{c}\right) &= \frac{\mathrm{d}}{\mathrm{d}|\boldsymbol{r}|}(\boldsymbol{r}-\boldsymbol{r}')\cdot\frac{\boldsymbol{v}}{c} + \frac{(\boldsymbol{r}-\boldsymbol{r}')}{c}\cdot\frac{\mathrm{d}\boldsymbol{v}}{\mathrm{d}|\boldsymbol{r}|}\\
&= \left(\frac{\mathrm{d}\boldsymbol{r}}{\mathrm{d}|\boldsymbol{r}|} - \frac{\mathrm{d}\boldsymbol{r}'}{\mathrm{d}t'}\frac{\mathrm{d}t'}{\mathrm{d}|\boldsymbol{r}|}\right)\cdot\frac{\boldsymbol{v}}{c} + (\boldsymbol{r}-\boldsymbol{r}')\cdot\frac{\dot{\boldsymbol{v}}}{c}\frac{\mathrm{d}t'}{\mathrm{d}|\boldsymbol{r}|}\\
&= \left(\frac{\mathrm{d}\boldsymbol{r}}{\mathrm{d}|\boldsymbol{r}|} - \boldsymbol{v}\frac{\mathrm{d}t'}{\mathrm{d}|\boldsymbol{r}|}\right)\cdot\frac{\boldsymbol{v}}{c} + \boldsymbol{R}\cdot\frac{\dot{\boldsymbol{v}}}{c}\frac{\mathrm{d}t'}{\mathrm{d}|\boldsymbol{r}|}.
\end{aligned}
\tag{4.52}
$$

Examination of eqns (4.52) reveals that every term on the right-hand side, except one, contains the factor $\mathrm{d}t'/\mathrm{d}|\boldsymbol{r}|$ which relates a small change in the spatial position of P to a change in the retarded time. If we can express $\mathrm{d}t'/\mathrm{d}|\boldsymbol{r}|$ as a factor multiplying the unit vector $\mathrm{d}\boldsymbol{r}/\mathrm{d}|\boldsymbol{r}|$ then we can solve eqn (4.51).

To do this, we consider the quantity $\mathrm{d}\boldsymbol{R}/\mathrm{d}|\boldsymbol{r}|$ which is equal to $\mathrm{d}(\boldsymbol{r}-\boldsymbol{r}')/\mathrm{d}|\boldsymbol{r}|$:

$$
\begin{aligned}
\frac{\mathrm{d}\boldsymbol{R}}{\mathrm{d}|\boldsymbol{r}|} &= \frac{\mathrm{d}\boldsymbol{R}}{\mathrm{d}t'}\frac{\mathrm{d}t'}{\mathrm{d}|\boldsymbol{r}|} = \frac{\mathrm{d}\boldsymbol{r}}{\mathrm{d}|\boldsymbol{r}|} - \frac{\mathrm{d}\boldsymbol{r}'}{\mathrm{d}|\boldsymbol{r}|}\\
&= \frac{\mathrm{d}\boldsymbol{r}}{\mathrm{d}|\boldsymbol{r}|} - \frac{\mathrm{d}\boldsymbol{r}'}{\mathrm{d}t'}\frac{\mathrm{d}t'}{\mathrm{d}|\boldsymbol{r}|} = \frac{\mathrm{d}\boldsymbol{r}}{\mathrm{d}|\boldsymbol{r}|} - \boldsymbol{v}\frac{\mathrm{d}t'}{\mathrm{d}|\boldsymbol{r}|}.
\end{aligned}
$$

Now form the scalar product of the unit vector $n = (r - r')/|r - r'|$ with each term in eqn (4.61) and remember that $n \cdot dR/dt'$ is just the component of the vector dR/dt' in the direction of R, which is $d|R|/dt' = -c$, so that

$$n \cdot \frac{dR}{dt'} \frac{dt'}{d|r|} = n \cdot \frac{dr}{d|r|} - n \cdot v \frac{dt'}{d|r|}$$

and

$$\frac{dt'}{d|r|} = -\frac{1}{c(1 - n \cdot v/c)} n \cdot \frac{dr}{d|r|} \tag{4.53}$$

which is the relationship we need so that we can bring together the expressions in eqns (4.51) to obtain

$$\frac{dr}{d|r|} \cdot \nabla \left[|r - r'| \left(1 - \frac{n \cdot v}{c} \right) \right]$$

$$= \left[\frac{n \cdot dr/d|r|}{(1 - n \cdot v/c)} \right] - \left[\left(\frac{dr}{d|r|} - \frac{v}{c} \frac{n \cdot dr/d|r|}{(1 - n \cdot v/c)} \right) \cdot \frac{v}{c} \right]$$

$$+ \left[(r - r') \cdot \frac{\dot{v}}{c} \frac{n \cdot dr/d|r|}{c(1 - n \cdot v/c)} \right]$$

so that

$$\nabla \left[|r - r'| \left(1 - \frac{n \cdot v}{c} \right) \right]$$

$$= \left[\frac{1}{c(1 - n \cdot v/c)} \right] \left[\left(1 - \frac{v^2}{c^2} \right) n - c \left(1 - \frac{n \cdot v}{c} \right) \frac{v}{c} + |r - r'| \frac{n \cdot \dot{v}}{c} n \right],$$

where $r - r'$ has been replaced by $|r - r'| n$. A bit more algebra leads to the expression for ∇V from eqn (4.43):

$$\nabla V(r, t) = \frac{q}{4\pi\varepsilon_0} \nabla \left\{ \frac{1}{[|r - r'|][1 - n \cdot v/c]} \right\}$$

$$= -\frac{q}{4\pi\varepsilon_0} \left\{ \frac{\nabla[|r - r'|(1 - n \cdot v/c)]}{[|r - r'|][1 - [n \cdot v]/c]^2} \right\}$$

$$= -\frac{q}{4\pi\varepsilon_0 c} \frac{1}{[1 - [n \cdot v]/c]^3}$$

$$\times \left\{ \left[\frac{c(1 - v^2/c^2) n - c(1 - n \cdot v/c) v/c}{|r - r'|^2} + \frac{(n \cdot \dot{v}/c) n}{|r - r'|} \right] \right\}. \tag{4.54}$$

We note, in passing, that eqn (4.53) is also a relationship between the change in any spatial vector at the retarded time through the definition $v = dr'/dt$, so that

$$\frac{dr'}{d|r|} = \frac{dr'}{dt'} \frac{dt'}{d|r|} = -\frac{v}{c(1 - n \cdot v/c)} n \cdot \frac{dr}{d|r|}.$$

To obtain the expression for the electric field component of the electromagnetic field, we now combine eqns (4.49) and (4.54) in accordance with eqn (4.28) and make use of the fact that $c^2 = 1/(\varepsilon_0\mu_0)$. We make use of the expression for the vector triple product to simplify the second term on the right-hand side and arrive, at last, with

$$\boldsymbol{E}(\boldsymbol{r},t) = \frac{q}{4\pi\varepsilon_0}\frac{1}{[1-\boldsymbol{n}\cdot\boldsymbol{v}/c]^3}\left\{\left[\frac{\left(1-v^2/c^2\right)(\boldsymbol{n}-\boldsymbol{v}/c)}{|\boldsymbol{r}-\boldsymbol{r}'|^2}\right] + \frac{1}{c}\left[\frac{\boldsymbol{n}\times\{(\boldsymbol{n}-\boldsymbol{v}/c)\times\dot{\boldsymbol{v}}/c\}}{|\boldsymbol{r}-\boldsymbol{r}'|}\right]\right\}. \tag{4.55}$$

A similar procedure can be followed to obtain the expression for the magnetic field component. In this case we must obtain curl $\boldsymbol{v}$ or $\nabla\times\boldsymbol{v}$ which is equal to $\nabla\times(\partial\boldsymbol{r}'/\partial t')$. We can obtain this quantity by expanding the curl operator in terms of its three components. It is convenient to use the determinant notation (see Appendix).

$$\nabla\times\boldsymbol{v} = \begin{vmatrix} \boldsymbol{i} & \boldsymbol{j} & \boldsymbol{k} \\ \dfrac{\partial}{\partial x} & \dfrac{\partial}{\partial y} & \dfrac{\partial}{\partial z} \\ v_x & v_y & v_z \end{vmatrix} = \begin{vmatrix} \boldsymbol{i} & \boldsymbol{j} & \boldsymbol{k} \\ \dfrac{\partial}{\partial t'}\dfrac{\partial t'}{\partial x} & \dfrac{\partial}{\partial t'}\dfrac{\partial t'}{\partial y} & \dfrac{\partial}{\partial t'}\dfrac{\partial t'}{\partial z} \\ v_x & v_y & v_z \end{vmatrix}$$

$$= \begin{vmatrix} \boldsymbol{i} & \boldsymbol{j} & \boldsymbol{k} \\ \dfrac{\partial t'}{\partial x} & \dfrac{\partial t'}{\partial y} & \dfrac{\partial t'}{\partial z} \\ \dot{v}_x & \dot{v}_y & \dot{v}_z \end{vmatrix} = (\nabla t')\times\dot{\boldsymbol{v}}. \tag{4.56}$$

Notice that we can use the rules for the evaluation of determinants to factor out the common multiplier $(\partial/\partial t')$ from the second row and apply it to the third row without changing the value of the result.

Next we obtain $\nabla t'$ by using the definition of the gradient operator given in eqn (4.50) to write

$$\frac{\mathrm{d}\boldsymbol{r}}{\mathrm{d}|\boldsymbol{r}|}\cdot\nabla t' = \frac{\mathrm{d}t'}{\mathrm{d}|\boldsymbol{r}|},$$

which, from eqn (4.53) is

$$\frac{\mathrm{d}\boldsymbol{r}}{\mathrm{d}|\boldsymbol{r}|}\cdot\nabla t' = -\frac{1}{c\,(1-\boldsymbol{n}\cdot\boldsymbol{v}/c)}\boldsymbol{n}\cdot\frac{\mathrm{d}\boldsymbol{r}}{\mathrm{d}|\boldsymbol{r}|},$$

so that

$$\nabla t' = -\frac{\boldsymbol{n}}{c\,(1-\boldsymbol{n}\cdot\boldsymbol{v}/c)}$$

and, from (4.56),

$$\nabla\times\boldsymbol{v} = -\frac{\boldsymbol{n}\times\dot{\boldsymbol{v}}}{c\,(1-\boldsymbol{n}\cdot\boldsymbol{v}/c)}. \tag{4.57}$$

We now have the ingredients we need to use eqns (4.26) and (4.43b) for evaluating the magnetic field $\boldsymbol{B}$ generated by the moving charge. We can write

$$\begin{aligned}\boldsymbol{B} &= \nabla \times \boldsymbol{A} = \frac{\mu_0 q}{4\pi}\left[\nabla \times \left(\frac{\boldsymbol{v}}{|\boldsymbol{r}-\boldsymbol{r}'|\,(1-[\boldsymbol{n}\cdot\boldsymbol{v}]/c)}\right)\right] \\ &= \frac{\mu_0 q}{4\pi}\left[\nabla \times \left(\frac{1}{|\boldsymbol{r}-\boldsymbol{r}'|\,(1-[\boldsymbol{n}\cdot\boldsymbol{v}]/c)}\right) + \frac{\nabla \times \boldsymbol{v}}{|\boldsymbol{r}-\boldsymbol{r}'|\,(1-[\boldsymbol{n}\cdot\boldsymbol{v}]/c)}\right].\end{aligned}$$

The first term in the square brackets has already been calculated [eqn (4.54)] and $\nabla \times \boldsymbol{v}$ is given in eqn (4.57) so that working through the algebra yields

$$\begin{aligned}\boldsymbol{B}(\boldsymbol{r},t) &= -\frac{\mu_0 q}{4\pi}\frac{1}{[1-\boldsymbol{n}\cdot\boldsymbol{v}/c]^3} \\ &\times\left\{\left[\frac{\left(1-v^2/c^2\right)(\boldsymbol{n}\times\boldsymbol{v}/c)}{|\boldsymbol{r}-\boldsymbol{r}'|^2}\right]+\frac{1}{c}\left[\frac{\boldsymbol{n}\times\{\boldsymbol{n}\times(\boldsymbol{v}/c\times(\dot{\boldsymbol{v}}/c))\}}{|\boldsymbol{r}-\boldsymbol{r}'|}\right]\right\},\end{aligned} \tag{4.58}$$

if, in the terms in the curly brackets, we replace $\boldsymbol{n} \times (\boldsymbol{v}/c)$ by $-\boldsymbol{n} \times (\boldsymbol{n} - (\boldsymbol{v}/c))$ which are equal because $\boldsymbol{n} \times \boldsymbol{n} = 0$, eqn (4.58) reduces to

$$\boldsymbol{B}(\boldsymbol{r},t) = \frac{1}{c}[\boldsymbol{n} \times \boldsymbol{E}(\boldsymbol{r},t)]. \tag{4.59}$$

As usual in these expressions the square brackets indicate that the quantities within them are to be evaluated at the retarded time t' as defined previously.

It is obvious that the expressions (4.55) and (4.58) for $\boldsymbol{E}$ and $\boldsymbol{B}$ contain two components. The first, known as the near field component, decreases as the inverse square of the distance from the charge and reduces to the electrostatic expression for the field when $\boldsymbol{v} = 0$. The second term, the far field component, decreases linearly with distance and corresponds to the electromagnetic wave whose intensity, proportional to $\boldsymbol{E} \times \boldsymbol{B}$ or E^2, decreases as the inverse square. This expression depends on the acceleration, $\mathrm{d}\boldsymbol{v}/\mathrm{d}t = \dot{\boldsymbol{v}}$, so that, if the acceleration of the charge is zero, then there is no radiation. At large distances, the electric and magnetic fields are dominated by the far component. The near field component can be neglected.

Equation (4.59) indicates that the magnetic field vector, generated by the moving charge, is directed at right angles to the electric vector. This is the general case already noted in connection with the plane wave solutions of Maxwell's equations in the absence of charges and currents.

The conclusion of this chapter is that accelerating electric charges can produce electric and magnetic fields which can be observed at large distances from the point of generation. These fields are coupled together and are detected as electromagnetic radiation. In the next chapter we will show how this radiative process leads to the observed properties of synchrotron radiation.

5

Synchrotron radiation from dipole magnets

Properties of circular motion

The simplest situation in which synchrotron radiation is produced is when an electron (or a positron, which is a positively charged electron) is forced to move in a circular path by the action of a magnetic field which is uniform (has the same value) along the entire path of the electron.

Let us first establish some elementary properties of circular motion. Figure 5.1 shows a right-handed coordinate system in which a small object with position vector $\boldsymbol{R}$ is moving with velocity $\boldsymbol{v}$ whose magnitude is constant and equal to $\mathrm{d}s/\mathrm{d}t$, where $\mathrm{d}s$ is a small distance along the trajectory (see Fig. 5.2).

If the trajectory is part of a circle, so that $|\boldsymbol{R}| = R$ is constant, we can write $\boldsymbol{R} \cdot \boldsymbol{R} =$ constant so that $\boldsymbol{R} \cdot \mathrm{d}\boldsymbol{R}/\mathrm{d}t = 0$. Now $\mathrm{d}\boldsymbol{R}/\mathrm{d}t$ is equal to $(\mathrm{d}\boldsymbol{R}/\mathrm{d}s)(\mathrm{d}s/\mathrm{d}t)$ and $\mathrm{d}\boldsymbol{R}/\mathrm{d}s$ is a unit vector along the tangent to the trajectory at the point $\boldsymbol{R}$ (the unit tangent, see Fig. 5.2) so that $\mathrm{d}\boldsymbol{R}/\mathrm{d}t = \boldsymbol{v}$, $\boldsymbol{R} \cdot \boldsymbol{v} = 0$, and $\boldsymbol{R}$ is always at right angles to $\boldsymbol{v}$. Now the time derivative of $\boldsymbol{R} \cdot \boldsymbol{v} = 0$ is

$$\boldsymbol{R} \cdot \frac{\mathrm{d}\boldsymbol{v}}{\mathrm{d}t} + \boldsymbol{v} \cdot \boldsymbol{v} = 0$$

so that, if $\boldsymbol{f} = \mathrm{d}\boldsymbol{v}/\mathrm{d}t$ denotes the acceleration, then

$$\boldsymbol{R} \cdot \boldsymbol{f} = -\boldsymbol{v} \cdot \boldsymbol{v} = -v^2, \tag{5.1}$$

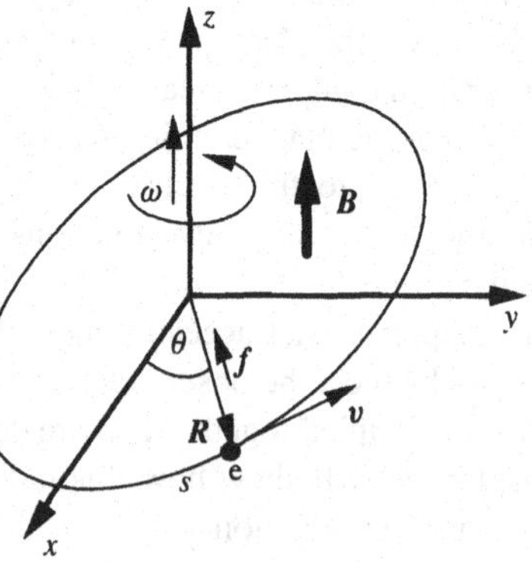

Fig. 5.1 Circular motion.

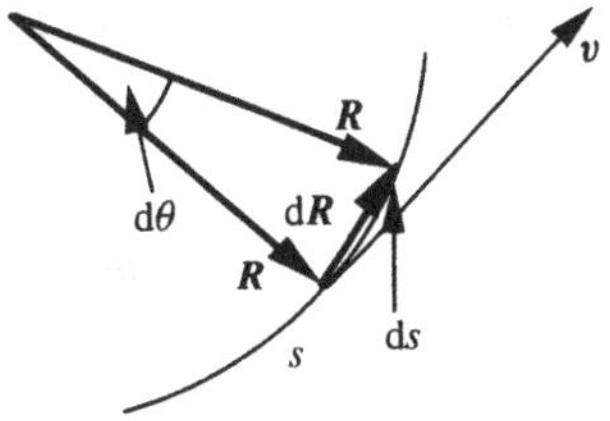

Fig. 5.2 Unit tangent $\mathrm{d}\boldsymbol{R}/\mathrm{d}s$.

and the magnitude of $\boldsymbol{f}$ is $-v^2/R$. What is the direction of $\boldsymbol{f}$? Because $\boldsymbol{v} \cdot \boldsymbol{v} = \text{constant}$, $\boldsymbol{v} \cdot \mathrm{d}\boldsymbol{v}/\mathrm{d}t = \boldsymbol{v} \cdot \boldsymbol{f} = 0$ so that $\boldsymbol{f}$ is always at right angles to $\boldsymbol{v}$. The minus sign indicates that $\boldsymbol{f}$ points in the direction opposite to $\boldsymbol{R}$, i.e. towards the centre of the circular orbit.

If ω is the angular velocity of the object, whose magnitude (equal to $\mathrm{d}\theta/\mathrm{d}t$) and direction are constant, then ω is related to $\boldsymbol{v}$ and $\boldsymbol{R}$ through the vector equation $\boldsymbol{v} = \omega \times \boldsymbol{R}$. The quantity

$$\omega \times \boldsymbol{v} = \omega \times (\omega \times \boldsymbol{R}) = -\omega^2 \boldsymbol{R} = \boldsymbol{f}$$

(by expansion of the triple scalar product, remembering that $\omega \cdot \boldsymbol{R} = 0$ because ω and $\boldsymbol{R}$ are always perpendicular to each other) is another expression for the acceleration vector whose magnitude is $-\omega^2 R$, equal to $-v^2/R$ as before.

Motion of a moving charge in a constant magnetic field

In order to generate this circular trajectory, a force must act on the object to produce a constant acceleration along the radius vector. If the object is an electrically charged particle with charge e, moving in electric and magnetic fields $\boldsymbol{E}$ and $\boldsymbol{B}$ respectively, then the force $\boldsymbol{F}$ which acts on the charge is given by eqn (2.6):

$$\boldsymbol{F} = e(\boldsymbol{E} + \boldsymbol{v} \times \boldsymbol{B}).$$

The equations of motion, which describe how the momentum $\boldsymbol{p}$ and energy U of the particle carrying the charge change with time, are

$$\begin{aligned} \frac{\mathrm{d}\boldsymbol{p}}{\mathrm{d}t} &= e(\boldsymbol{E} + \boldsymbol{v} \times \boldsymbol{B}), \\ \frac{\mathrm{d}U}{\mathrm{d}t} &= e\boldsymbol{v} \cdot (\boldsymbol{E} + \boldsymbol{v} \times \boldsymbol{B}) = e\boldsymbol{v} \cdot \boldsymbol{E}. \end{aligned} \tag{5.2}$$

The constant magnetic field produces a force $\boldsymbol{F}$ acting on the charge which is always at right angles to its direction of motion v, and so cannot change the energy of the charge. When the electric field $\boldsymbol{E}$ is zero the equations of motion reduce to

$$\frac{\mathrm{d}\boldsymbol{p}}{\mathrm{d}t} = e\boldsymbol{v} \times \boldsymbol{B}, \qquad \frac{\mathrm{d}U}{\mathrm{d}t} = 0, \tag{5.3}$$

so that U is constant as expected.

Because we are dealing with a particle which may be moving at a speed close to that of light, we must use the relativistic expressions for the particle's energy and momentum, so that

$$\boldsymbol{p} = \gamma m_0 \boldsymbol{v}, \qquad U = \gamma m_0 c^2, \tag{5.4}$$

and the relationship between momentum and energy,

$$U^2 - p^2c^2 = m_0^2 c^4. \tag{5.5}$$

As usual, m_0 is the mass of the particle, measured in a reference frame in which the particle is at rest and γ is the relativistic Lorentz factor

$$\gamma = \frac{1}{\sqrt{1 - v^2/c^2}}.$$

We can now apply these equations to the motion of the particle in a magnetic field. The quantity $\boldsymbol{v} \times \boldsymbol{B}$ in eqn (5.3) reduces to vB, because the field is everywhere at right angles to the velocity of the particle and $\mathrm{d}\boldsymbol{p}/\mathrm{d}t$ becomes $\gamma m_0 v^2/R$, from eqns (5.1) and (5.4). Equating these quantities gives the relationship between the radius of the trajectory and the momentum of the particle:

$$p = BeR. \tag{5.6}$$

These results can be applied to any electrically charged particle. When the particle is an electron, and when we are dealing with electron energies which are much larger than the electron rest energy so that the term on the right-hand side of eqn (5.5) can be equated to zero, we may also write

$$pc = U = BeRc. \tag{5.7}$$

When the quantities in eqns (5.6) and (5.7) are measured in MKS (metre kilogram second) or SI (Système Internationale) units, the energy of the electron is measured in joules. However, it is useful to measure the energy of the electron in electron volts, where 1 eV is the energy gained by an electron when it is accelerated by a potential difference of 1 V. The corresponding unit for the momentum is eV/c. Because the charge q of the electron is numerically 1.6×10^{-19} C, 1 eV is 1.6×10^{-19} J and 1 MeV is 1.6×10^{-13} J. If we substitute these values, along with the velocity of light, into eqns (5.6) and (5.7), we find that

$$p\,(\mathrm{MeV}/c) = U\,(\mathrm{MeV}) = 300BR\,(\mathrm{T\,m})$$

and, for example, an electron with an energy of 2 GeV, moving in a magnetic field of 1.2 T, would describe a trajectory with a radius of 5.56 m. Because the mass of an electron at rest is 0.511 MeV, it is valid to take the approximation that $pc = U$ in this case.

Radiation of energy by a moving charge in a constant magnetic field

Now let us consider how this electron radiates energy as it moves in a constant magnetic field. The electron is being subjected to an acceleration, $\dot{\boldsymbol{v}}$, of magnitude v^2/R so that a distant observer detects the electric and magnetic fields in the form of an electromagnetic wave [eqns (4.55) and (4.58)] from the radiating electron. The total amount of energy transported by the wave to the observer, per second, per unit area, is given by the value of the Poynting vector $\boldsymbol{S}$ [eqn (4.16)] at the observation point,

$$\boldsymbol{S} = \boldsymbol{E} \times \frac{\boldsymbol{B}}{\mu_0} \quad (\text{W/m}^2),$$

and since [eqn (4.59)]

$$\boldsymbol{B} = \frac{1}{c}\boldsymbol{n} \times \boldsymbol{E},$$

where $\boldsymbol{n}$ is a unit vector along the line from the point at which the radiation is emitted to the observation point, we can write

$$\boldsymbol{S} = \varepsilon_0 c \boldsymbol{E} \times (\boldsymbol{n} \times \boldsymbol{E}) = \varepsilon_0 c\{\boldsymbol{n}E^2 - (\boldsymbol{n} \cdot \boldsymbol{E})\boldsymbol{E}\}.$$

In the far field, from eqn (4.55)

$$\boldsymbol{E}(\boldsymbol{r}, t) = \frac{q}{4\pi\varepsilon_0 c}\left[\frac{\boldsymbol{n} \times \{(\boldsymbol{n} - \boldsymbol{v}/c) \times \dot{\boldsymbol{v}}/c\}}{|\boldsymbol{r} - \boldsymbol{r}'|(1 - (\boldsymbol{n} \cdot \boldsymbol{v})/c)^3}\right], \tag{5.8}$$

so that $\boldsymbol{n} \cdot \boldsymbol{E} = 0$, and

$$\boldsymbol{S} = \varepsilon_0 c E^2 \boldsymbol{n}. \tag{5.9}$$

This was shown to be true for plane electromagnetic waves in eqn (4.18). Here it is seen to be true in general.

Suppose the radiation is being generated by an electron, which, at the moment of emission, is located at the origin of the coordinate system, then we can describe what is happening using spherical coordinates as in Fig. 5.3.

Consider first, the case when the instantaneous velocity of the electron at the origin is very much less than the speed of light. In this case we can forget about the movement of the electron during the time of observation $\mathrm{d}t$, so that the power, $\mathrm{d}U/\mathrm{d}t$, radiated into a solid angle $\mathrm{d}\Omega$ in J/(sr s) is given by

$$\mathrm{d}U/\mathrm{d}t = \boldsymbol{n} \cdot \boldsymbol{S}\,\mathrm{d}A = \varepsilon_0 c E^2 r^2\,\mathrm{d}\Omega\,\boldsymbol{n} \cdot \boldsymbol{n} = \varepsilon_0 c E^2 r^2\,\mathrm{d}\Omega \tag{5.10}$$

from eqn (5.9). In this case, when $|v/c| \ll 1$, and $|\boldsymbol{r} - \boldsymbol{r}'| = r$, eqn (5.8) reduces to

$$\boldsymbol{E}(\boldsymbol{r}, t) = \frac{q}{4\pi\varepsilon_0 c}\left(\frac{\boldsymbol{n} \times (\boldsymbol{n} \times \dot{\boldsymbol{v}}/c)}{r}\right), \tag{5.11}$$

and inserting this into (5.10), we find that

$$\frac{d^2U}{d\Omega\,dt} = \frac{q^2}{(4\pi)^2\varepsilon_0 c^3}|\boldsymbol{n} \times (\boldsymbol{n} \times \dot{\boldsymbol{v}})|^2. \tag{5.12}$$

If θ is the angle between the direction of observation, $\boldsymbol{n}$, and the direction of acceleration, $\dot{\boldsymbol{v}}$, then $\boldsymbol{n} \times \boldsymbol{n} \times \dot{\boldsymbol{v}} = (\boldsymbol{n} \times \boldsymbol{m})\dot{v} \sin\theta$, where $\boldsymbol{m}$ is a unit vector perpendicular to $\boldsymbol{n}$ (and $\dot{\boldsymbol{v}}$) (see Fig. 5.4), so that $|\boldsymbol{n} \times \boldsymbol{m}|^2 = 1$ and

$$\frac{d^2U}{d\Omega\,dt} = \frac{q^2}{(4\pi)^2\varepsilon_0 c^3}\dot{v}^2 \sin^2\theta. \tag{5.13}$$

This angular distribution is a maximum in the plane defined by the tangent to the electron orbit and the normal to the plane of the orbit. In this plane $\sin^2\theta$ has its maximum value of 1. Figure 5.4 shows the shape of this angular distribution.

The electric vector, $\boldsymbol{E}$, of the radiation points in a direction perpendicular to both $\boldsymbol{n}$ and $\boldsymbol{m}$ [from eqn (5.12)] and is therefore plane polarized with the electric vector

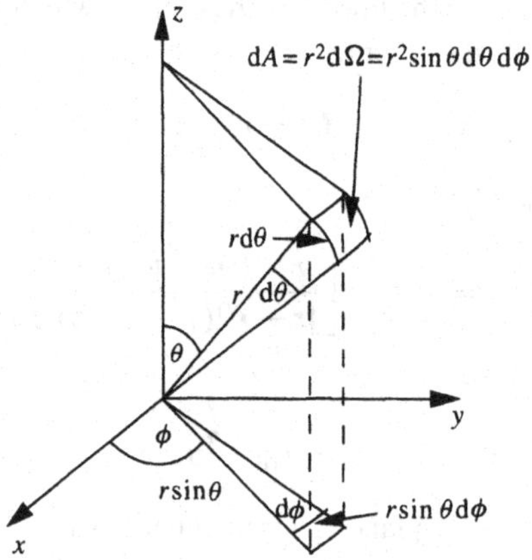

Fig. 5.3 Spherical coordinate system.

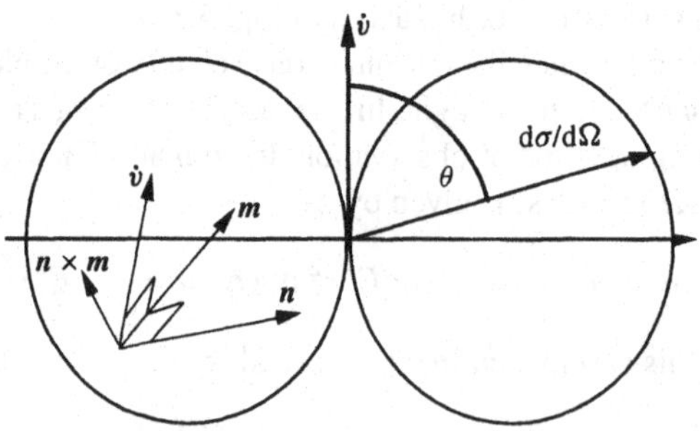

Fig. 5.4 The $\sin^2\theta$ radiation distribution.

lying in the plane containing both $\boldsymbol{n}$ and $\dot{\boldsymbol{v}}$. The directions of these vectors are shown, as an inset, in Fig. 5.4. An observer looking along the direction of acceleration of the electron ($\theta = 0$) sees no radiation. An observer looking at right angles to this direction detects maximum radiated energy. Moreover, this radiation is polarized parallel to the direction of $\dot{\boldsymbol{v}}$. The ϕ distribution of the radiation can be visualized in Fig. 5.4 by rotation of the $\sin^2\theta$ around the $\dot{\boldsymbol{v}}$ axis to form a doughtnut-shaped surface in three dimensions.

Equation (5.13) can be integrated over the whole of the 4π solid angle to give the total amount of power radiated by a non-relativistic charge in unit time. The element of solid angle $\mathrm{d}\Omega = \sin\theta\,\mathrm{d}\theta\,\mathrm{d}\phi$ so that, by symmetry

$$\frac{\mathrm{d}U}{\mathrm{d}t} = \frac{8q^2\dot{v}^2}{(4\pi)^2\varepsilon_0 c^3}\int_0^{\pi/2}\sin^2\theta\sin\theta\,\mathrm{d}\theta\int_0^{\pi/2}\mathrm{d}\phi$$

and, after carrying out the integration

$$\frac{\mathrm{d}U}{\mathrm{d}t} = \frac{2}{3}\frac{q^2\dot{v}^2}{4\pi\varepsilon_0 c^3}\ \text{(W)}, \tag{5.14}$$

which is often called the Larmor formula after Sir Joseph Larmor (1857–1942).

In this case, the electron is moving in a circular orbit, with constant angular frequency $\nu = \omega/2\pi$ and radius R, so that $|\dot{\boldsymbol{v}}| = R\omega^2$ and $\mathrm{d}U/\mathrm{d}t$ is then given by

$$\frac{\mathrm{d}U}{\mathrm{d}t} = \frac{2}{3}\frac{q^2R^2\omega^4}{4\pi\varepsilon_0 c^3}. \tag{5.15}$$

We must now generalize this result to include all electrons, whatever their velocities. We can do this either by returning to the full expression for $\boldsymbol{E}(\boldsymbol{r}, t)$ [eqn (5.8)] or by expressing eqn (5.14) in a Lorentz invariant form. It is instructive to follow this latter route first. The quantity $\mathrm{d}U/\mathrm{d}t$ in the Larmor formula [eqn (5.14)] is a Lorentz invariant quantity even though the expression in eqn (5.15) is only a non-relativistic approximation. In other words, the rate at which energy is radiated is independent of the reference frame of the observer. This is by no means obvious, because an amount of energy, ΔU radiated in a time interval Δt in one reference frame will become an amount $\Delta U'$, radiated in a time $\gamma\Delta t$ in a second frame related to it by the Lorentz factor γ. The question is whether $\Delta U' = \gamma\Delta U$. An indication of the proof that this assertion is correct is given by Jackson.[1] We will assume it to be so.

With this assumption we can express the left-hand side as a Lorentz invariant by substituting the four-acceleration $\boldsymbol{A}$ for the three-acceleration $\dot{\boldsymbol{v}}$, so that

$$\frac{\mathrm{d}U}{\mathrm{d}t} = -\frac{2}{3}\frac{q^2}{4\pi\varepsilon_0 c^3}|A|^2. \tag{5.16}$$

The minus sign is needed to make $\mathrm{d}U/\mathrm{d}t$ a rate of energy loss. In a rigorous treatment eqn (5.16), with the minus sign, comes from the Lorentz invariant energy–momentum tensor.

Consider now the moment when the electron is instantaneously at rest from the point of view of the observer so that the acceleration is at right angles to the direction of motion of the electron along the x-axis. The components of the acceleration four-vector $\boldsymbol{A}$ are obtained from eqn (3.29) which, since $\boldsymbol{v} \cdot \dot{\boldsymbol{v}} = 0$, gives

$$\boldsymbol{A} = (0, 0, \gamma^2 \dot{\boldsymbol{v}}, 0)$$

and

$$\frac{\mathrm{d}U}{\mathrm{d}t} = \frac{2}{3}\frac{q^2}{4\pi\varepsilon_0 c^3}\gamma^4 \dot{v}^2. \tag{5.17}$$

Now, since the energy and the magnitude of the momentum of the electron are unchanged by the magnetic field, we can write, from the definition of momentum in eqn (5.4),

$$\dot{\boldsymbol{v}} = \frac{1}{\gamma m_0}\frac{\mathrm{d}\boldsymbol{p}}{\mathrm{d}t} \tag{5.18}$$

and $\mathrm{d}\boldsymbol{p}/\mathrm{d}t = \boldsymbol{p}\,\mathrm{d}\theta/\mathrm{d}t = \boldsymbol{p}\omega$ so that $|\mathrm{d}\boldsymbol{p}/\mathrm{d}t| = pv/R$. Substitution of this expression into eqn (5.18) and inserting eqn (5.18) into eqn (5.17) gives the upgraded Larmor formula which is valid for electrons of any energy:

$$\frac{\mathrm{d}U}{\mathrm{d}t} = \frac{2}{3}\frac{q^2}{4\pi\varepsilon_0 c^3}\frac{\gamma^2 p^2 v^2}{m_0^2 c^2 R^2}. \tag{5.19}$$

As we would expect, in the low energy limit, when $\gamma = 1$, eqn (5.19) reduces to (5.14). In the high energy limit, $E \gg m_0 c^2$, $p^2c^2 = E^2$ and

$$\frac{\mathrm{d}U}{\mathrm{d}t} = \frac{2}{3}\frac{q^2 c}{4\pi\varepsilon_0}\frac{\gamma^4}{R^2} = \frac{2}{3}\frac{q^2 c}{4\pi\varepsilon_0}\frac{E^4}{(m_0c^2)^4 R^2}. \tag{5.20}$$

One particularly significant feature of eqn (5.20) is that the total power radiated by the electron in unit time is proportional to the fourth power of the electron energy and inversely proportional to the fourth power of the electron (or positron) mass. An important consequence of this power law is that charged particles heavier than the electron are very poor radiators by comparison. For the same energy and orbit radius, a proton radiates energy at a rate reduced by a factor $(1/1836)^4$, compared with the electron.

Suppose that the electron is being accelerated in the direction of its motion, for example, in a linear accelerator or in the radio-frequency cavity of a circular accelerator. In this case, $\dot{\boldsymbol{v}}$ and $\boldsymbol{v}$ are parallel, along the x-axis and, from eqn (3.29) as before, with $|\boldsymbol{v}| = \beta c$,

$$\begin{aligned}\boldsymbol{A} &= \left(\gamma^4 \beta \dot{v}, \left(\gamma^2 \dot{v} + \gamma^4 \beta^2 \dot{v}\right), 0, 0\right) \\ &= \left(\gamma^4 \beta \dot{v}, \gamma^4 \dot{v}, 0, 0\right)\end{aligned}$$

since, from the definition of γ, $1+\gamma^2\beta^2 = \gamma^2$. We now evaluate $|\boldsymbol{A}|^2$, and remember that $|\boldsymbol{A}|^2 = \boldsymbol{A} \cdot \boldsymbol{A} = (A_0^2 - A \cdot A)$ and insert the result into eqn (5.16) to give

$$\frac{\mathrm{d}U}{\mathrm{d}t} = \frac{2}{3}\frac{q^2}{4\pi\varepsilon_0 c^3}\gamma^6\dot{v}^2. \tag{5.21}$$

At first sight, it appears that the rate of loss of energy in this case would be much greater than when the acceleration is perpendicular to the direction of motion, but in fact this is not so, because, from eqn (3.34),

$$\frac{\mathrm{d}E}{\mathrm{d}t} = m_0\gamma^3 v\dot{v}$$

so that

$$\gamma^3\dot{v} = \frac{1}{m_0}\frac{\mathrm{d}E}{\mathrm{d}x}$$

and

$$\frac{\mathrm{d}U}{\mathrm{d}t} = \frac{2}{3}\frac{q^2}{4\pi\varepsilon_0 c^3}\frac{1}{m_0^2}\left(\frac{\mathrm{d}E}{\mathrm{d}x}\right)^2 \tag{5.22}$$

which is independent of the energy of the electron so that there is no energy limit imposed by the need to make up for the synchrotron radiation energy loss. Of course, $\mathrm{d}E/\mathrm{d}x$, the energy gain per unit distance, is just the accelerating force acting on the electron in the direction of its motion. What electric field would be required to provide a force which would give the same loss of radiated power? From eqns (5.20) and (5.22), the condition for equal radiated power loss in the two cases will be

$$\frac{\mathrm{d}E}{\mathrm{d}x} = \frac{E^2}{m_0 c^2}\frac{1}{R}.$$

For example, the Daresbury SRS has $E = 2\,\mathrm{GeV}$ and $R = 5.57\,\mathrm{m}$, so, with $m_0c^2 = 0.511\,\mathrm{MeV}$, $\mathrm{d}E/\mathrm{d}x = 1.4 \times 10^6\,\mathrm{MeV/m}$. In order to provide such a force, a field gradient of 1400 MV/mm would be required, which is about 10^5 times that which can be reached in practice.

We can also examine the general case in the following way. The acceleration four-vector $\boldsymbol{A}$ is given by eqn (3.29) which we can write, with $\boldsymbol{v} = \beta c$ for convenience:

$$\boldsymbol{A} = \left(\gamma^4\beta \cdot \dot{\boldsymbol{v}}, (\gamma^2\dot{\boldsymbol{v}} + \gamma^4\beta(\beta \cdot \dot{\boldsymbol{v}}))\right).$$

From this we can form the length of this four-vector:

$$|\boldsymbol{A}|^2 = \gamma^8(\beta \cdot \dot{\boldsymbol{v}})^2 - (\gamma^2\dot{\boldsymbol{v}} + \gamma^4\beta(\beta \cdot \dot{\boldsymbol{v}}))^2.$$

After some elementary algebra and the use of the vector identity for the square of a vector product which in this case is

$$\beta^2\dot{\boldsymbol{v}}^2 - (\beta \cdot \dot{\boldsymbol{v}})^2 = (\beta \times \dot{\boldsymbol{v}})^2,$$

we obtain

$$|\boldsymbol{A}|^2 = \gamma^6 c^2[(\beta \times \dot{v})^2 - \dot{v}^2].$$

The Larmor formula for an electron travelling with any allowed velocity must reduce to the non-relativistic formula when β is small and γ is close to unity so that, from eqn (5.16),

$$\frac{dU}{dt} = \frac{2}{3}\frac{q^2}{4\pi\varepsilon_0 c^3}\gamma^6\left(\dot{v}^2 - \frac{1}{c^2}(v \times \dot{v})^2\right) \tag{5.23}$$

as obtained by Liénard (1898). For the two cases examined above, eqn (5.23) reduces to the same expressions as before.

We have obtained the above result [eqn (5.23)] using an intuitive approach which imposes the condition of Lorentz invariance on the expression for the radiated power given by the Larmor formula, eqn (5.14). For a rigorous treatment we must return to the expressions for the electric and magnetic fields and remove the condition that the speed of the electron is much less than the speed of light. If we are detecting the radiation at time t during a time interval dt, this radiation is emitted by the electron at time t', during a time interval dt' and $t' = t - |\boldsymbol{r}|/c$. The observer is stationary and the electron is moving so that r is a function of t' only and $dt/dt' = 1 - (\boldsymbol{v} \cdot \boldsymbol{n})/c$ [eqn (4.45)]. As usual, $\boldsymbol{v}$ is the velocity of the electron and $\boldsymbol{n}$ is a unit vector along the direction of the vector $\boldsymbol{r}$ linking the observer at time t and the electron at the retarded time t'. These vectors are shown in Fig. 5.5.

If we wish to compute the total amount of radiated energy U which will be detected at the position given by the vector $\boldsymbol{r}$ over a time interval Δt, we must calculate the integral of dU/dt over that time interval. But dU/dt is expressed as a function of quantities which must be calculated at the retarded time t' so we must write

$$\int_{t_0}^{t_0+\Delta t} \frac{dU}{dt} dt = \varepsilon_0 c \int_{t'_0}^{t'_0+\Delta t'} |r[\boldsymbol{E}(\boldsymbol{r}, t)]|^2 \frac{dt}{dt'}\, dt'.$$

Equation (5.8) gives the expression for $\boldsymbol{E}(\boldsymbol{r}, t)$ and $dt/dt' = 1 - (\boldsymbol{v} \cdot \boldsymbol{n})/c$, as noted above, so that, following the same treatment as in eqns (5.10)–(5.12), but with v/c taking any value <1,

$$\frac{d^2U}{d\Omega\, dt} = \frac{q^2}{(4\pi)^2\varepsilon_0 c}\frac{|\boldsymbol{n} \times \{(\boldsymbol{n} - \boldsymbol{v}/c) \times \dot{\boldsymbol{v}}/c\}|^2}{(1 - (\boldsymbol{n} \cdot \boldsymbol{v})/c)^5}. \tag{5.24}$$

Equation (5.24) is a completely general expression which applies to all situations. If we consider first the case where $\boldsymbol{v}$ and $\dot{\boldsymbol{v}}$ are parallel to each other then $\boldsymbol{v} \times \dot{\boldsymbol{v}} = 0$, and

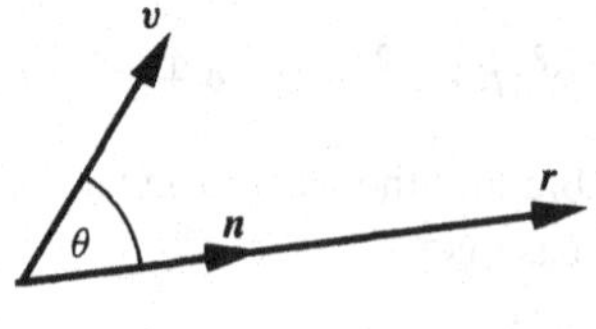

Fig. 5.5 Relation between $\boldsymbol{v}$, $\boldsymbol{n}$, and $\boldsymbol{r}$.

the triple vector product $\boldsymbol{n} \times (\boldsymbol{n} \times \dot{\boldsymbol{v}}) = (\boldsymbol{n} \cdot \dot{\boldsymbol{v}})\boldsymbol{n} - (\boldsymbol{n} \cdot \boldsymbol{n})\dot{\boldsymbol{v}}$. We can write $\boldsymbol{n} \cdot \dot{\boldsymbol{v}} = |\dot{\boldsymbol{v}}| \cos\theta$ and $\boldsymbol{n} \cdot \boldsymbol{v} = |\boldsymbol{v}| \cos\theta$, so that

$$\left|\boldsymbol{n} \times \left\{\left(\boldsymbol{n} - \frac{\boldsymbol{v}}{c}\right) \times \frac{\dot{\boldsymbol{v}}}{c}\right\}\right|^2 = \left|\boldsymbol{n}\frac{|\dot{\boldsymbol{v}}|}{c}\cos\theta - \frac{\dot{\boldsymbol{v}}}{c}\right|^2 = \frac{|\dot{\boldsymbol{v}}|^2}{c^2}\sin^2\theta,$$

and putting this into eqn (5.24) yields

$$\frac{\mathrm{d}^2 U}{\mathrm{d}\Omega\,\mathrm{d}t} = \frac{q^2}{4\pi^2\varepsilon_0 c^3}\frac{\sin^2\theta}{(1-(v/c)\cos\theta)^5}|\dot{\boldsymbol{v}}|^2.$$

This expression, apart from the factor $1/(1-(v/c)\cos\theta)^5$, has the same form as that obtained for the non-relativistic case [eqn (5.13)]. Integration of this expression over the whole solid angle, with $\mathrm{d}\Omega = 2\pi \sin\theta\,\mathrm{d}\theta$ and noting that

$$\int \frac{\sin\theta\,\mathrm{d}\theta}{(1-a\cos\theta)^n} = -\frac{1}{a(n-1)(1-a\cos\theta)^{n-1}}$$

gives the result of eqn (5.21) for the total radiated power as we would expect.

In the case where the acceleration is at right angles to the direction of motion of the electron, $\boldsymbol{v} \cdot \dot{\boldsymbol{v}} = 0$, which is the only case of practical interest for the production of synchrotron radiation, to evaluate eqn (5.24), we must expand the numerator:

$$\left|\boldsymbol{n} \times \left\{\left(\boldsymbol{n} - \frac{\boldsymbol{v}}{c}\right) \times \frac{\dot{\boldsymbol{v}}}{c}\right\}\right|^2 = \left|\left(\boldsymbol{n} \cdot \frac{\dot{\boldsymbol{v}}}{c}\right)\left(\boldsymbol{n} - \frac{\boldsymbol{v}}{c}\right) - \left(\boldsymbol{n} \cdot \left(\boldsymbol{n} - \frac{\boldsymbol{v}}{c}\right)\right) \cdot \left(\frac{\dot{\boldsymbol{v}}}{c}\right)\right|^2$$

and, referring to the coordinate system shown in Fig. 5.6, $\boldsymbol{n} \cdot \dot{\boldsymbol{v}} = |\dot{\boldsymbol{v}}| \sin\theta\cos\phi$, and $\boldsymbol{n} \cdot \boldsymbol{v} = |\boldsymbol{v}| \cos\theta$ as before, so that

$$\frac{\mathrm{d}^2 U}{\mathrm{d}\Omega\,\mathrm{d}t} = \frac{q^2}{4\pi^2\varepsilon_0 c^3}\frac{|\dot{\boldsymbol{v}}|^2}{(1-(v/c)\cos\theta)^3}\left(1 - \frac{\sin^2\theta\cos^2\phi}{\gamma^2(1-(v/c)\cos\theta)^2}\right) \tag{5.25}$$

and integration gives eqn (5.17).

It is very useful to express eqn (5.25) in the small angle approximation, for which we write $(1-\beta\cos\theta) \approx (1/2\gamma^2)(1+\gamma^2\theta^2)$, so that

$$\frac{\mathrm{d}^2 U}{\mathrm{d}\Omega\,\mathrm{d}t} = \frac{q^2}{4\pi^2\varepsilon_0 c^3}\frac{\gamma^6|\dot{\boldsymbol{v}}|^2}{(1+\gamma^2\theta^2)^3}\left(1 - \frac{4\gamma^2\theta^2\cos^2\phi}{(1+\gamma^2\theta^2)^2}\right),$$

which shows that the important scaling parameter for the angular distribution is not θ but $(\gamma\theta)$.

Figure 5.7 shows the shape of the distribution, plotted against $\gamma\theta$ in the y–z plane ($\phi = 90°$), in the plane where $\phi = 45°$ and the plane containing $\boldsymbol{v}$ and $\dot{\boldsymbol{v}}$ (the x–z plane, where $\phi = 0°$). It is clear that the distribution is strongly peaked in the direction

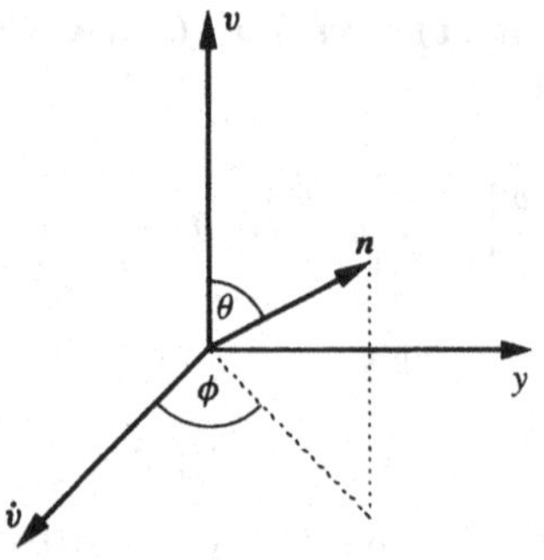

Fig. 5.6 Coordinate system for synchrotron radiation production.

of motion of the electron. The mean square value $\langle\theta\rangle^2$ is a measure of the width of the angular distribution which is shown to be equal to $1/\gamma^2$ by evaluating the rather formidable expression of eqn (5.26):

$$\langle\theta\rangle^2 = \frac{\displaystyle\int_0^\infty \frac{\theta^2\theta\,\mathrm{d}\theta}{(1+\gamma^2\theta^2)^3}\int_0^{\pi/2}\mathrm{d}\phi - \int_0^\infty \frac{4\gamma^2\theta^2\theta^2\theta\,\mathrm{d}\theta}{(1+\gamma^2\theta^2)^5}\int_0^{\pi/2}\cos^2\phi\,\mathrm{d}\phi}{\displaystyle\int_0^\infty \frac{\theta\,\mathrm{d}\theta}{(1+\gamma^2\theta^2)^3}\int_0^{\pi/2}\mathrm{d}\phi - \int_0^\infty \frac{4\gamma^2\theta^2\theta\,\mathrm{d}\theta}{(1+\gamma^2\theta^2)^5}\int_0^{\pi/2}\cos^2\phi\,\mathrm{d}\phi}. \tag{5.26}$$

Integration over ϕ reduces eqn (5.26) to

$$\langle\theta\rangle^2 = \frac{\displaystyle\int_0^\infty \frac{\theta^3\,\mathrm{d}\theta}{2(1+\gamma^2\theta^2)^3} - \int_0^\infty \frac{\gamma^2\theta^2\theta^3\,\mathrm{d}\theta}{(1+\gamma^2\theta^2)^5}}{\displaystyle\int_0^\infty \frac{\theta\,\mathrm{d}\theta}{2(1+\gamma^2\theta^2)^3} - \int_0^\infty \frac{\gamma^2\theta^2\theta\,\mathrm{d}\theta}{(1+\gamma^2\theta^2)^5}}.$$

Because the angular distribution is confined to small values of θ, there is no appreciable loss of accuracy when the upper limit of the integral over θ is extended to infinity. Doing this means that the substitution $\gamma\theta = \tan\alpha$ can be used, with α running from 0 to $\pi/2$, and the expression reduces further to

$$\langle\theta\rangle^2 = \frac{1}{\gamma^2}\frac{\int_0^{\pi/2}\sin^3\alpha\cos\alpha\,\mathrm{d}\alpha - \int_0^{\pi/2}\sin^5\alpha\cos^3\alpha\,\mathrm{d}\alpha}{\int_0^{\pi/2}\sin\alpha\cos^3\alpha\,\mathrm{d}\alpha - \int_0^{\pi/2}\sin^3\alpha\cos^5\alpha\,\mathrm{d}\alpha}. \tag{5.27}$$

The integrals in eqn (5.27) can be evaluated in terms of factorial functions (gamma functions) but it is easy to show, using integration by parts, and without evaluation, that

$$\int_0^{\pi/2}\sin^{n-1}\alpha\cos^{n+1}\alpha\,\mathrm{d}\alpha = \int_0^{\pi/2}\sin^{n+1}\alpha\cos^{n-1}\alpha\,\mathrm{d}\alpha.$$

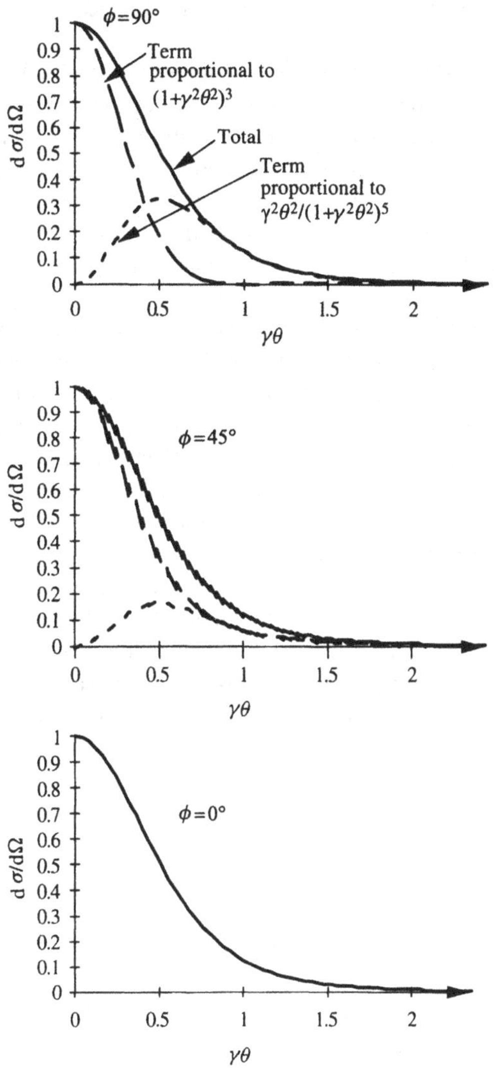

Fig. 5.7 Angular distribution of synchrotron radiation from a point on the circular orbit of an electron in a magnetic field.

Either way, eqn (5.27) reduces to

$$\langle \theta \rangle^2 = \frac{1}{\gamma^2}, \tag{5.28}$$

which is the characteristic opening angle of synchrotron radiation.

Dependence of radiation on frequency and angle

Throughout our discussion so far we have considered the radiation to be emitted by the electron at the moment when the electron passes through the origin of the coordinate system. This tells what an observer would see during an infinitely short length of time. We must now relax that assumption in order to determine how the radiation from a single electron appears to an observer whose detection system collects information over a long time period. In other words, we need to know how the radiation from the electron changes with time as seen by a stationary observer.

Qualitatively, an observer viewing a single electron is presented with a cone of radiation with cone angle of order $1/\gamma^2$. Figure 5.8 shows part of the trajectory of an electron travelling along an arc of a circle of radius R with a velocity βc. A detector at a distance D from the origin of the coordinate system at O registers the arrival of a photon at a time $t_0 = D/c$. The first photons to be registered at the detector, produced within a cone angle θ are emitted as the electron passes the point A, to arrive at the detector at time t_a and the last to be registered arrive at a time t_b, having been emitted at point B on the trajectory. The pulse length of the photon pulse, or, more precisely, the length of the electric field pulse at the detector is $t_b - t_a$. In order to generate a pulse of this duration, the electric field must rise and fall again in the time $t_b - t_a$, so that the wavelength of the electric field must be in the region $c(t_b - t_a)$ and the frequency $1/(t_b - t_a)$. The frequency can be lower than this, but it cannot be much higher without a corresponding decrease in the cone angle of the radiation. The frequency of the radiation is determined by the electron energy $E = \gamma m_0 c^2$.

The pulse length is equal to the time taken for the electron to travel along the arc AOB, reduced by the time taken for the radiation to travel directly from A to B, so that

$$t_b - t_a = \frac{2R\theta}{\beta c} - \frac{2R\sin\theta}{c},$$

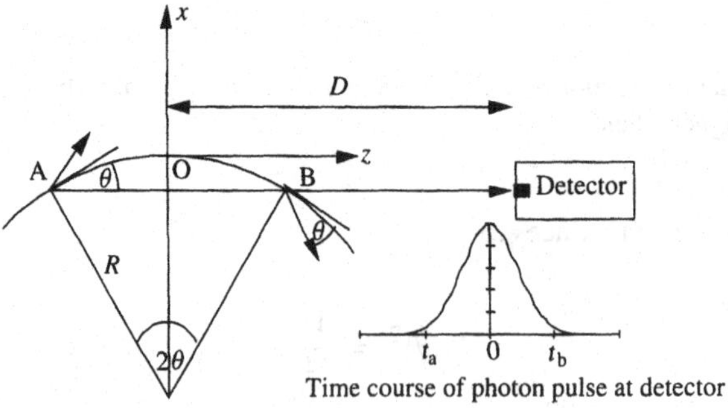

Fig. 5.8 Time variation of synchrotron radiation pulse.

and when θ is given by eqn (5.28) and is sufficiently small so that $\sin\theta$ can be approximated by $\sin\theta = \theta - \theta^3/6$ then

$$t_b - t_a = \frac{2R}{c}\left(\frac{1}{\gamma\beta} - \frac{1}{\gamma} + \frac{1}{6\gamma^3}\right),$$

which reduces to

$$t_b - t_a = \frac{4R}{3c\gamma^3}$$

with the approximation $1/\gamma\beta = 1/\gamma + 1/2\gamma^3$, so that the corresponding frequency of the radiation is given by

$$\nu = \frac{3c\gamma^3}{4R}$$

and the wavelength by

$$\lambda = \frac{4R}{3\gamma^3}. \tag{5.29}$$

The argument which has just been given is meant to show that for the case of an ultra-relativistic radiator, the cone angle of the radiator is so small that an observer sees a pulse of radiation whose duration in time determines the order of magnitude of the frequency present in the radiation. Figure 5.8 assumes that the electron is moving in a magnetic field which extends over the trajectory AOB. However, if the field covers a shorter distance, the pulse duration at the detector is shorter as well and the minimum frequency is even higher. This may seem strange. After all, the colour of a car headlight as the car sweeps round a bend in the road does not depend on the radius of the bend. In that case the car is moving slowly and the cone angle of the radiation produced is large so that $c(t_b - t_a)$ is very long compared with the wavelength of the light produced by the car headlight and so does not restrict the radiation observed at wavelengths in the visible region of the spectrum. On the other hand, for the case of an electron, moving at a speed close to the speed of light, $c(t_b - t_a)$ is comparable, in the X-ray region for ~2 GeV electrons, to the wavelength response of the detector and so exerts a strong constraint on the wavelengths that can be observed.

This intuitive argument indicates that the frequency spectrum may be obtained by calculating the Fourier transform (see Chapter 1) of the time distribution of the electric field at the observer. When we follow this procedure we shall see that it is convenient to use a wavelength parameter which is π times that given by eqn (5.29). This parameter is often called the critical or characteristic wavelength λ_c given by

$$\lambda_c = \frac{4\pi R}{3\gamma^3}. \tag{5.30}$$

What value of λ_c can we expect? In the example given above, $R = 5.56$ m, $E = 2$ GeV so that γ is 3914; then λ_c would be about 0.4 nm, which is in the X-ray region of the electromagnetic spectrum. Also, λ_c varies inversely as γ^3 so that doubling the electron energy reduces the characteristic wavelength by a factor of 8, moving it further into the X-ray region.

We must now proceed to the calculation of the frequency spectrum of the electromagnetic radiation produced by an object in circular motion. We know from eqns (5.9) and (5.10) that, in general, for any kind of motion

$$\frac{\mathrm{d}^2U}{\mathrm{d}\Omega\,\mathrm{d}t} = \varepsilon_0 c E^2 r^2. \tag{5.31}$$

In eqn (5.31), E^2 is the square of the modulus of the electromagnetic field vector at the observation point. We can write eqn (5.31), showing the time dependence explicitly, as

$$\frac{\mathrm{d}^2U}{\mathrm{d}\Omega\,\mathrm{d}t} = |\boldsymbol{G}(t)|^2, \tag{5.32}$$

where $\boldsymbol{G}(t) = \sqrt{\varepsilon_0 c}[rE]$, calculated at the retarded time at which the radiation observed is emitted from the electron. The stationary observer who is detecting the radiation emitted into the solid angle $\mathrm{d}\Omega$ measures the total energy received, in J/sr at the detector from the moving charge during a time interval $\mathrm{d}t$ as

$$\frac{\mathrm{d}U}{\mathrm{d}\Omega} = \int_{-\infty}^{+\infty} |\boldsymbol{G}(t)|^2\,\mathrm{d}t.$$

From the discussion above, we know that the radiation pulse is emitted over a very short period of time so that the only finite contribution to this integral comes from times close to $t = 0$. Extension of the integral to infinite times is a mathematical convenience which does not affect the physical result.

$\boldsymbol{G}(t)$ is the time course of the electric field vector but what we need is $\boldsymbol{G}(\omega)$, the electric field vector as a function of frequency. These two quantities are related by a Fourier transformation, which means that we expand the continuous function $\boldsymbol{G}(t)$ as a Fourier integral, so that $\boldsymbol{G}(t)$ is related to $\boldsymbol{G}(\omega)$ by

$$\boldsymbol{G}(t) = \frac{1}{\sqrt{2\pi}}\int_{-\infty}^{+\infty} \boldsymbol{G}(\omega)\exp(-\mathrm{i}\omega t)\,\mathrm{d}\omega$$

and $\boldsymbol{G}(\omega)$ to $\boldsymbol{G}(t)$ by the inverse transform

$$\boldsymbol{G}(\omega) = \frac{1}{\sqrt{2\pi}}\int_{-\infty}^{+\infty} \boldsymbol{G}(t)\exp(+\mathrm{i}\omega t)\,\mathrm{d}t. \tag{5.33}$$

Using this procedure, we can calculate $\boldsymbol{G}(\omega)$ using eqn (5.33). There is need for special care at this point because although we understand what negative times mean physically and we can, if we wish, make all times positive by a simple change to the coordinate system, a negative frequency has no physical meaning and we must obtain the energy radiated into unit solid angle by integrating the frequency spectrum over positive frequencies only so that

$$\frac{\mathrm{d}U}{\mathrm{d}\Omega} = \int_0^\infty \frac{\mathrm{d}^2 I(\omega, \boldsymbol{n})}{\mathrm{d}\Omega\,\mathrm{d}\omega}\,\mathrm{d}\omega. \tag{5.34}$$

On the other hand, the formalism demands that there is actual power being radiated into frequencies which lie between zero and $-\infty$, so, to make sure we include that power, we must write

$$\frac{\mathrm{d}^2 I(\omega, \boldsymbol{n})}{\mathrm{d}\Omega\,\mathrm{d}\omega} = |\boldsymbol{G}(\omega)|^2 + |\boldsymbol{G}(-\omega)|^2,$$

where ω is always positive, as in eqn (5.34). However, from eqn (5.33)

$$\boldsymbol{G}(-\omega) = \frac{1}{\sqrt{2\pi}} \int_{-\infty}^{+\infty} \boldsymbol{G}(t) \exp(-\mathrm{i}\omega t)\,\mathrm{d}t$$

and the complex conjugate $\boldsymbol{G}^*(\omega)$ (see Chapter 1) is related to the complex conjugate of $\boldsymbol{G}(t)$, which is $\boldsymbol{G}^*(t)$ by

$$\boldsymbol{G}^*(\omega) = \frac{1}{\sqrt{2\pi}} \int_{-\infty}^{+\infty} \boldsymbol{G}^*(t) \exp(-\mathrm{i}\omega t)\,\mathrm{d}t$$

but, since $\boldsymbol{G}(t)$ must be real, $\boldsymbol{G}(t) = \boldsymbol{G}^*(t)$ so that $\boldsymbol{G}(-\omega) = \boldsymbol{G}^*(\omega)$, so that

$$\frac{\mathrm{d}^2 I(\omega, \boldsymbol{n})}{\mathrm{d}\Omega\,\mathrm{d}\omega} = 2|\boldsymbol{G}(\omega)|^2 \tag{5.35}$$

and

$$\frac{\mathrm{d}U}{\mathrm{d}\Omega} = \int_0^{+\infty} 2|\boldsymbol{G}(\omega)^2|\,\mathrm{d}\omega. \tag{5.36}$$

Fourier transform of the electric field

From the definition of $\boldsymbol{G}(t) = \sqrt{\varepsilon_0 c}[r\boldsymbol{E}]$, given at eqn (5.32), we write [from eqn (5.8)]

$$\boldsymbol{G}(t) = \frac{q}{4\pi\sqrt{\varepsilon_0 c}} \left[\frac{\boldsymbol{n} \times \{(\boldsymbol{n} - \boldsymbol{v}/c) \times \dot{\boldsymbol{v}}/c\}}{(1 - (\boldsymbol{n}\cdot\boldsymbol{v})/c)^3}\right].$$

The quantity in square brackets must be evaluated at the retarded time $t' = t - r(t')/c$ so that we must write eqn (5.33) with t' as the continuous variable instead of t, so that

$$\boldsymbol{G}(\omega) = \frac{q}{\sqrt{32\pi^3 \varepsilon_0 c}} \int_{-\infty}^{+\infty} \left[\frac{\boldsymbol{n} \times \{(\boldsymbol{n} - \boldsymbol{v}/c) \times \dot{\boldsymbol{v}}/c\}}{(1 - (\boldsymbol{n}\cdot\boldsymbol{v})/c)^2}\right] \exp\left(+\mathrm{i}\omega\left(t' + \frac{r(t')}{c}\right)\right)\mathrm{d}t'$$

because $\mathrm{d}t = \mathrm{d}t'(1 - \boldsymbol{n}\cdot\boldsymbol{v}/c)$, from eqn (4.45).

Figure 5.9 shows the relationship between an observer at a fixed point P whose coordinates are $(\boldsymbol{x}, t)$, $\boldsymbol{x}$ being the position vector of P and the radiating electron at $(\boldsymbol{R}, t')$. Since the distance $r(t')$ is much larger than $|\boldsymbol{R}|$, the zero order approximation would make $|\boldsymbol{r}(t')| = |\boldsymbol{x}|$. However, because the electron is moving we must take the next approximation and write $|\boldsymbol{r}(t')| = |\boldsymbol{x}| - \boldsymbol{n}\cdot\boldsymbol{R}(t')$ so that we can write the exponential factor in eqn (5.36) as

$$\exp\left(+\mathrm{i}\omega\frac{x}{c}\right)\exp\left(+\mathrm{i}\omega\left(t'-\frac{\boldsymbol{n}\cdot\boldsymbol{R}(t')}{c}\right)\right).$$

The factor $\exp(\mathrm{i}\omega x/c)$ is a constant phase angle which can be placed outside the integral and when, at the end of the calculation, we take $|\boldsymbol{G}|^2$ this multiplies out to unity. Physically, this corresponds to a constant phase shift of the electric field vector which is of no interest to the observer and eqn (5.36) can be written as

$$\boldsymbol{G}(\omega) = \frac{q}{\sqrt{32\pi^3\varepsilon_0 c}}\int_{-\infty}^{+\infty}\left[\frac{\boldsymbol{n}\times\{(\boldsymbol{n}-\boldsymbol{v}/c)\times\dot{\boldsymbol{v}}/c\}}{(1-(\boldsymbol{n}\cdot\boldsymbol{v})/c)^2}\right] \times\exp\left(+\mathrm{i}\omega\left(t'-\frac{\boldsymbol{n}\cdot\boldsymbol{R}(t')}{c}\right)\right)\mathrm{d}t'.$$

In this form $\boldsymbol{G}(\omega)$ can be evaluated if we note that

$$\frac{\mathrm{d}}{\mathrm{d}t'}\left[\frac{\boldsymbol{n}\times\{\boldsymbol{n}\times\dot{\boldsymbol{v}}/c\}}{1-(\boldsymbol{n}\cdot\boldsymbol{v})/c}\right] = \left[\frac{\boldsymbol{n}\times\{(\boldsymbol{n}-\boldsymbol{v}/c)\times\dot{\boldsymbol{v}}/c\}}{(1-(\boldsymbol{n}\cdot\boldsymbol{v})/c)^2}\right]$$

and

$$\frac{\mathrm{d}}{\mathrm{d}t'}\left(\exp\left(+\mathrm{i}\omega\left(t'-\frac{\boldsymbol{n}\cdot\boldsymbol{R}(t')}{c}\right)\right)\right) = \mathrm{i}\omega\left(1-\frac{\boldsymbol{n}\cdot\boldsymbol{v}}{c}\right)\exp\left(+\mathrm{i}\omega\left(t'-\frac{\boldsymbol{n}\cdot\boldsymbol{R}(t')}{c}\right)\right). \quad (5.37)$$

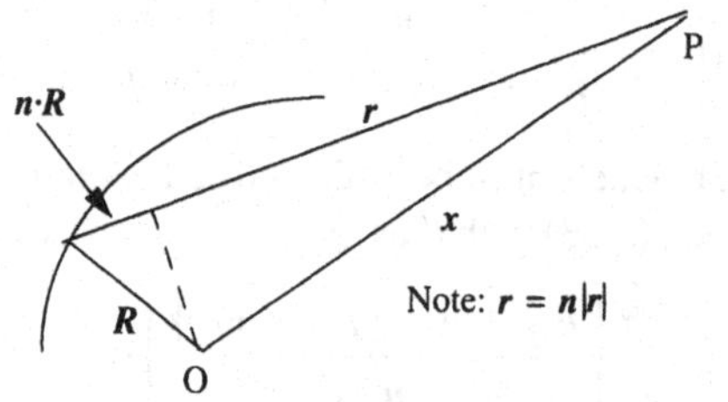

Fig. 5.9 Position vectors for evaluation of integral.

Integration by parts yields

$$\begin{aligned}
&\int_{-\infty}^{+\infty}\left[\frac{\boldsymbol{n}\times\{(\boldsymbol{n}-\boldsymbol{v}/c)\times\dot{\boldsymbol{v}}/c\}}{(1-(\boldsymbol{n}\cdot\boldsymbol{v})/c)^2}\right]\exp\left(+\mathrm{i}\omega\left(t'-\frac{\boldsymbol{n}\cdot\boldsymbol{R}(t')}{c}\right)\right)\mathrm{d}t' \\
&= \left|\frac{\boldsymbol{n}\times(\boldsymbol{n}\times(\boldsymbol{v}/c))}{1-(\boldsymbol{n}\cdot\boldsymbol{v})/c}\exp\left(+\mathrm{i}\omega\left(t'-\frac{\boldsymbol{n}\cdot\boldsymbol{R}(t')}{c}\right)\right)\right|_{t'=-\infty}^{t'=+\infty} \\
&\quad -\mathrm{i}\omega\int_{-\infty}^{+\infty}\boldsymbol{n}\times\left(\boldsymbol{n}\times\frac{\boldsymbol{v}}{c}\right)\exp\left(+\mathrm{i}\omega\left(t'-\frac{\boldsymbol{n}\cdot\boldsymbol{R}(t')}{c}\right)\right)\mathrm{d}t' \\
&= -\mathrm{i}\omega\int_{-\infty}^{+\infty}\boldsymbol{n}\times\left(\boldsymbol{n}\times\frac{\boldsymbol{v}}{c}\right)\exp\left(+\mathrm{i}\omega\left(t'-\frac{\boldsymbol{n}\cdot\boldsymbol{R}(t')}{c}\right)\right)\mathrm{d}t'.
\end{aligned}$$

Because the radiation is confined to a forward cone close to $t' = 0$, the contribution from the first term, which is to be evaluated at $t' = \pm\infty$, must be zero and we can write, from eqn (5.35)

$$\begin{aligned}
&\frac{\mathrm{d}^2 I(\omega,\boldsymbol{n})}{\mathrm{d}\Omega\,\mathrm{d}\omega} \\
&= \frac{q^2}{16\pi^3\varepsilon_0 c}\left|\int_{-\infty}^{+\infty}\left[\frac{\boldsymbol{n}\times\{(\boldsymbol{n}-\boldsymbol{v}/c)\times\dot{\boldsymbol{v}}/c\}}{(1-(\boldsymbol{n}\cdot\boldsymbol{v})/c)^2}\right]\exp\left(+\mathrm{i}\omega\left(t'-\frac{\boldsymbol{n}\cdot\boldsymbol{R}(t')}{c}\right)\right)\mathrm{d}t'\right|^2 \\
&= \frac{q^2\omega^2}{16\pi^3\varepsilon_0 c}\left|\int_{-\infty}^{+\infty}\left(\boldsymbol{n}\times\left(\boldsymbol{n}\times\frac{\boldsymbol{v}}{c}\right)\right)\exp\left(+\mathrm{i}\omega\left(t'-\frac{\boldsymbol{n}\cdot\boldsymbol{R}(t')}{c}\right)\right)\mathrm{d}t'\right|^2 \qquad (5.38)
\end{aligned}$$

Radiation from an electron moving along the arc of a circle

We can now use the formalism developed above to calculate the frequency spectrum of the radiation from a relativistic electron which is moving along the arc of a circle with radius R. The geometry is shown in Fig. 5.10.

In Fig 5.10, the observer is located at P, in the y–z plane, and the line from P to the origin of the coordinate system, the direction defined by the unit vector $\boldsymbol{n}$, makes an angle ψ with the z-direction. At a time t' the radius vector $\boldsymbol{R}$ makes an angle θ with the x-axis, equal to $\omega_0 t'$, and $\omega_0 = |\boldsymbol{v}|/R$, the angular frequency of the electron in its circular orbit. Because the radiation is produced within a narrow cone around the tangent to the orbit only a short segment of the electron trajectory on either side of the origin of the coordinate system actually contributes to the integral in eqn (5.38). We can now write down the components of the vectors $\boldsymbol{n}$, $\boldsymbol{R}(t')$, and $\boldsymbol{v}$ in terms of the angles shown:

$$\begin{aligned}
\boldsymbol{n} &= (0, \sin\psi, \cos\psi), \\
\boldsymbol{R}(t') &= |\boldsymbol{R}|(\cos(\omega_0 t'), 0, \sin(\omega_0 t')), \qquad (5.39)\\
\boldsymbol{v}(t') &= |\boldsymbol{v}|(\sin(\omega_0 t'), 0, \cos(\omega_0 t')).
\end{aligned}$$

We can also define two polarization vectors which indicate the two possible directions of the electric field. From the properties of the radiation field established earlier, we

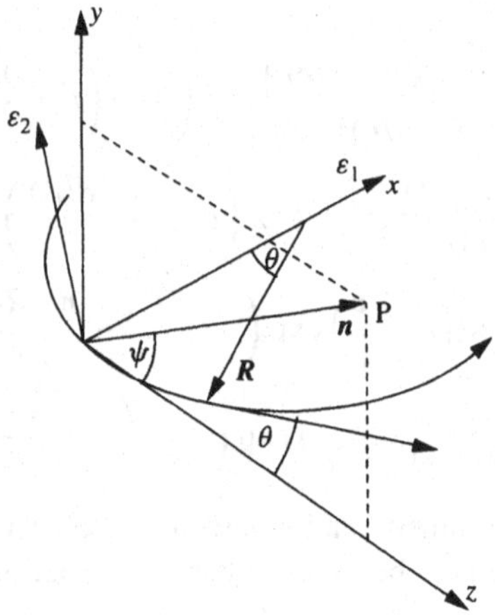

Fig. 5.10 Geometry for synchrotron radiation production from circular motion.

know that these directions must be at right angles to the direction of observation. From this it follows that the horizontal polarization vector ε_1 must be a unit vector along the x-axis. The 'vertical' direction of polarization of the radiation field must be at right angles to both the x-axis and the direction of observation and so must be a unit vector $\varepsilon_2 = \boldsymbol{n} \times \varepsilon_1$ which has components $(0, \cos\psi, \sin\psi)$. Using these unit vectors and the definition of $\boldsymbol{v}$ in eqn (5.39), we can write, for the vector product in eqn (5.38),

$$\begin{aligned}
\boldsymbol{v} &= |\boldsymbol{v}|(\varepsilon_1 \sin\omega_0 t' + (\boldsymbol{n}\cos\psi - \varepsilon_2 \sin\psi)\cos\omega_0 t'),\\
\boldsymbol{n} \times \boldsymbol{v} &= |\boldsymbol{v}|(\varepsilon_2 \sin\omega_0 t' + \varepsilon_1 \sin\psi\cos\omega_0 t'),\\
\boldsymbol{n} \times (\boldsymbol{n} \times \boldsymbol{v}) &= |\boldsymbol{v}|(-\varepsilon_1 \sin\omega_0 t' + \varepsilon_2 \sin\psi\cos\omega_0 t').
\end{aligned}$$

The definitions of $\boldsymbol{n}$ and $\boldsymbol{R}$ in eqn (5.39) can be used to compute the scalar product in the exponential term in eqn (5.38) so that

$$\boldsymbol{n} \cdot \boldsymbol{R} = R\cos\psi \sin\omega_0 t' \tag{5.40}$$

so that the exponential factor becomes

$$\exp\left(\mathrm{i}\omega\left(t' - \frac{R}{c}\cos\psi \sin\omega_0 t'\right)\right).$$

Because the angles in eqn (5.40) are very small and $v \approx c$, we can replace ψ and $\omega_0 t'$ by the first two terms in the series expansions for the trigonometric functions so that

$$\cos\psi \approx 1 - \frac{\psi^2}{2}, \qquad \sin\omega_0 t' \approx \omega_0 t' - \frac{{\omega_0}^3 t'^3}{6}.$$

It is also convenient to replace ω_0 by v/R, $(1 - v/c)$ by $1/2\gamma^2$, where γ has its usual definition, and set $\beta = 1$ when higher order terms can be neglected. If we then follow through the algebra we find that

$$t' - \frac{R}{c}\cos\psi\sin\omega_0 t' = \frac{t'}{2\gamma^2}\left(1 + \gamma^2\psi^2\right) + \frac{c^2 t'^3}{6R^2}.$$

We can now write $\boldsymbol{G}(\omega)$ from eqn (5.37) as

$$\begin{aligned}\boldsymbol{G}(\omega) &= \frac{q}{\sqrt{32\pi^3\varepsilon_0 c}}\left\{\varepsilon_1\left[\mathrm{i}\omega\int_{-\infty}^{+\infty}\omega_0 t' \exp\left(\mathrm{i}\omega\left(\frac{t'}{2\gamma^2}\left(1+\gamma^2\psi^2\right)+\frac{c^2 t'^3}{6R^2}\right)\right)\mathrm{d}t'\right]\right.\\ &\left.\quad -\varepsilon_2\left[\mathrm{i}\omega\int_{-\infty}^{+\infty}\psi\exp\left(\mathrm{i}\omega\left(\frac{t'}{2\gamma^2}\left(1+\gamma^2\psi^2\right)+\frac{c^2 t'^3}{6R^2}\right)\right)\mathrm{d}t'\right]\right\}.\end{aligned} \tag{5.41}$$

It is immediately obvious that an observer in the plane of the electron trajectory ($\psi = 0$) observes only a horizontal component of the radiated electric field and the radiation for such an observer is polarized horizontally.

The next stage is to evaluate, as far as possible, the integrals in eqn (5.41). For this we need a change of variables as follows:

$$x = \frac{\gamma\omega_0}{\left(1+\gamma^2\psi^2\right)^{1/2}}t', \qquad \xi = \frac{1}{2}\frac{\omega}{\omega_c}\left(1+\gamma^2\psi^2\right)^{3/2},$$

where $\omega_c = \frac{3}{2}\gamma^3\omega_0$, so that, with these substitutions, the polarization components of $\boldsymbol{G}(\omega)$ simplify to

$$\boldsymbol{G}_1(\omega) = \mathrm{i}\frac{q}{\sqrt{32\pi^3\varepsilon_0 c}}\frac{\left(1+\gamma^2\psi^2\right)}{\gamma^2\omega_0}\int_{-\infty}^{+\infty} x\exp\left(\mathrm{i}\frac{3}{2}\xi\left(x+\frac{1}{3}x^3\right)\right)\mathrm{d}x,$$

$$\boldsymbol{G}_2(\omega) = -\mathrm{i}\psi\frac{q}{\sqrt{32\pi^3\varepsilon_0 c}}\frac{\left(1+\gamma^2\psi^2\right)^{1/2}}{\gamma\omega_0}\int_{-\infty}^{+\infty}\exp\left(\mathrm{i}\frac{3}{2}\xi\left(x+\frac{1}{3}x^3\right)\right)\mathrm{d}x,$$

where $\boldsymbol{G}(\omega) = \varepsilon_1\boldsymbol{G}_1(\omega) + \varepsilon_2\boldsymbol{G}_2(\omega)$. The exponential function under the integral can be expanded using De Moivre's theorem so that

$$\begin{aligned}&\int_{-\infty}^{+\infty} x\exp\left(\mathrm{i}\frac{3}{2}\xi\left(x+\frac{1}{3}x^3\right)\right)\mathrm{d}x\\ &\quad = \int_{-\infty}^{+\infty} x\cos\left(\frac{3}{2}\xi\left(x+\frac{1}{3}x^3\right)\right)\mathrm{d}x + \mathrm{i}\int_{-\infty}^{+\infty} x\sin\left(\frac{3}{2}\xi\left(x+\frac{1}{3}x^3\right)\right)\mathrm{d}x,\end{aligned}$$

$$\begin{aligned}&\int_{-\infty}^{+\infty}\exp\left(\mathrm{i}\frac{3}{2}\xi\left(x+\frac{1}{3}x^3\right)\right)\mathrm{d}x\\ &\quad = \int_{-\infty}^{+\infty}\cos\left(\frac{3}{2}\xi\left(x+\frac{1}{3}x^3\right)\right)\mathrm{d}x + \mathrm{i}\int_{-\infty}^{+\infty}\sin\left(\frac{3}{2}\xi\left(x+\frac{1}{3}x^3\right)\right)\mathrm{d}x.\end{aligned}$$

Now

$$\int_{-\infty}^{+\infty} x\cos\left(\frac{3}{2}\xi\left(x+\frac{1}{3}x^3\right)\right)\mathrm{d}x \quad \text{and} \quad \int_{-\infty}^{+\infty} \sin\left(\frac{3}{2}\xi\left(x+\frac{1}{3}x^3\right)\right)\mathrm{d}x$$

are both zero because, in each case, the integrand changes sign when the sign of x is changed from plus to minus so that the portion of each integral from $-\infty$ to 0 exactly cancels the portion from 0 to $+\infty$ and so the expressions for $\boldsymbol{G}(\omega)$ reduce to

$$\begin{aligned} \boldsymbol{G}_1(\omega) &= -\frac{q}{\sqrt{32\pi^3\varepsilon_0 c}}\frac{(1+\gamma^2\psi^2)}{\gamma^2\omega_0}\int_{-\infty}^{+\infty} x\sin\left(\frac{3}{2}\xi\left(x+\frac{1}{3}x^3\right)\right)\mathrm{d}x, \\ \boldsymbol{G}_2(\omega) &= -\mathrm{i}\frac{q\psi}{\sqrt{32\pi^3\varepsilon_0 c}}\frac{(1+\gamma^2\psi^2)^{1/2}}{\gamma\omega_0}\int_{-\infty}^{+\infty} \cos\left(\frac{3}{2}\xi\left(x+\frac{1}{3}x^3\right)\right)\mathrm{d}x. \end{aligned} \tag{5.42}$$

This expression can be simplified further because the integrals in eqn (5.42) can be expressed in terms of the modified Bessel functions $K_{2/3}$ and $K_{1/3}$:

$$\int_0^{+\infty} x\sin\left(\frac{3}{2}\xi\left(x+\frac{1}{3}x^3\right)\right)\mathrm{d}x = \frac{1}{\sqrt{3}}K_{2/3}(\xi),$$

$$\int_0^{+\infty} \cos\left(\frac{3}{2}\xi\left(x+\frac{1}{3}x^3\right)\right)\mathrm{d}x = \frac{1}{\sqrt{3}}K_{1/3}(\xi),$$

so that

$$\begin{aligned} \boldsymbol{G}_1(\omega) &= -\frac{q}{\sqrt{32\pi^3\varepsilon_0 c}}\frac{(1+\gamma^2\psi^2)}{\gamma^2\omega_0}\frac{2}{\sqrt{3}}K_{2/3}(\xi), \\ \boldsymbol{G}_2(\omega) &= -\mathrm{i}\psi\frac{q}{\sqrt{32\pi^3\varepsilon_0 c}}\frac{(1+\gamma^2\psi^2)^{1/2}}{\gamma\omega_0}\frac{2}{\sqrt{3}}K_{1/3}(\xi), \end{aligned} \tag{5.43}$$

and from (5.35), since $\boldsymbol{G}(\omega)^2 = \boldsymbol{G}_1(\omega)^2 + \boldsymbol{G}_2(\omega)^2$, and inserting the definition of ω_c, we have

$$\begin{aligned} \frac{\mathrm{d}^2 I(\omega,\boldsymbol{n})}{\mathrm{d}\Omega\,\mathrm{d}\omega} &= \frac{3}{4}\frac{q^2}{4\pi^3\varepsilon_0 c}\gamma^2\left(\frac{\omega}{\omega_c}\right)^2\left(1+\gamma^2\psi^2\right)^2 \\ &\quad\times\left[K_{2/3}^2(\xi)+\frac{\gamma^2\psi^2}{1+\gamma^2\psi^2}K_{1/3}^2(\xi)\right]. \end{aligned} \tag{5.44}$$

Equation (5.44) is the required expression for the frequency spectrum of the radiation from an electron moving along a trajectory which is an arc of a circle, measured in J/sr radiated into unit frequency interval.

In the next chapter we will discuss the physical meaning of this expression.

Reference

1. J. D. Jackson, *Classical electrodynamics* (2nd edn), pp. 659 and 660. John Wiley & Sons, New York (USA) (1962).

6

The spectral distribution of synchrotron radiation

Properties of the modified Bessel functions

Equation (5.44) is an expression for the angular distribution of the synchrotron radiation which contains the modified Bessel functions, $K_{1/3}$ and $K_{2/3}$. Their behaviour, as a function of the general variable x, is shown in Fig. 6.1.

These functions are the solutions of the modified Bessel equation

$$x^2\frac{d^2y}{dx^2} + x\frac{dy}{dx} - (x^2 + n^2)y = 0. \tag{6.1}$$

The parameter n is called the order of the Bessel function, which, in this case, is either 1/3 or 2/3.

Before we go into more detail about the properties of these functions, we start by considering a much simpler equation, (6.2), below, whose solution leads to functions which we already know about,

$$\frac{d^2y}{dx^2} + n^2y = 0. \tag{6.2}$$

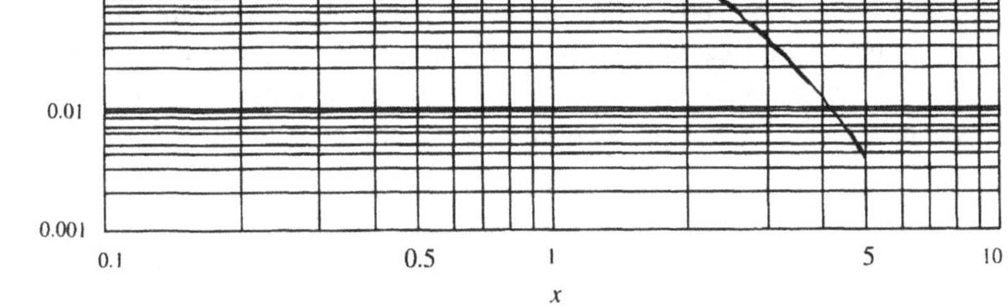

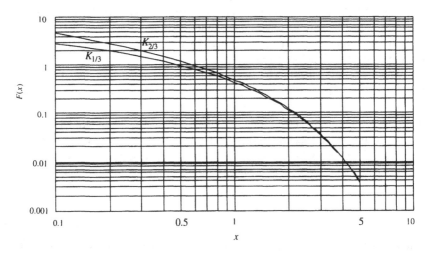

Fig. 6.1 Modified Bessel functions $K_{1/3}$ and $K_{2/3}$.

Equation (6.2) is a simple form of the wave equation in one dimension which we have met before. Its solutions, where we express y as a function of x, can be written as linear combinations of the trigonometric functions (also called circular functions), $\sin(nx)$ and $\cos(nx)$, so that, for example, $y = A\cos(nx) + B\sin(nx)$ is a solution, as can be seen by calculating the double differential, $\mathrm{d}^2y/\mathrm{d}x^2$, and noting that it is equal to n^2y. Here A and B can be any numbers, real or complex, depending on the precise physical situation in which the wave motion is taking place.

These circular functions are found throughout the mathematical description of physical phenomena. They have well-known mathematical properties, and, as such, develop a life of their own, far beyond their elementary definition in terms of the ratios of sides of a right-angled triangle. In particular, we can expand the sine and cosine functions as power series in x:

$$\begin{aligned} \sin x &= x - \frac{x^3}{3!} + \frac{x^5}{5!} - \frac{x^7}{7!} + \cdots = \sum_{k=0}^{\infty}(-1)^k \frac{x^{2k+1}}{\Gamma(2k+2)}, \\ \cos x &= 1 - \frac{x^2}{2!} + \frac{x^4}{4!} - \frac{x^6}{6!} + \cdots = \sum_{k=0}^{\infty}(-1)^k \frac{x^{2k}}{\Gamma(2k+1)}. \end{aligned} \tag{6.3}$$

In eqn (6.3), Γ is called the factorial function, and is defined so that $\Gamma(a+1) = a! = a \times (a-1) \times (a-2) \times \cdots \times 1$. The use of the gamma function in eqn (6.3) is not really necessary because the summation index k is restricted to integer values. For example, $\Gamma(1) = 0!$, which is equal to 1, $\Gamma(2) = 1! = 1$, $\Gamma(3) = 2! = 2 \times 1 = 2$, $\Gamma(4) = 3! = 3 \times 2 \times 1 = 6$, etc. Physically, factorial k (written $k!$) means the number of possible arrangements of a number k of non-identical objects. If there are four objects, for example, the first can be chosen in four ways, the second in three ways, the third in two ways and the fourth in only one way. The total number of ways is the product of these, namely twelve, which is 4!, or $\Gamma(5)$. If there are no objects, the number of ways of arranging them is still one, so $\Gamma(1) = 0! = 1$. It is, perhaps, unfortunate that the gamma function is defined so that, for example, $\Gamma(5) = 4!$, but this will make more sense when, later on, we have to make use of the gamma function when the argument is no longer an integer.

The power series, equations (6.3), can be used to calculate the value of the circular function for any value of x. The terms in the series get numerically smaller and smaller as x gets larger and larger (the series is said to converge) so we can obtain the value of $\sin x$ [or $\sin(nx)$, by substituting nx for x in the series], and similarly $\cos x$, or $\cos(nx)$, to any desired accuracy. This is how the values of the functions are prepared for tabulation in textbooks or calculated by means of a computer algorithm.

If we change the sign in eqn (6.2), we obtain a new equation, (6.4), below:

$$\frac{\mathrm{d}^2y}{\mathrm{d}x^2} - n^2y = 0 \tag{6.4}$$

which is just as simple as eqn (6.2) and its solutions can be written in terms of the hyperbolic functions $\sinh(nx)$ and $\cosh(nx)$. The corresponding series expansions defining these functions are:

$$\sinh x = x + \frac{x^3}{3!} + \frac{x^5}{5!} + \frac{x^7}{7!} + \cdots = \sum_{k=0}^{\infty} \frac{x^{2k+1}}{\Gamma(2k+2)},$$
$$\cosh x = 1 + \frac{x^2}{2!} + \frac{x^4}{4!} + \frac{x^6}{6!} + \cdots = \sum_{k=0}^{\infty} \frac{x^{2k}}{\Gamma(2k+1)}. \tag{6.5}$$

A quick glance at eqns (6.4) and (6.5) shows that each hyperbolic function reduces to the corresponding circular function when n is complex so that, in the particular case when $n^2 = -1$, the hyperbolic functions are related to the circular functions by

$$\sinh x = -\mathrm{i} \sin \mathrm{i}x, \qquad \cosh x = \cos \mathrm{i}x.$$

These are not the only ways the solutions can be written. We already know that the exponential functions $\exp(nx)$ and $\exp(-nx)$ are also solutions of eqns (6.2) and (6.4). The solutions in terms of exponential functions are not independent of the solutions in terms of sine and cosine functions so we can write the exponential solutions as combinations of the circular functions. In particular,

$$\cos nx = \frac{\exp(\mathrm{i}nx) + \exp(-\mathrm{i}nx)}{2},$$
$$\sin nx = \frac{\exp(\mathrm{i}nx) - \exp(-\mathrm{i}nx)}{2\mathrm{i}}. \tag{6.6}$$

The factor 1/2 (or 1/2i) is there to normalize the function, that is to make sure that $\cos(nx) = 1$ when $nx = 0$ and $\sin(nx) = 1$ when $nx = \pi/2$. The factor 1/i in the expression for the sine function is there to make sure that when nx is real, $\sin(nx)$ is real too.

The corresponding expansion for the hyperbolic functions is given by eqn (6.7) below:

$$\cosh nx = \frac{\exp(nx) + \exp(-nx)}{2},$$
$$\sinh nx = \frac{\exp(nx) - \exp(-nx)}{2}. \tag{6.7}$$

It is simply a matter of convenience what functions we use to express the solutions, so, in practice, we select those functions which make it easiest for us to carry out any subsequent mathematical manipulations.

Figure 6.2 shows the behaviour of these functions. The behaviour of the sine and cosine functions is well known, their values oscillate between $+1$ and -1 and pass through zero when nx takes odd values of $\pi/2$ (in the case of the cosine function), and when nx takes even values of $\pi/2$ (including zero) for the sine function. The hyperbolic functions, on the other hand, increase as x increases, as we would expect

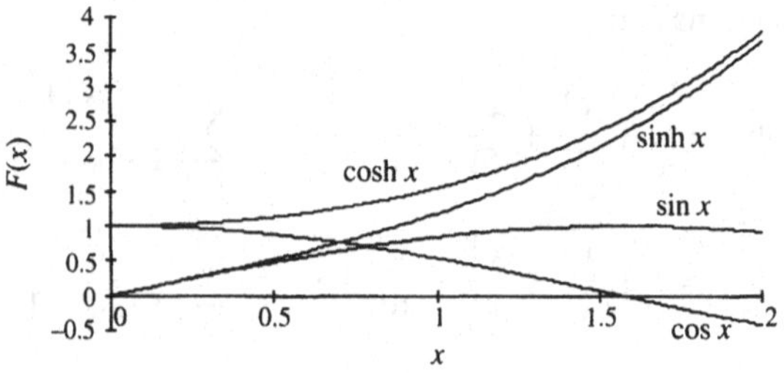

Fig. 6.2 Trigonometric and hyperbolic functions.

from their definition in terms of the exponential functions in eqn (6.7). The hyperbolic function $\cosh(nx)$ is never zero; $\sinh(nx)$ is zero only at $x = 0$. Physically, $\cosh(nx)$ is the shape taken up by a perfectly flexible chain, hung from two points at the same height and a distance $2n$ apart.

This brief description of the trigonometric and hyperbolic functions has been given because the Bessel functions, which are required for the description of synchrotron radiation, are analogous to them. For example, eqn (6.1) is the modified form of the standard Bessel equation

$$x^2\frac{d^2y}{dx^2} + x\frac{dy}{dx} + (x^2 - n^2)y = 0, \tag{6.8}$$

which is analogous to eqn (6.2). The solutions to eqn (6.8) can be written as a series expansion, just like (6.2). In this case the solutions are called Bessel functions, J_{+n} and J_{-n}. The general solution (under certain conditions) is a linear combination of these, just as the solution to (6.2) is a combination of sine and cosine functions. The Bessel functions are given by:

$$\begin{aligned} J_{+n}(x) &= \sum_{k=0}^{\infty} \frac{(-1)^k}{k!\Gamma(k+n+1)}\left(\frac{x}{2}\right)^{2k+n}, \\ J_{-n}(x) &= \sum_{k=0}^{\infty} \frac{(-1)^k}{k!\Gamma(k-n+1)}\left(\frac{x}{2}\right)^{2k-n}. \end{aligned} \tag{6.9}$$

In eqn (6.9), n is called the order of the Bessel function and can take integer or non-integer values. Strictly speaking, the solutions of eqn (6.9) are valid only when n is not an integer. Figures 6.3 and 6.4 show the behaviour of the Bessel functions J with order 1/3 and 2/3. The shapes of the curves are reminiscent of the cosine and sine functions, but the values of the J_+ functions are infinite at $x = 0$.

When x is replaced by ix, so that x^2 is replaced by $-x^2$ in eqn (6.8), we obtain the modified Bessel equation, eqn (6.1) and its solutions (for non-integral order n) are

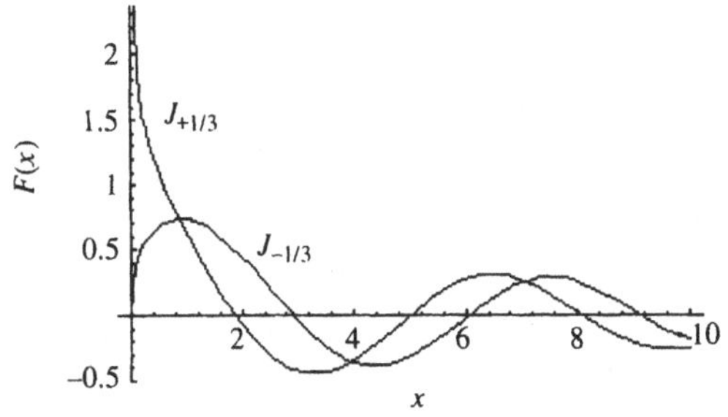

Fig. 6.3 Bessel functions $J_{+1/3}(x)$ and $J_{-1/3}(x)$.

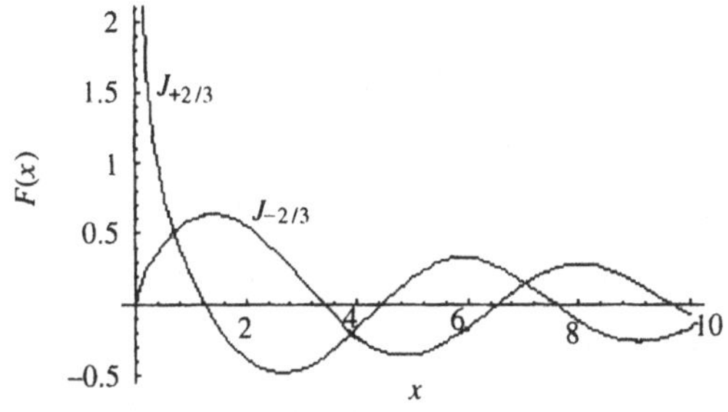

Fig. 6.4 Bessel functions $J_{+2/3}(x)$ and $J_{-2/3}(x)$.

written as

$$
\begin{aligned}
I_{+n}(x) &= \sum_{k=0}^{\infty} \frac{1}{k!\Gamma(k+n+1)} \left(\frac{x}{2}\right)^{2k+n}, \\
I_{-n}(x) &= \sum_{k=0}^{\infty} \frac{1}{k!\Gamma(k-n+1)} \left(\frac{x}{2}\right)^{2k-n},
\end{aligned}
\tag{6.10}
$$

which are equivalent to the hyperbolic solutions of the modified wave equation (6.4). A similar relation holds between the solutions I and J of the two Bessel equations and the circular and hyperbolic functions, namely,

$$I_n(x) = \mathrm{i}^{-n} J(\mathrm{i}x).$$

Graphs of these functions for $n = \pm 1/3$ and $\pm 2/3$ are shown in Figs 6.5 and 6.6. Their behaviour is similar to that of the hyperbolic functions shown in Fig. 6.2 except that $I_+(x)$ is infinite at $x = 0$.

The functions K, which occur in eqn (5.44), are also solutions of eqn (6.1). They are formed from the solutions I_{+n} and I_{-n} in eqn (6.10):

$$K_n(x) = \frac{1}{2} \frac{\pi}{\sin(n\pi)} (I_{-n}(x) - I_{+n}(x)). \tag{6.11}$$

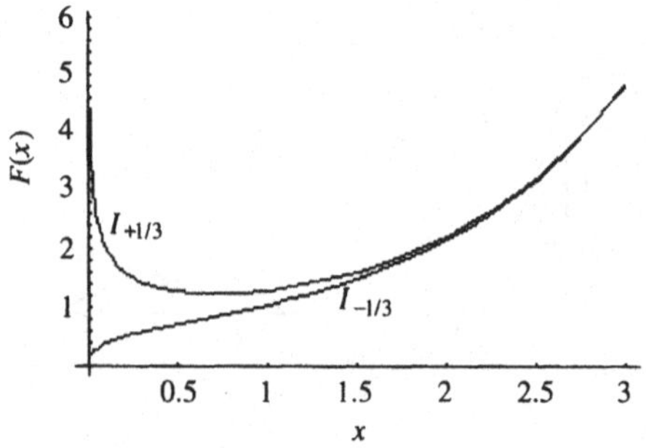

Fig. 6.5 Bessel functions $I_{+1/3}(x)$ and $I_{-1/3}(x)$.

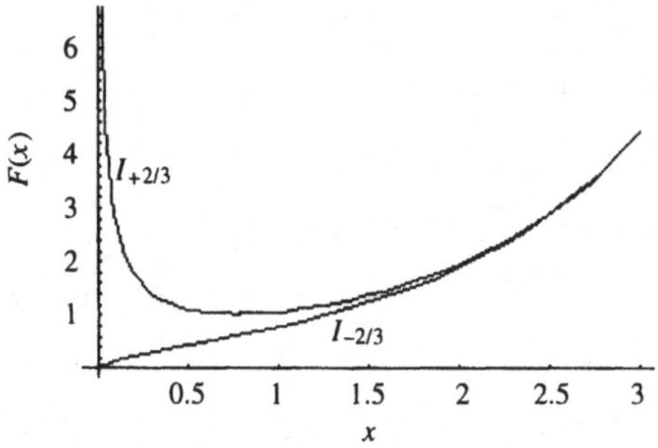

Fig. 6.6 Bessel functions $I_{+2/3}(x)$ and $I_{-2/3}(x)$.

The factor $\pi/\sin(n\pi)$ is a useful way of writing $\Gamma(p)\Gamma(p-1)$.[1] This is analogous to the expression of the solutions of the wave equation as combinations of exponential functions in eqns (6.6) and (6.7).

Now we are in a position to calculate the numerical values of the $K_{1/3}$ and $K_{2/3}$ functions by making use of the series expansion for the function I, with the corresponding order, given in eqn (6.10). The required gamma function is obtained using the recurrence relation $\Gamma(n+1) = n\Gamma(n)$, starting with $\Gamma(1/3) = 2.6789385$ and $\Gamma(2/3) = 1.3541179$. Values of the I function can also be obtained from published tables (see, for example, ref. 1). Values of the K function are obtained from the I functions by using eqn (6.11). A selection of calculated values of $K_{1/3}(x)$ and $K_{2/3}(x)$ are given in Table 6.1.

Armed with this knowledge we can investigate the properties of the synchrotron radiation angular distribution calculated in the previous chapter.

Photon distribution as a function of energy and angle

Imagine a beam of relativistic electrons which is passing through a deflecting magnet as shown in Fig. 6.7.

TABLE 6.1 Values of the functions $K_{1/3}(x)$, $K_{2/3}(x)$, and $K_{5/3}(x)$

x	$K_{1/3}(x)$	$K_{2/3}(x)$	$K_{5/3}(x)$
0.00001	78.2979	2315.51	3.09E+08
0.00002	62.1289	1458.68	9.72E+07
0.00003	54.2627	1113.18	4.95E+07
0.00004	49.2913	918.909	3.06E+07
0.00005	45.7498	791.891	2.11E+07
0.00006	43.045	701.258	1.56E+07
0.00007	40.8826	632.771	1.21E+07
0.00008	39.0969	578.875	9.65E+06
0.00009	37.5862	535.159	7.93E+06
0.0001	36.284	498.859	6.65E+06
0.0002	28.7637	314.259	2.10E+06
0.0003	25.1019	239.822	1.07E+06
0.0004	22.7858	197.966	659911
0.0005	21.1348	170.6	454955
0.0006	19.873	151.073	335737
0.0007	18.8637	136.317	259670
0.0008	18.0297	124.704	207859
0.0009	17.3238	115.285	170810
0.001	16.715	107.464	143302
0.002	13.1916	67.686	45137.2
0.003	11.4689	51.6433	22964
0.004	10.3755	42.6207	14217.3
0.005	9.59371	36.7201	9801.62
0.006	8.99454	32.5087	7233.15
0.007	8.51396	29.3255	5594.32
0.008	8.1159	26.8198	4478.08
0.009	7.77816	24.7867	3679.89
0.010	7.48622	23.0981	3087.23
0.020	5.78056	14.4979	972.308
0.030	4.9321	11.0171	494.581
0.040	4.3861	9.05206	306.121
0.050	3.99102	7.76193	210.976
0.060	3.68507	6.83745	155.628
0.070	3.43741	6.13585	120.311
0.080	3.23055	5.58135	96.2531
0.090	3.05372	5.12965	79.0485
0.100	2.89983	4.75296	66.2727
0.200	1.97934	2.80179	20.6579
0.300	1.50911	1.98662	10.3385
0.400	1.20576	1.51713	6.26287
0.500	0.989031	1.20593	4.20485
0.600	0.825094	0.98283	3.00916
0.700	0.69653	0.814784	2.2485
0.800	0.59318	0.683871	1.73297
0.900	0.508597	0.579388	1.36695
1.000	0.438431	0.494475	1.09773
2.000	0.116545	0.124839	0.199771
3.000	0.0353059	0.0370571	0.0517757
4.000	0.0112999	0.0117308	0.0152102

TABLE 6.1 (*Continued*)

x	$K_{1/3}(x)$	$K_{2/3}(x)$	$K_{5/3}(x)$
5.000	0.00372888	0.00384442	0.00475405
6.000	0.00125474	0.00128749	0.00154085
7.000	0.00042797	0.00043762	0.00051133
8.000	0.00014744	0.00015036	0.0001725
9.000	5.1181E−05	5.2089E−05	5.8897E−05
10.000	1.7875E−05	1.8161E−05	2.0296E−05

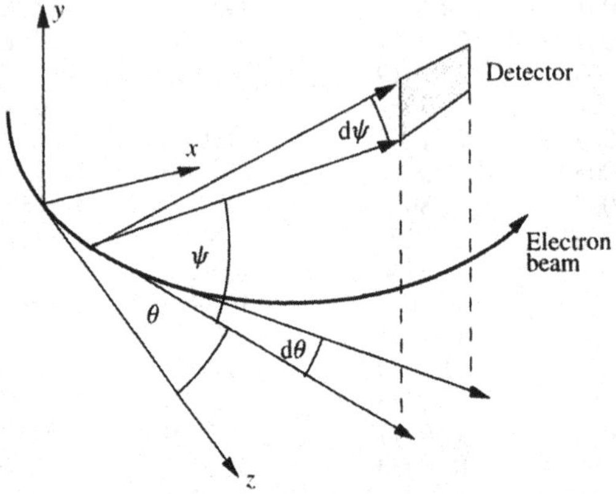

Fig. 6.7 Photons reaching a detector at angles θ and ψ.

A detector placed, for example, above the plane of the electron beam, observes the photons radiated by the beam from that portion of the electron beam trajectory lying between the horizontal angles θ and $\theta + \mathrm{d}\theta$ into the angular range between ψ and $\psi + \mathrm{d}\psi$. Equation (5.44) tells us the radiant energy in joules falling on the detector in a bandwidth of 1 Hz from the passage of one electron. We can convert this quantity into the number of photons in the same angular interval by dividing the power by the amount of energy carried by one photon which is $\hbar\omega$. Further, if I is the electric current carried by the beam, then the number of electrons passing the observation point in 1 s is I/e so that we may write for the number of photons/(s mrad^2) into a fractional frequency bandwidth $\Delta\omega/\omega$,

$$\frac{\mathrm{d}^2 N(\omega, \boldsymbol{n})}{\mathrm{d}\theta\, \mathrm{d}\psi} = \frac{3}{4}\frac{\alpha}{\pi^2}\gamma^2 \frac{I}{e}\frac{\Delta\omega}{\omega}\left(\frac{\omega}{\omega_c}\right)^2 \left(1 + \gamma^2\psi^2\right) \times \left[K_{2/3}^2(\xi) + \frac{\gamma^2\psi^2}{1+\gamma^2\psi^2} K_{1/3}^2(\xi)\right]. \tag{6.12}$$

In eqn (6.12),[2] α, the fine structure constant, and ξ are given by

$$\alpha = \frac{e^2}{4\pi\varepsilon_0\hbar c} = \frac{1}{137},$$
$$\xi = \frac{1}{2}\frac{\omega}{\omega_c}\left(1+\gamma^2\psi^2\right)^{3/2}. \tag{6.13}$$

If we divide eqn (6.13) by γ^2, we can write eqn (6.12) as a function of (ω/ω_c) and plot a series of universal curves showing the number of photons/(mrad^2 s) into unit fractional bandwidth and for one electron ($I/e = 1$) in Figs 6.8–6.12. In these figures, as well as the total photon flux, the photon fluxes resulting from the two terms in square brackets in eqn (6.12) are shown separately.

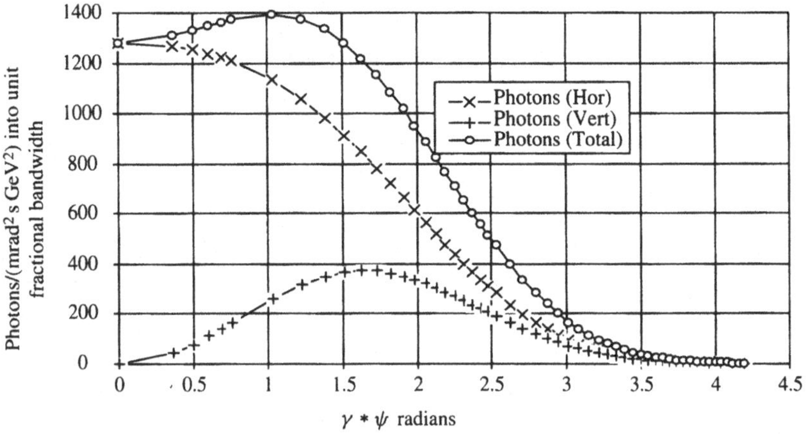

Fig. 6.8 Photon angular distribution for $\omega/\omega_c = 0.1$.

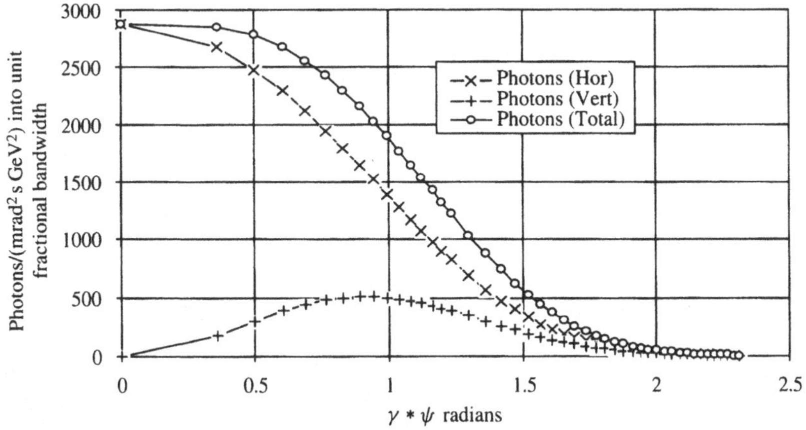

Fig. 6.9 Photon angular distribution for $\omega/\omega_c = 0.5$.

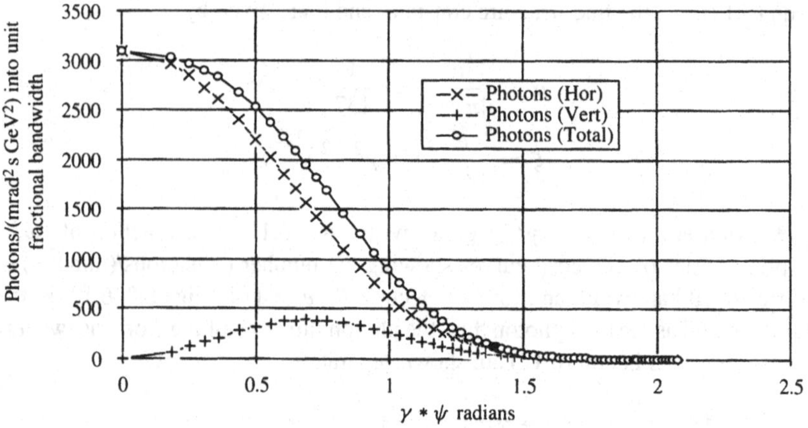

Fig. 6.10 Photon angular distribution for $\omega/\omega_c = 1.0$.

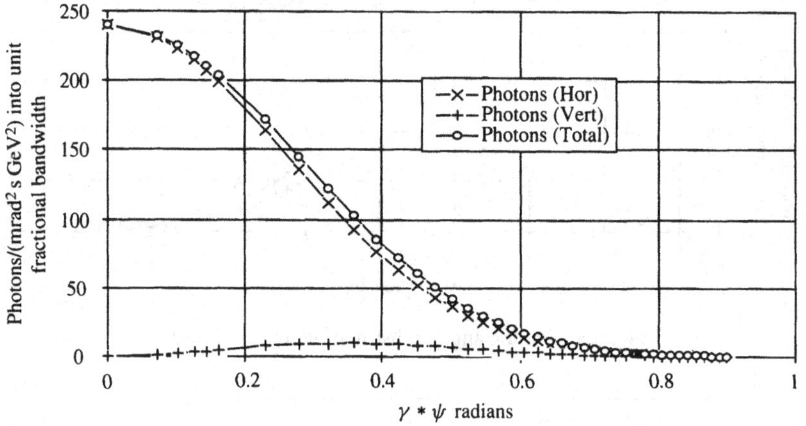

Fig. 6.11 Photon angular distribution for $\omega/\omega_c = 5.0$.

Figures 6.8–6.12 show that the angular distribution is dominated by the $K_{2/3}$ term in eqn (6.12). The effect of the $K_{1/3}$ term is only significant when $\omega/\omega_c \ll 1$.

Photon polarization as a function of energy and angle

If you imagine yourself to be an observer at a point P viewing the approaching electron as it sweeps round the curve defined by the magnetic field (Fig. 6.7) you will be interested to know in which direction the electric field vector $\boldsymbol{E}$ is pointing at your position of observation. In other words, you need to know the state of polarization of

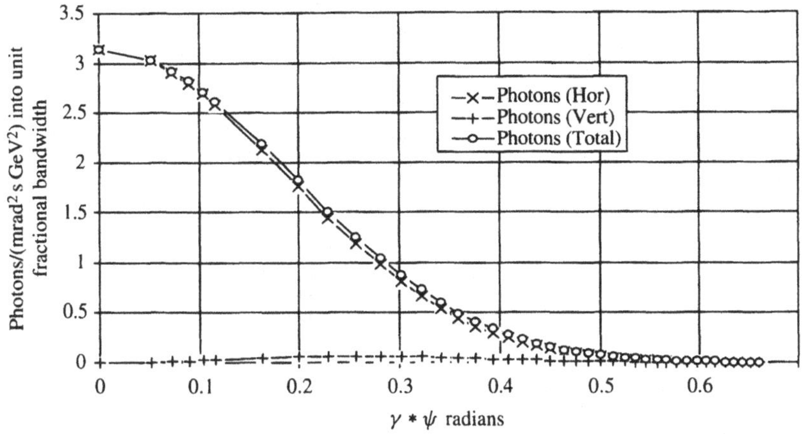

Fig. 6.12 Photon angular distribution for $\omega/\omega_c = 10.0$.

the radiation field because the force exerted by the field on a charge e placed at P is given, in magnitude and direction, by $e\boldsymbol{E}$. The polarization state has important physical consequences. For example, if a single crystal is placed at P, the intensities of the Bragg (Sir William Lawrence Bragg, 1890–1971) reflections are affected according to the orientation of the crystal planes with respect to the polarization direction of the incident beam.[3]

In general, $\boldsymbol{E}(\omega)$ is given by the relationship

$$\boldsymbol{E}(\omega) = \frac{1}{\sqrt{\varepsilon_0 c}} \frac{\boldsymbol{G}(\omega)}{[r]}$$

from (5.31) and (5.32). Because we are talking about the polarization at a fixed point at a distance r from the radiating electron, it is sufficient to consider the vector $\boldsymbol{G}(\omega)$ which differs from $\boldsymbol{E}(\omega)$ only by a multiplying constant and, as before

$$\boldsymbol{G}(\omega) = \varepsilon_1 G_1(\omega) + \varepsilon_2 G_2(\omega). \tag{6.14}$$

So from eqn (5.43)

$$\begin{aligned} G_1(\omega) &= -\frac{q}{\sqrt{32\pi^3 \varepsilon_0 c}} \frac{\left(1+\gamma^2\psi^2\right)}{\gamma^2 \omega_0} \frac{2}{\sqrt{3}} K_{2/3}(\xi), \\ G_2(\omega) &= -\mathrm{i}\psi \frac{q}{\sqrt{32\pi^3 \varepsilon_0 c}} \frac{\left(1+\gamma^2\psi^2\right)^{1/2}}{\gamma \omega_0} \frac{2}{\sqrt{3}} K_{1/3}(\xi); \end{aligned} \tag{6.15}$$

hence we can write the two components of the electric field vector, in terms of the unit vectors ε_1 and ε_2 as

$$\boldsymbol{E}_1(\omega) = \kappa\varepsilon_1 K_{2/3}(\xi),$$
$$\boldsymbol{E}_2(\omega) = \kappa\varepsilon_2 \frac{\mathrm{i}\gamma\psi}{\sqrt{1+\gamma^2\psi^2}} K_{1/3}(\xi),$$

where

$$\kappa = -\frac{q}{4\pi\varepsilon_0[r]}\sqrt{\frac{3}{2\pi}}\frac{\gamma\left(1+\gamma^2\psi^2\right)}{\omega_c},$$
$$\omega_c = \frac{3}{2}\gamma^3\omega_0,$$

and

$$\xi = \frac{1}{2}\frac{\omega}{\omega_c}\left(1+\gamma^2\psi^2\right)^{3/2}.$$

At the moment when the electron radiates a photon, the observer at P is located at a distance $[r]$ from the electron and the directions of the two unit vectors are as shown in Fig. 5.10. Because the angle ψ is very small, ε_1 and ε_2 can be regarded as pointing in the horizontal and vertical directions, respectively. The linear polarization of the radiation is defined by

$$P_{\text{linear}} = \frac{N_x - N_y}{N_x + N_y} = \frac{|\boldsymbol{E}_1|^2 - |\boldsymbol{E}_2|^2}{|\boldsymbol{E}_1|^2 + |\boldsymbol{E}_2|^2}$$

which can be written as

$$P_{\text{linear}} = \frac{K_{2/3}^2(\xi) - \left[\gamma^2\psi^2/(1+\gamma^2\psi^2)\right]K_{1/3}^2(\xi)}{K_{2/3}^2(\xi) + \left[\gamma^2\psi^2/(1+\gamma^2\psi^2)\right]K_{1/3}^2(\xi)}.$$

It is worth being a little more careful in our definition because the observer at P, of course, sees an electric field which changes with time so that

$$\boldsymbol{E}_1(\omega, t) = \varepsilon_1 E_{0,1}\exp\left(-\mathrm{i}(\omega t + \phi_1)\right),$$
$$\boldsymbol{E}_2(\omega, t) = \varepsilon_2 E_{0,2}\exp\left(-\mathrm{i}(\omega t + \phi_2)\right),$$

and the plane of polarization is that defined by the direction of the vectors ε_1 (or ε_2) and the product $\varepsilon_1 \times \varepsilon_2$.

The introduction of the phases ϕ_1 and ϕ_2 allows for the possibility that $\boldsymbol{E}_1$ and $\boldsymbol{E}_2$ may differ in phase by an amount $\phi_1 - \phi_2$. In the particular case of the electric fields generated by an electron moving in the field of a dipole magnet, given by eqn (6.15), the phase difference $\phi_1 - \phi_2$, between $\boldsymbol{E}_1$ and $\boldsymbol{E}_2$, is $\pi/2$, as can be seen by noting that in eqn (6.15), $\boldsymbol{G}_2(\omega)$ is multiplied by the factor $\exp(\mathrm{i}\pi/2) = \mathrm{i}$ and the wave form of $\boldsymbol{E}_2$ is shifted by 90° compared with $\boldsymbol{E}_1$ as is shown in Fig. 6.13.

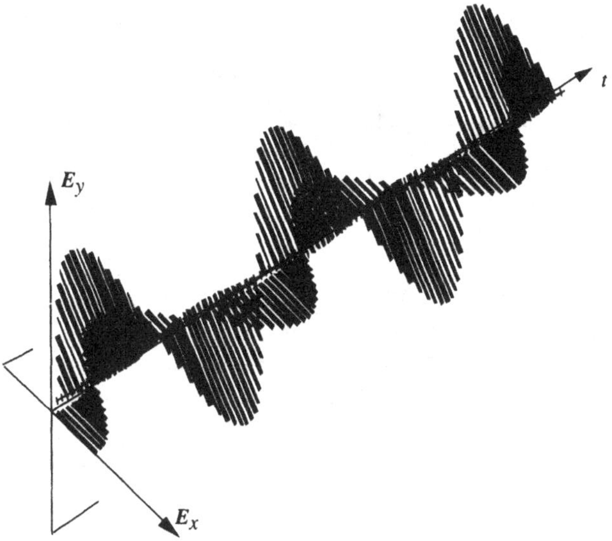

Fig. 6.13 Behaviour of $\boldsymbol{E}_x$ and $\boldsymbol{E}_y$ as a function of time.

An observer detects an electric field which varies with time in such a way that when $\boldsymbol{E}_x$ is a maximum, $\boldsymbol{E}_y$ is a minimum and vice versa. So we can write

$$\boldsymbol{E}_x = \boldsymbol{E}_{0,x} \cos \omega t, \quad \text{and} \quad \boldsymbol{E}_y = \boldsymbol{E}_{0,y} \sin \omega t$$

so that

$$\frac{|\boldsymbol{E}_x|^2}{|\boldsymbol{E}_{0,x}|^2} + \frac{|\boldsymbol{E}_y|^2}{|\boldsymbol{E}_{0,y}|^2} = \cos^2 \omega t + \sin^2 \omega t = 1$$

which is the equation of an ellipse with major axis $\boldsymbol{E}_x$ and minor axis $\boldsymbol{E}_y$. When the observer is in the horizontal plane, $\boldsymbol{E}_y$ is always zero so that the ellipse reduces to the special case of a straight line and the radiation is 100% linearly polarized. At the other extreme, at large values of ψ, when $\boldsymbol{E}_x = \boldsymbol{E}_y$ the radiation is circularly polarized. In fact, because ψ must be large for this condition to apply, the state of circularly polarized radiation is reached only when the photon intensity is very low.

How is this vertical component of the electric field generated? Refer to Fig. 6.14. We know that the acceleration of the electron, $\dot{\beta}$, directed towards the centre of its orbit, generates a component of the electric field which always lies in the horizontal plane. We say horizontal because the dipole magnets, which generate the curved trajectory of the electron, are almost always arranged with the magnetic field pointing in the vertical direction so that the plane of the electron orbit, defined by the curvature of its motion, must lie at right angles to the field direction and therefore be horizontal. The acceleration vector of the electron lies in this plane and an observer in this plane, viewing the electron as it passes the point O, observes an electric field, pointing in the direction of $\dot{\beta}$ as shown in Fig. 6.14.

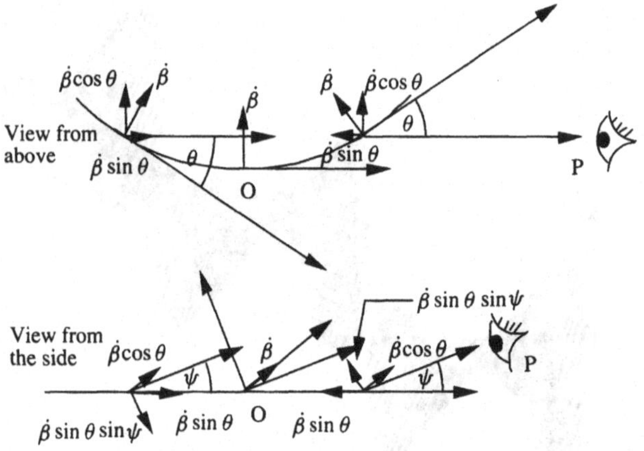

Fig. 6.14 Directions of acceleration vector components as seen by an observer at P, drawn in plane view (top) and elevation (bottom).

When the trajectory of the electron has turned through an angle θ, $\dot{\beta}$ can be resolved into two components as shown: $\dot{\beta}\cos\theta$, at right angles to the line of sight which generates the horizontally polarized component $\boldsymbol{E}_1$; and $\dot{\beta}\sin\theta$ which points along the line of sight. When P is in the horizontal plane, this component cannot contribute to the radiation field at P, and the observer sees only the field generated by the component $\dot{\beta}\cos\theta$ but as soon as P moves out of the horizontal plane to an angle ψ above or below the plane, the observer sees an electric field at right angles to the line of sight, generated by the acceleration component $\dot{\beta}\sin\theta\sin\psi$ and pointing predominantly upwards or downwards depending on whether the electron is moving towards O or away from O. It follows that, over the time period of the pulse, the observer sees an electric vector generated by the resultant of the acceleration vectors $\dot{\beta}\cos\theta$ and $\dot{\beta}\sin\theta\sin\psi$ which appears to rotate through 2π radians during the duration of the pulse. Of course, this is exactly what we would expect from eqn (6.14), where the quantity $\boldsymbol{G}_2$ depends on ψ, the angle of observation. It is also clear from eqn (6.14) (and from Fig. 6.14) that the direction of rotation of polarization depends on the sign of ψ so that an observer above the median plane detects right-handed elliptical polarization whereas an observer below the plane observes left-handed polarization.

Let us calculate the degree of elliptical polarization. To do this, we separate the elliptically polarized radiation into two new vector components, one which is rotating, with frequency ω in the clockwise direction (as viewed by the observer) and the other which is rotating, with the same frequency, in the opposite (anticlockwise) direction. These components are often called right-handed and left-handed, respectively and will be denoted by $\boldsymbol{E}_R$ and $\boldsymbol{E}_L$. Figure 6.15 shows a vector diagram in which the total electric field vector $\boldsymbol{E}$ is shown resolved into its components $\boldsymbol{E}_x$ and $\boldsymbol{E}_y$ or $\boldsymbol{E}_R$ and $\boldsymbol{E}_L$.

In terms of these vectors, we can write

$$\begin{aligned} \boldsymbol{E}_x &= \boldsymbol{E}_{0,\mathrm{R}} \cos \omega t + \boldsymbol{E}_{0,\mathrm{L}} \cos \omega t = \boldsymbol{E}_{0,x} \cos \omega t, \\ \boldsymbol{E}_y &= \boldsymbol{E}_{0,\mathrm{R}} \sin \omega t - \boldsymbol{E}_{0,\mathrm{L}} \sin \omega t = \boldsymbol{E}_{0,y} \sin \omega t \end{aligned} \tag{6.16}$$

so that, from eqn (6.16),

$$\boldsymbol{E}_{0,x} = \boldsymbol{E}_{0,\mathrm{R}} + \boldsymbol{E}_{0,\mathrm{L}}, \qquad \boldsymbol{E}_{0,y} = \boldsymbol{E}_{0,\mathrm{R}} - \boldsymbol{E}_{0,\mathrm{L}} \tag{6.17}$$

and, from eqn (6.17)

$$\boldsymbol{E}_{0,\mathrm{R}} = \tfrac{1}{2}\left(\boldsymbol{E}_{0,x} + \boldsymbol{E}_{0,y}\right), \qquad \boldsymbol{E}_{0,\mathrm{L}} = \tfrac{1}{2}\left(\boldsymbol{E}_{0,x} - \boldsymbol{E}_{0,y}\right).$$

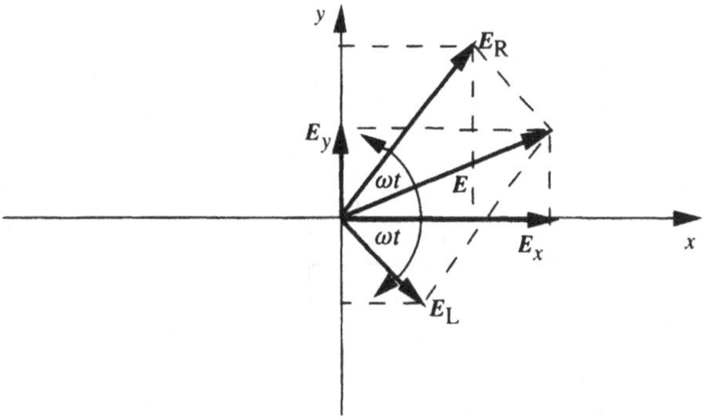

Fig. 6.15 Time-varying electric vector $\boldsymbol{E}$ expressed in terms of the vectors $\boldsymbol{E}_x$ and $\boldsymbol{E}_y$ or $\boldsymbol{E}_\mathrm{R}$ and $\boldsymbol{E}_\mathrm{L}$.

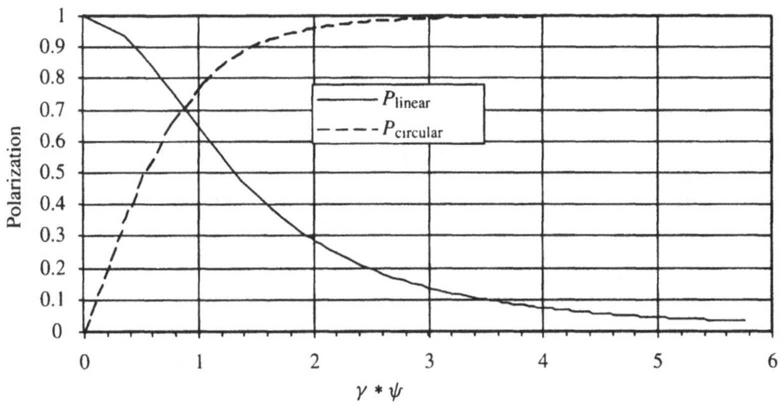

Fig. 6.16 Linear and circular polarization for $\omega/\omega_\mathrm{c} = 0.1$.

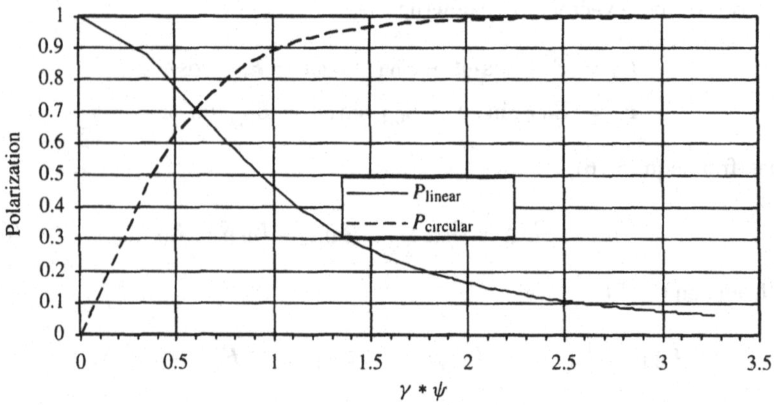

Fig. 6.17 Linear and circular polarization for $\omega/\omega_c = 0.5$.

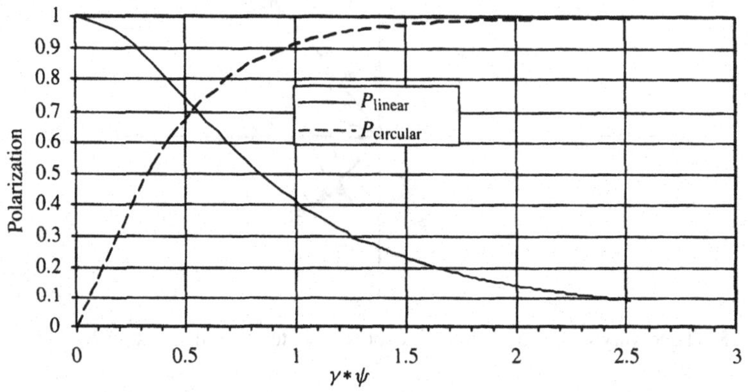

Fig. 6.18 Linear and circular polarization for $\omega/\omega_c = 1.0$.

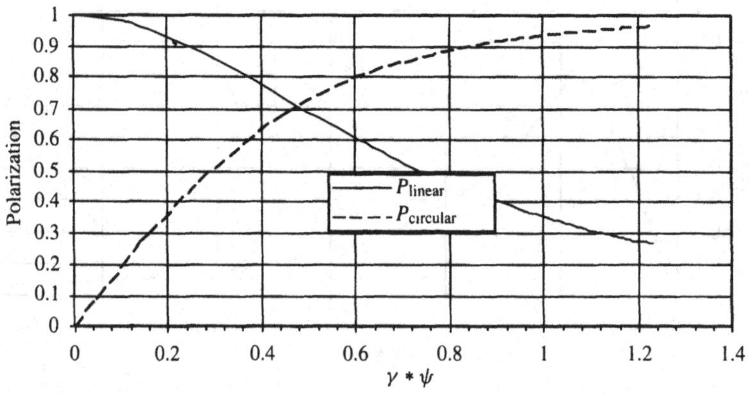

Fig. 6.19 Linear and circular polarization for $\omega/\omega_c = 5.0$.

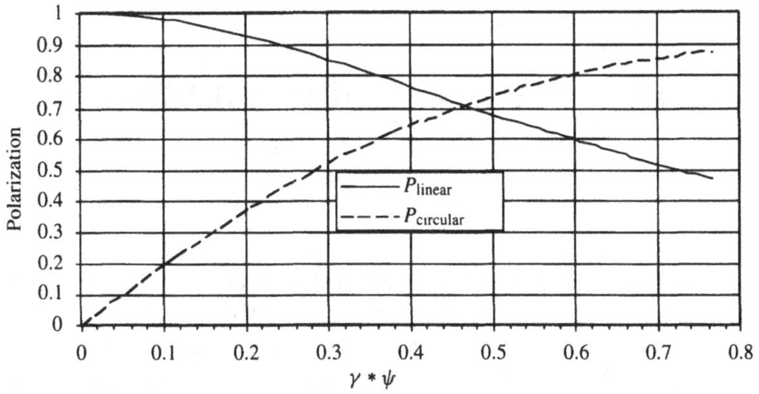

Fig. 6.20 Linear and circular polarization for $\omega/\omega_c = 10.0$.

So that, defining

$$P_{\text{circular}} = \frac{|\boldsymbol{E}_{0,\text{R}}|^2 - |\boldsymbol{E}_{0,\text{L}}|^2}{|\boldsymbol{E}_{0,\text{R}}|^2 + |\boldsymbol{E}_{0,\text{L}}|^2}$$

we have the required expression for P_{circular}:

$$\begin{aligned} P_{\text{circular}} &= \frac{2E_{0,x}E_{0,y}}{|E_{0,x}|^2 + |E_{0,y}|^2} \\ &= \frac{2K_{2/3}(\xi)K_{1/3}(\xi)\left(\gamma\psi/\sqrt{1+\gamma^2\psi^2}\right)}{K_{2/3}^2(\xi) + \left(\gamma^2\psi^2/(1+\gamma^2\psi^2)\right)K_{1/3}^2(\xi)}. \end{aligned}$$

We note again that P_{circular} is zero when $\psi = 0$ and its sign (which corresponds to the direction of rotation), depends on the sign of ψ.

Figures 6.16–6.20 show the result of a polarization calculation for various values of ω/ω_c.

References

1. Mary L. Boas, *Mathematical methods in the physical sciences* (2nd edn). Chapter 11, p. 462. John Wiley & Sons, New York, NY, USA (1983).
2. J. Schwinger, *Physical Review* **75**, 1912 (1949).
3. J. R. Helliwell, *Macromolecular crystallography and synchrotron radiation*. Cambridge University Press, Cambridge, England (1992).

7

Photon spectral distribution integrated over vertical angles

The fact that the radiation from the accelerated electron is emitted into a narrow forward cone means that quite often an observer is interested in the photon energy (or the total number of photons) emitted over all vertical angles. This is the case when the target being irradiated subtends a vertical angle which is significantly larger than the opening angle of the forward cone from the emission point. The energy radiated by the passage of one electron into unit solid angle, located at an angle ψ above the plane of the electron orbit, and unit frequency interval is given as usual by eqn (7.1) which is identical to (5.44):[1]

$$\frac{\mathrm{d}^2 I(\omega, \boldsymbol{n})}{\mathrm{d}\Omega\,\mathrm{d}\omega} = \frac{3}{4}\frac{q^2}{4\pi^3\varepsilon_0 c}\gamma^2\left(\frac{\omega}{\omega_\mathrm{c}}\right)^2\left(1+\gamma^2\psi^2\right)^2 \times\left[K_{2/3}^2(\xi)+\frac{\gamma^2\psi^2}{1+\gamma^2\psi^2}K_{1/3}^2(\xi)\right] \tag{7.1}$$

where $\xi = \frac{1}{2}(\omega/\omega_\mathrm{c})\left(1+\gamma^2\psi^2\right)^{3/2}$ as before and the units are J s/rad^2 into unit frequency bandwidth. When we use eqn (7.1) to calculate the energy emitted by the electron we multiply the result by the frequency bandwidth of the detector to obtain the energy (in J/rad^2) incident on the detector resulting from that one electron.

We can simplify the equation and make calculations easier by noting that the fine structure constant, which is a dimensionless number approximately equal to 1/137 is given by

$$\alpha = \frac{q^2}{4\pi\varepsilon_0\hbar c}$$

so we can rewrite eqn (7.1) in the form

$$\frac{\mathrm{d}^2 I(\omega, \boldsymbol{n})}{\mathrm{d}\Omega\,\mathrm{d}\omega} = \frac{3\alpha\hbar}{4\pi^2}\gamma^2\left(\frac{\omega}{\omega_\mathrm{c}}\right)^2\left(1+\gamma^2\psi^2\right)^2 \times\left[K_{2/3}^2(\xi)+\frac{\gamma^2\psi^2}{1+\gamma^2\psi^2}K_{1/3}^2(\xi)\right]$$

and $\hbar$ which is the Planck constant divided by 2π is equal to 1.054×10^{-34} J s or 6.582×10^{-22} MeV s.

Integration of eqn (7.1), over the angle ψ (the observation angle above or below the median plane), gives the required result for both states of polarization. The integration is not straightforward because of the dependence of the parameter ξ on ψ. The treatment given here follows that presented by Wiedemann.[2] In order to simplify the problem, we note that the solid angle $d\Omega = d\theta\, d\psi$ and we separate the ψ dependence from the other quantities in an obvious way [eqn (7.2)]:

$$\frac{d^2 I(\omega)}{d\theta\, d\omega} = \frac{3}{4}\frac{q^2}{4\pi^3\varepsilon_0 c}\gamma^2\left(\frac{\omega}{\omega_c}\right)^2 \times \int_{-\infty}^{+\infty}\left[\left(1+\gamma^2\psi^2\right)^2 K_{2/3}^2(\xi) + \gamma^2\psi^2\left(1+\gamma^2\psi^2\right) K_{1/3}^2(\xi)\right] d\psi \tag{7.2}$$

and replace the modified Bessel functions with Airy functions defined as

$$\mathrm{Ai}(x) = \frac{1}{\pi}\sqrt{\frac{x}{3}}K_{1/3}\left(\frac{2x^{3/2}}{3}\right) \tag{7.3a}$$

and its derivative is

$$\mathrm{Ai}'(x) = -\frac{1}{\pi}\frac{x}{\sqrt{3}}K_{2/3}\left(\frac{2x^{3/2}}{3}\right) \tag{7.3b}$$

so that making this replacement transforms the dependence of the Bessel function from the 3/2 power of $(1+\gamma^2\psi^2)$ to a linear dependence. Inspection of eqns (7.2) and (7.3) shows that $\xi = \frac{2}{3}x^{2/3}$, so that

$$x = \left(\frac{3\omega}{4\omega_c}\right)^{2/3}\left(1+\gamma^2\psi^2\right). \tag{7.4}$$

With this substitution we can rewrite eqn (7.2) in terms of the Airy functions:

$$\begin{aligned}\frac{d^2 I(\omega)}{d\theta\, d\omega} &= \frac{4q^2}{4\pi^3\varepsilon_0 c}\gamma^2 \\ &\times \int_{-\infty}^{+\infty}\left(\frac{3\omega}{4\omega_c}\right)^{2/3}\mathrm{Ai}'^2\left(\left(\frac{3\omega}{4\omega_c}\right)^{2/3}\left(1+\gamma^2\psi^2\right)\right) d\psi \\ &+ \int_{-\infty}^{+\infty}\left(\frac{3\omega}{4\omega_c}\right)^{4/3}\gamma^2\psi^2\mathrm{Ai}^2\left(\left(\frac{3\omega}{4\omega_c}\right)^{2/3}\left(1+\gamma^2\psi^2\right)\right) d\psi.\end{aligned} \tag{7.5}$$

Equation (7.5) is an equivalent statement of the radiation spectrum to eqn (7.2) but is in a more suitable form for integration. We concentrate first on the integral over the square of the Airy function, multiplied by ψ^2, which is the second term of eqn (7.5) and to make a start on this we note that the Airy function itself can be written as[3]

$$\mathrm{Ai}(x) = \frac{1}{\pi}\int_0^\infty \cos\left(\frac{u^3}{3} + xu\right) \mathrm{d}u$$

so that

$$\mathrm{Ai}^2(x) = \frac{1}{\pi^2}\int_0^\infty \cos\left(\frac{u^3}{3} + xu\right) \mathrm{d}u \int_0^\infty \cos\left(\frac{v^3}{3} + xv\right) \mathrm{d}v \qquad (7.6)$$

and make the substitution

$$s = u + v, \qquad t = u - v \qquad (7.7)$$

so that

$$\cos\left(\frac{u^3}{3} + xu\right) = \cos\left[\frac{1}{2}\left\{\left(\frac{s^3}{12} + \frac{3st^2}{12} + xs\right) + \left(\frac{t^3}{12} + \frac{3ts^2}{12} + xt\right)\right\}\right]$$

and

$$\cos\left(\frac{v^3}{3} + xv\right) = \cos\left[\frac{1}{2}\left\{\left(\frac{s^3}{12} + \frac{3st^2}{12} + xs\right) - \left(\frac{t^3}{12} + \frac{3ts^2}{12} + xt\right)\right\}\right].$$

To complete the substitution we need the area element $\mathrm{d}u\,\mathrm{d}v$ and we can obtain this by noting that eqns (7.7) can be written as a rotation of the (s, t) coordinate system, scaled by a factor $1/\sqrt{2}$, through an angle of 45° [eqns (7.8)], i.e.

$$\frac{s}{\sqrt{2}} = u\cos\theta + v\sin\theta, \qquad \frac{t}{\sqrt{2}} = u\cos\theta - v\sin\theta$$
$$\text{where } \theta = 45^\circ \text{ and } \cos\theta = \sin\theta = \frac{1}{\sqrt{2}} \qquad (7.8)$$

so that $\mathrm{d}u\,\mathrm{d}v = (\mathrm{d}s/\sqrt{2})(\mathrm{d}t/\sqrt{2})$ and eqn (7.6) becomes

$$\mathrm{Ai}^2(x) = \frac{1}{4\pi^2}\int_0^\infty\int_0^\infty \left\{\cos\left(\frac{s^3}{12} + \frac{3st^2}{12} + xs\right) + \cos\left(\frac{t^3}{12} + \frac{3ts^2}{12} + xt\right)\right\} \mathrm{d}s\,\mathrm{d}t \qquad (7.9)$$

where we have made use of the identity

$$2\cos\frac{a+b}{2}\cos\frac{a-b}{2} = \cos a + \cos b. \qquad (7.10)$$

Equation (7.9) is symmetric in s and t so the integration over the cosine function is the same for both terms and we can simplify the integral further to give

$$\mathrm{Ai}^2(x) = \frac{1}{2\pi^2}\int_0^\infty\int_0^\infty \cos\left(\frac{1}{12}\left(s^3 + 3st^2\right) + xs\right) \mathrm{d}s\,\mathrm{d}t \qquad (7.11)$$

for the square of the Airy function.

We now insert the ψ dependence explicitly and obtain [from the definition of x in eqn (7.4)], the form of the second integral in eqn (7.5):

$$\int_{-\infty}^{+\infty} \psi^2 \mathrm{Ai}^2\left(\left(\frac{3\omega}{4\omega_c}\right)^{2/3}\left(1+\gamma^2\psi^2\right)\right)\mathrm{d}\psi$$
$$= \frac{1}{\pi^2}\int_0^{\infty}\int_0^{\infty}\int_0^{\infty} \psi^2 \cos\left(\frac{1}{12}\left(s^3+3st^2\right)\right.$$
$$\left. + s\left(\frac{3\omega}{4\omega_c}\right)^{2/3}\left(1+\gamma^2\psi^2\right)\right)\mathrm{d}s\,\mathrm{d}t\,\mathrm{d}\psi. \quad (7.12)$$

Because the integral depends on ψ^2, we need to integrate over positive values of ψ only.

Next we rearrange eqn (7.12) by collecting terms depending on the variable of integration s so that

$$\int_{-\infty}^{+\infty} \psi^2 \mathrm{Ai}^2\left(\left(\frac{3\omega}{4\omega_c}\right)^{2/3}\left(1+\gamma^2\psi^2\right)\right)\mathrm{d}\psi$$
$$= \frac{1}{\pi^2}\int_0^{\infty}\int_0^{\infty}\int_0^{\infty} \psi^2 \cos\left(\frac{1}{12}s^3 + s\left(\frac{1}{4}t^2+\gamma^2\psi^2\left(\frac{3\omega}{4\omega_c}\right)^{2/3}\right.\right.$$
$$\left.\left. + \left(\frac{3\omega}{4\omega_c}\right)^{2/3}\right)\right)\mathrm{d}s\,\mathrm{d}t\,\mathrm{d}\psi. \quad (7.13)$$

Now make the substitution

$$\frac{t}{2} = r\cos\chi, \qquad \left(\frac{3\omega}{4\omega_c}\right)^{1/3}\gamma\psi = r\sin\chi. \quad (7.14)$$

With this substitution

$$r^2 = \frac{t^2}{4} + \gamma^2\psi^2\left(\frac{3\omega}{4\omega_c}\right)^{2/3} \quad (7.15)$$

and

$$\tan\chi = 2\left(\frac{3\omega}{4\omega_c}\right)^{1/3}\frac{\gamma\psi}{t} \quad (7.16)$$

so that the integration of ψ from 0 to ∞ is replaced by the integration of χ from 0 to $\pi/2$. The substitution defined in eqn (7.14) is formally equivalent to a transformation from rectangular to polar coordinates so that the area element $r\,\mathrm{d}\chi\,\mathrm{d}r$ is given by

$$\mathrm{d}\left(\frac{t}{2}\right)\mathrm{d}\left(\gamma\psi\left(\frac{3\omega}{4\omega_c}\right)^{1/3}\right) = r\mathrm{d}\chi\mathrm{d}r \quad (7.17)$$

so that

$$\mathrm{d}t\,\mathrm{d}\psi = \frac{2r\mathrm{d}\chi\mathrm{d}r}{\gamma\left(3\omega/4\omega_c\right)^{1/3}}. \quad (7.18)$$

Inserting all this into eqn (7.13) gives

$$\int_{-\infty}^{+\infty} \psi^2 \mathrm{Ai}^2\left(\left(\frac{3\omega}{4\omega_c}\right)^{2/3}\left(1+\gamma^2\psi^2\right)\right) \mathrm{d}\psi$$
$$= \frac{1}{\pi^2\gamma^3\,(3\omega/4\omega_c)} \int_0^{\infty} \sin^2\chi\,\mathrm{d}\chi$$
$$\times \int_0^{\infty}\int_0^{\infty} r^2 \cos\left(\frac{1}{12}s^3 + s\left(r^2 + \left(\frac{3\omega}{4\omega_c}\right)^{2/3}\right)\right) 2r\mathrm{d}r\mathrm{d}s \tag{7.19}$$

and we can integrate over χ [eqn (7.20)]:

$$\int_0^{\pi/2} \sin^2\chi\,\mathrm{d}\chi = \int_0^{\pi/2} \frac{1}{2}(1-\cos 2\chi)\,\mathrm{d}\chi = \left[\frac{1}{2}\left(\chi - \frac{\sin 2\chi}{2}\right)\right]_0^{\pi/2} = \frac{\pi}{4} \tag{7.20}$$

to obtain

$$\int_{-\infty}^{+\infty} \psi^2 \mathrm{Ai}^2\left(\left(\frac{3\omega}{4\omega_c}\right)^{2/3}\left(1+\gamma^2\psi^2\right)\right) \mathrm{d}\psi$$
$$= \frac{1}{2\pi\gamma^3\,(3\omega/4\omega_c)} \int_0^{\infty}\int_0^{\infty} r^2 \cos\left(\frac{1}{12}s^3 + s\left(r^2 + \left(\frac{3\omega}{4\omega_c}\right)^{2/3}\right)\right) r\,\mathrm{d}r\,\mathrm{d}s. \tag{7.21}$$

The final step is to transform the argument of the cosine function into a form resembling an Airy function which we can do by setting $w^3/3 = s^3/12$ so that $w = s/2^{2/3}$, $\mathrm{d}s = 2^{2/3}\mathrm{d}w$ and

$$x = 2^{2/3}\left(r^2 + \left(\frac{3\omega}{4\omega_c}\right)^{2/3}\right)$$

so that $\mathrm{d}x = 2^{5/3} r\,\mathrm{d}r$. The new limits of integration are: when $r = 0$, $x = x_0$ given by

$$x_0 = \left(\frac{3\omega}{2\omega_c}\right)^{2/3}, \tag{7.22}$$

and when $r = \infty$, $x = \infty$. Also

$$r^2 = \left(\frac{x}{2^{2/3}} - \left(\frac{3\omega}{4\omega_c}\right)^{2/3}\right) \tag{7.23}$$

and $2r\,\mathrm{d}r\,\mathrm{d}s = \mathrm{d}x\,\mathrm{d}w$. The integral now becomes

$$\int_{-\infty}^{+\infty} \psi^2 \mathrm{Ai}^2\left(\left(\frac{3\omega}{4\omega_c}\right)^{2/3}\left(1+\gamma^2\psi^2\right)\right)\mathrm{d}\psi$$

$$= \frac{1}{4\gamma^3(3\omega/4\omega_c)}\int_{x_0}^{\infty}\left(\frac{x}{2^{2/3}} - \left(\frac{3\omega}{4\omega_c}\right)^{2/3}\right)\frac{1}{\pi}\int_0^{\infty}\cos\left(\frac{w^3}{3}+xw\right)\mathrm{d}w\,\mathrm{d}x$$

$$= \frac{1}{4\gamma^3(3\omega/4\omega_c)}\int_{x_0}^{\infty}\left(\frac{x}{2^{2/3}} - \left(\frac{3\omega}{4\omega_c}\right)^{2/3}\right)\mathrm{Ai}(x)\,\mathrm{d}x,$$

$$\text{where } x_0 = \left(\frac{3\omega}{2\omega_c}\right)^{2/3}. \tag{7.24}$$

The Airy functions are solutions of the equation

$$\mathrm{Ai}''(x) - x\mathrm{Ai}(x) = 0 \tag{7.25}$$

so that

$$\int_{x_0}^{\infty} x\mathrm{Ai}\,(x)\,\mathrm{d}x = \int_{x_0}^{\infty}\mathrm{Ai}''(x)\,\mathrm{d}x = [\mathrm{Ai}'(x)]_{x_0}^{\infty} = -\mathrm{Ai}'(x_0) \tag{7.26}$$

and inserting this result into eqn (7.24) gives the final result for this integral:

$$\int_{-\infty}^{+\infty} \psi^2 \mathrm{Ai}^2\left(\left(\frac{3\omega}{4\omega_c}\right)^{2/3}\left(1+\gamma^2\psi^2\right)\right)\mathrm{d}\psi$$

$$= -\frac{1}{4\gamma^3(3\omega/4\omega_c)^{1/3}}\left(\frac{\mathrm{Ai}'(x_0)}{x_0} + \int_{x_0}^{\infty}\mathrm{Ai}(x)\,\mathrm{d}x\right) \tag{7.27}$$

with $x_0 = (3\omega/2\omega_c)^{2/3}$ as before.

In order to compute the integral over the first derivative of the Airy function in eqn (7.5) we use eqn (7.11) and write down the integral over the square of the Airy function itself, by analogy with eqn (7.12):

$$\int_{-\infty}^{+\infty} \mathrm{Ai}^2\left(\left(\frac{3\omega}{4\omega_c}\right)^{2/3}\left(1+\gamma^2\psi^2\right)\right)\mathrm{d}\psi$$

$$= \frac{1}{\pi^2}\int_0^{\infty}\int_0^{\infty}\int_0^{\infty}\cos\left(\frac{1}{12}\left(s^3+3st^2\right)+s\left(\frac{3\omega}{4\omega_c}\right)^{2/3}\left(1+\gamma^2\psi^2\right)\right)\mathrm{d}s\,\mathrm{d}t\,\mathrm{d}\psi$$

and use the same argument as before [eqns (7.13)–(7.19)] to reduce this integral to a manageable form. Now eqn (7.20) becomes

$$\int_0^{\pi/2}\mathrm{d}\chi = \frac{\pi}{2}$$

and, by making the same substitutions in the integral as detailed in eqns (7.21)–(7.26), we obtain

$$\int_{-\infty}^{+\infty} \mathrm{Ai}^2\left(\left(\frac{3\omega}{4\omega_c}\right)^{2/3}\left(1+\gamma^2\psi^2\right)\right)\mathrm{d}\psi = \frac{1}{2\gamma\,(3\omega/4\omega_c)^{1/3}}\int_{x_0}^{\infty}\mathrm{Ai}(x)\,\mathrm{d}x \tag{7.28}$$

with $x_0 = (3\omega/2\omega_c)^{2/3}$ as before.

Finally we can now obtain the integral over $\mathrm{Ai}'^2(x)$. The trick is to carry out a double differentiation of eqn (7.28) with respect to the variable $(3\omega/4\omega_c)^{2/3}$. To do this we need the rules for carrying out this operation which involves differentiation under an integral sign. The integral is of the general form

$$I = \int_{u(x)}^{v(x)} f(x,t)\,\mathrm{d}t = \left|F(x,t)\right|_{t=u(x)}^{t=v(x)} = F(x,v(x)) - F(x,u(x)) \tag{7.29}$$

and we require the total differential $\mathrm{d}I/\mathrm{d}x$. We note first of all that if the small incremental quantity δI, which is some function of x, u, and v, is imagined as a vector in a multidimensional space with components $\delta x, \delta v$, and δu along the x-, v-, and u-directions with the direction cosines $\partial x/\partial I, \partial v/\partial I$, and $\partial u/\partial I$, then the magnitude of δI can be written as the sum over each component divided by the corresponding direction cosine, so that

$$\delta I = \frac{\partial I}{\partial x}\delta x + \frac{\partial I}{\partial v}\delta v + \frac{\partial I}{\partial u}\delta u$$

and, in the limit, when δI, δx, etc. tend to zero, we can write

$$\frac{\mathrm{d}I}{\mathrm{d}x} = \frac{\partial I}{\partial x} + \frac{\partial I}{\partial v}\frac{\mathrm{d}v}{\mathrm{d}x} + \frac{\partial I}{\partial u}\frac{\mathrm{d}u}{\mathrm{d}x}. \tag{7.30}$$

We remember that the partial derivative $\partial I/\partial x$ is the derivative of I with respect to x, all other variables being held constant, and likewise for the other partial derivatives, and with this in mind we can apply eqn (7.30) to eqn (7.29) with the result

$$\begin{aligned}\frac{\mathrm{d}I}{\mathrm{d}x} &= \frac{\mathrm{d}}{\mathrm{d}x}\int_{u(x)}^{v(x)} f(x,t)\,\mathrm{d}t \\ &= \int_{u(x)}^{v(x)} \frac{\partial f(x,t)}{\partial x}\mathrm{d}t + f(x,v(x))\frac{\partial v(x)}{\partial x} - f(x,u(x))\frac{\partial u(x)}{\partial x}\end{aligned} \tag{7.31}$$

which is often known as Leibnitz's rule (Gottfried Wilhelm Leibnitz, 1646–1716).[4]

Now apply Leibnitz's rule to eqn (7.28). To clarify the algebra we first express eqn (7.28) in terms of two new variables p and q, where

$$p = \left(\frac{3\omega}{4\omega_c}\right)^{2/3}, \qquad q = \gamma\left(\frac{3\omega}{4\omega_c}\right)^{1/3}$$

so we can rewrite eqn (7.28) as

$$\int_{-\infty}^{+\infty} \mathrm{Ai}^2\left(p + q^2\psi^2\right)\mathrm{d}\psi = \frac{1}{2q}\int_{2^{2/3}p}^{\infty}\mathrm{Ai}(x)\,\mathrm{d}x. \tag{7.32}$$

We now differentiate eqn (7.32) twice with respect to the variable p. Although q is itself a function of p, we can ignore this because we are applying the same operation to both sides of the equation. The operation is being applied, not to determine the rate of change of each side consequent upon a change in the underlying variable $3\omega/4\omega_c$ but as an operation which generates for us the function $\mathrm{Ai}'^2(x)$. In other words, we write

$$\frac{\mathrm{d}^2}{\mathrm{d}p^2}\left\{\int_{-\infty}^{+\infty}\mathrm{Ai}^2\left(p+q^2\psi^2\right)\mathrm{d}\psi\right\}=\frac{\mathrm{d}^2}{\mathrm{d}p^2}\left\{\frac{1}{2q}\int_{2^{2/3}p}^{\infty}\mathrm{Ai}(x)\,\mathrm{d}x\right\}$$

and make use of Leibnitz's rule. The first differentiation gives

$$\frac{\mathrm{d}}{\mathrm{d}p}\left\{\int_{-\infty}^{+\infty}2\,\mathrm{Ai}\left(p+q^2\psi^2\right)\mathrm{Ai}'\left(p+q^2\psi^2\right)\mathrm{d}\psi\right\}=\frac{\mathrm{d}}{\mathrm{d}p}\left\{-\frac{1}{2q}2^{2/3}\mathrm{Ai}\left(2^{2/3}p\right)\right\}$$

and the second

$$2\int_{-\infty}^{+\infty}\left[\mathrm{Ai}'^2\left(p+q^2\psi^2\right)+\mathrm{Ai}\left(p+q^2\psi^2\right)\mathrm{Ai}'\left(p+q^2\psi^2\right)\right]\mathrm{d}\psi$$
$$=-\frac{2^{1/3}}{q}\mathrm{Ai}'\left(2^{2/3}p\right). \qquad (7.33)$$

Next, we make use of the identity of eqn (7.25) and rewrite eqn (7.33) as

$$2\int_{-\infty}^{+\infty}\left[\mathrm{Ai}'^2\left(p+q^2\psi^2\right)+\left(p+q^2\psi^2\right)\mathrm{Ai}^2\left(p+q^2\psi^2\right)\right]\mathrm{d}\psi$$
$$=-\frac{2^{1/3}}{q}\mathrm{Ai}'\left(2^{2/3}p\right). \qquad (7.34)$$

The integrals over the square of the Airy function are already known from eqns (7.27) and (7.28); so, inserting these in eqn (7.34) gives the result

$$\int_{-\infty}^{+\infty}\mathrm{Ai}'^2\left(\left(\frac{3\omega}{4\omega_c}\right)^{2/3}\left(1+\gamma^2\psi^2\right)\right)\mathrm{d}\psi$$
$$=-\frac{(3\omega/4\omega_c)^{1/3}}{4\gamma}\left[\frac{3\mathrm{Ai}'(x_0)}{x_0}+\int_{x_0}^{\infty}\mathrm{Ai}(x)\,\mathrm{d}x\right] \qquad (7.35)$$

with $x_0=(3\omega/2\omega_c)^{2/3}$ as usual.

We now substitute these integrals [eqns (7.27) and (7.35)] in eqn (7.5) to give an expression for the energy spectrum of the synchrotron radiation, integrated over the whole range of vertical angles:

$$\frac{\mathrm{d}^2 I(\omega)}{\mathrm{d}\theta\,\mathrm{d}\omega}=\frac{3\gamma}{4}\frac{4q^2}{4\pi^3\varepsilon_0 c}\left(\frac{\omega}{\omega_c}\right)$$
$$\times\left\{-\left[\frac{3\mathrm{Ai}'(x_0)}{x_0}+\int_{x_0}^{\infty}\mathrm{Ai}(x)\,\mathrm{d}x\right]-\left[\frac{\mathrm{Ai}'(x_0)}{x_0}+\int_{x_0}^{\infty}\mathrm{Ai}(x)\,\mathrm{d}x\right]\right\}.$$

The last stage is to transform this expression into the form involving Bessel functions. From eqn (7.3) and the definition of x_0 we have

$$\frac{\mathrm{Ai}'(x_0)}{x_0} = -\frac{1}{\pi\sqrt{3}} K_{2/3}\left(\frac{\omega}{\omega_c}\right) \tag{7.36a}$$

and

$$\int_{x_0}^{\infty} \mathrm{Ai}(x)\,\mathrm{d}x = \frac{1}{\pi\sqrt{3}} \int_{x_0}^{\infty} \sqrt{x} K_{1/3}\left(\frac{2}{3}x^{3/2}\right) \mathrm{d}x \tag{7.36b}$$

if the variable of integration y is given by $y = \frac{2}{3}x^{3/2}$ so that $\mathrm{d}y = \sqrt{x}\,\mathrm{d}x$ then, when $x = x_0$, $y_0 = \omega/\omega_c$, the Airy integral [eqn (7.36)] can be rewritten as

$$\int_{x_0}^{\infty} \mathrm{Ai}(x)\,\mathrm{d}x = \frac{1}{\pi\sqrt{3}} \int_{\omega/\omega_c}^{\infty} K_{1/3}(y)\,\mathrm{d}y. \tag{7.37}$$

We now make use of the recurrence relation for the Bessel functions:[5]

$$\begin{aligned} K_{\nu-1}(y) - K_{\nu+1}(y) &= -2K'_{\nu}(y), \\ K_{-\nu}(y) &= K_{\nu}(y); \end{aligned} \tag{7.38}$$

so, using eqn (7.38) we may write the Airy integral in eqn (7.37) as

$$\int_{x_0}^{\infty} \mathrm{Ai}(x)\,\mathrm{d}x = -\frac{2}{\pi\sqrt{3}} \int_{\omega/\omega_c}^{\infty} K'_{2/3}(y)\,\mathrm{d}y - \frac{1}{\pi\sqrt{3}} \int_{\omega/\omega_c}^{\infty} K_{5/3}(y)\,\mathrm{d}y. \tag{7.39}$$

The first term in eqn (7.39) can be integrated if we remember that $K_{2/3}(y) \to 0$ as $y \to \infty$, so that

$$\int_{\omega/\omega_c}^{\infty} K'_{2/3}(y)\,\mathrm{d}y = \left|K_{2/3}(y)\right|_{\omega/\omega_c}^{\infty} = -K_{2/3}\left(\frac{\omega}{\omega_c}\right)$$

and we reach the result for the photon spectrum integrated over all vertical angles:

$$\begin{aligned} \frac{\mathrm{d}^2 I}{\mathrm{d}\theta\,\mathrm{d}\omega} = \frac{\sqrt{3}}{4\pi}\gamma \left(\frac{q^2}{4\pi\varepsilon_0 c}\right)\left(\frac{\omega}{\omega_c}\right) \\ \times \left\{\left[K_{2/3}\left(\frac{\omega}{\omega_c}\right) + \int_{\omega/\omega_c}^{\infty} K_{5/3}(y)\,\mathrm{d}y\right]\right. \\ \left. + \left[-K_{2/3}\left(\frac{\omega}{\omega_c}\right) + \int_{\omega/\omega_c}^{\infty} K_{5/3}(y)\mathrm{d}y\right]\right\}. \end{aligned} \tag{7.40}$$

In eqn (7.40) the first term describes the spectrum of the horizontally polarized radiation and the second term the vertically polarized radiation. The integral in eqn (7.40) must be calculated numerically. Values of the integral itself and the integral multiplied by ω/ω_c calculated using Mathematica[6] are tabulated in Table 7.1.

TABLE 7.1 Values of $\int_{\omega/\omega_c}^{\infty} K_{5/3}(y)\,dy$ and $(\omega/\omega_c)\int_{\omega/\omega_c}^{\infty} K_{5/3}(y)\,dy$ as a function of ω/ω_c and are shown graphically in Fig. 7.1

ω/ω_c	$\int_{\omega/\omega_c}^{\infty} K_{5/3}(y)\,dy$	$\omega/\omega_c \int_{\omega/\omega_c}^{\infty} K_{5/3}(y)\,dy$
0.001	0.213139	0.00021314
0.002	0.267196	0.00053439
0.003	0.304575	0.00091373
0.004	0.333962	0.00133585
0.005	0.358497	0.00179249
0.006	0.379715	0.00227829
0.007	0.398497	0.00278948
0.008	0.4154	0.0033232
0.009	0.430802	0.00387722
0.01	0.444973	0.00444973
0.02	0.547239	0.01094478
0.03	0.613607	0.01840821
0.04	0.662796	0.02651184
0.05	0.701572	0.0350786
0.06	0.733248	0.04399488
0.07	0.759722	0.05318054
0.08	0.782199	0.06257592
0.09	0.801493	0.07213437
0.1	0.818186	0.0818186
0.2	0.903386	0.1806772
0.3	0.917705	0.2753115
0.4	0.901937	0.3607748
0.5	0.870819	0.4354095
0.6	0.831475	0.498885
0.7	0.787875	0.5515125
0.8	0.742413	0.5939304
0.9	0.696603	0.6269427
1	0.651423	0.651423
2	0.301636	0.603272
3	0.128566	0.385698
4	0.052827	0.211308
5	0.021248	0.10624
6	0.008426	0.050556
7	0.003307	0.023149
8	0.001288	0.010304
9	0.000498	0.004482
10	0.00019224	0.0019224

The spectrum summed over both states of polarization is given by eqn (7.41)

$$\frac{d^2 I}{d\theta\, d\omega} = \frac{\sqrt{3}}{2\pi}\gamma\left(\frac{q^2}{4\pi\varepsilon_0 c}\right)\left(\frac{\omega}{\omega_c}\right)\int_{\omega/\omega_c}^{\infty} K_{5/3}(y)\,dy \tag{7.41}$$

where the units are now J s/rad into unit frequency bandwidth.

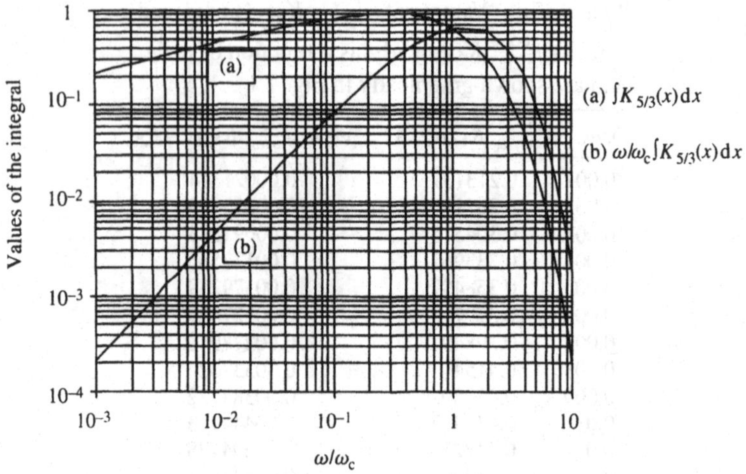

Fig. 7.1 Values of (a) $\int_{\omega/\omega_c}^{\infty} K_{5/3}(x)\,\mathrm{d}x$ and (b) $(\omega/\omega_c)\int_{\omega/\omega_c}^{\infty} K_{5/3}(x)\,\mathrm{d}x$ as a function of ω/ω_c.

Each photon with frequency ω (rad/s) carries an energy $\hbar\omega$ so that we can write

$$\frac{\mathrm{d}N_\gamma}{\mathrm{d}\theta} = \frac{\sqrt{3}}{2\pi}\alpha\gamma\left(\frac{\Delta\omega}{\omega}\right)\left(\frac{\omega}{\omega_c}\right)\int_{\omega/\omega_c}^{\infty} K_{5/3}(y)\,\mathrm{d}y \tag{7.42}$$

for the number of photons in the frequency range between ω and $\omega + \Delta\omega$ emitted by one electron in 1 rad of horizontal trajectory angle, summed over all vertical angles.

We can integrate eqn (7.41) over 2π rad and over all photon frequencies to obtain the total energy lost by one electron in one turn around the storage ring. We write

$$I = \sqrt{3}\gamma\left(\frac{q^2}{4\pi\varepsilon_0 c}\right)\int_0^{\infty}\left(\frac{\omega}{\omega_c}\right)\left\{\int_{\omega/\omega_c}^{\infty} K_{5/3}(y)\mathrm{d}y\right\}\mathrm{d}\omega$$

and integrate by parts:

$$\int_0^{\infty}\left(\frac{\omega}{\omega_c}\right)\left\{\int_{\omega/\omega_c}^{\infty} K_{5/3}(y)\,\mathrm{d}y\right\}\mathrm{d}\omega = \left|\frac{1}{2}\left(\frac{\omega^2}{\omega_c}\right)\int_{\omega/\omega_c}^{\infty} K_{5/3}(y)\,\mathrm{d}y\right|_0^{\infty} - \int_0^{\infty}\frac{1}{2}\left(\frac{\omega^2}{\omega_c}\right)\frac{\mathrm{d}}{\mathrm{d}\omega}\left\{\int_{\omega/\omega_c}^{\infty} K_{5/3}(y)\,\mathrm{d}y\right\}\mathrm{d}\omega.$$

The first term in this expression reduces to zero when the limits of integration are inserted and the second term can be evaluated further by applying the Leibnitz rule [eqn (7.31)]. The function whose integral is to be differentiated is independent of ω, as is the upper limit of integration, so we obtain

$$\frac{\mathrm{d}}{\mathrm{d}\omega}\left\{\int_{\omega/\omega_c}^{\infty} K_{5/3}(y)\,\mathrm{d}y\right\} = -\frac{1}{\omega_c} K_{5/3}\left(\frac{\omega}{\omega_c}\right)$$

and

$$\int_0^{\infty}\left(\frac{\omega}{\omega_c}\right)\left\{\int_{\omega/\omega_c}^{\infty} K_{5/3}(y)\,\mathrm{d}y\right\}\mathrm{d}\omega = \frac{1}{2}\int_0^{\infty}\left(\frac{\omega}{\omega_c}\right)^2 K_{5/3}\left(\frac{\omega}{\omega_c}\right)\mathrm{d}\omega.$$

This integral is evaluated using the general form given by Gradshteyn and Ryzhik:[7]

$$\int_0^{\infty} x^a K_\nu(x)\,\mathrm{d}x = 2^{a-1}\Gamma\left(\frac{a+\nu+1}{2}\right)\Gamma\left(\frac{a-\nu+1}{2}\right)$$

in which Γ denotes the gamma function whose properties are described e.g. in Boas.[8] In this case,

$$\begin{aligned}\int_0^{\infty}\left(\frac{\omega}{\omega_c}\right)\left\{\int_{\omega/\omega_c}^{\infty} K_{5/3}(y)\,\mathrm{d}y\right\}\mathrm{d}\omega &= \frac{\omega_c}{2}\int_0^{\infty}\left(\frac{\omega}{\omega_c}\right)^2 K_{5/3}\left(\frac{\omega}{\omega_c}\right)\mathrm{d}\left(\frac{\omega}{\omega_c}\right)\\ &= \frac{\omega_c}{2}\left[2\Gamma\left(2+\frac{1}{3}\right)\Gamma\left(\frac{2}{3}\right)\right]\end{aligned}$$

and we use the recurrence relation $\Gamma(p+1) = p\Gamma(p)$ and the product relation

$$\Gamma(p)\Gamma(p-1) = \frac{\pi}{\sin \pi p}$$

to evaluate the gamma functions to give

$$I = \frac{8\pi}{9}\gamma\left(\frac{q^2}{4\pi\varepsilon_0 c}\right)\omega_c = \frac{8\pi}{9}\alpha\gamma\varepsilon_c \tag{7.43}$$

for the energy radiated (in joules) by a single electron in one turn around a storage ring. This energy is radiated in a time $2\pi R/c$, so inserting the definition of $\omega_c = 3\gamma^3 c/2R$, we find for P_γ, the power (in watts) radiated,

$$P_\gamma = \frac{2}{3}\left(\frac{q^2 c}{4\pi\varepsilon_0}\right)\frac{\gamma^4}{R^2} = \frac{2}{3}\alpha\hbar c^2\frac{\gamma^4}{R^2} \tag{7.44}$$

which is identical to eqn (5.14) obtained by a different route and which confirms the correctness of the mathematical treatment.

Exactly the same method can be used to determine the total number of photons radiated by a single electron in one turn around the ring. In this case we integrate eqn (7.42) to obtain

$$N_\gamma = \frac{5\pi}{\sqrt{3}}\alpha\gamma. \tag{7.45}$$

This result is independent of the characteristic frequency or the characteristic energy of the spectrum and depends only on the fine structure constant and the energy of the electron.

Another useful quantity is the average photon energy, $\langle \varepsilon \rangle$, which is the total energy radiated [eqn (7.43)] divided by the total number of photons [eqn (7.45)], which gives

$$\langle \varepsilon \rangle = \frac{8}{15\sqrt{3}} \varepsilon_c. \tag{7.46}$$

Similarly, the mean square photon energy is defined as

$$\langle \varepsilon^2 \rangle = \frac{\int_0^\infty \varepsilon^2 N_\gamma(\omega)\, d\omega}{N_\gamma}, \tag{7.47}$$

where $N_\gamma(\omega)$ is the number of photons with frequency ω and energy ε obtained from eqn (7.42) and N_γ is obtained from eqn (7.45) as before. The integration follows the same pattern as above giving

$$\langle \varepsilon^2 \rangle = \frac{11}{27} \varepsilon_c^2. \tag{7.48}$$

References

1. J. Schwinger, *Phys. Rev.* **75**, 1912 (1949). D. Ivanenko and A. A. Sokolov, *Dokladii Akademiya Nauk (USSR)*, **59**, 1551 (1972).
2. Helmut Wiedemann, *Particle accelerator physics*, Vol. II, Chapter 7. Springer-Verlag, Berlin, Germany (1996).
3. J. D. Jackson, *Classical electrodynamics* (2nd edn). Chapter 7, p. 323. John Wiley & Sons, New York (1975).
4. Mary L. Boas, *Mathematical methods in the physical sciences* (2nd edn). Chapter 4, pp. 192–197. John Wiley & Sons, New York, NY, USA (1983).
5. G. N. Watson, *A treatise of the theory of Bessel functions* (2nd edn). Chapter III, § 3.71. Equations (2) and (8). p. 79. Cambridge University Press, Cambridge, England (1962).
6. Mathematica, A system for doing mathematics by computer. Version 2.2.1. Wolfram Research Inc. Champaign, Illinois, USA (1988–93).
7. I. S. Gradshteyn and I. M. Ryzhik (Alan Jeffrey, editor), *Table of integrals, series and products*, (5th edn). Academic Press, New York, NY, USA (1980). Equation 6.561–16, p. 708.
8. Mary L. Boas, *Mathematical methods in the physical sciences* (2nd edn). Chapter 11, p. 457 ff. John Wiley & Sons, New York, NY, USA (1983).

8

Introduction to electron storage rings

Principles of operation

The development of circular (or approximately circular) electron storage rings has been essential for the construction and utilization of synchrotron radiation sources. These machines have evolved from the first cyclotron, constructed by Ernest Orlando Lawrence (1901–58) at the Berkeley campus of the University of California in 1932, to the massive complexes operating or under construction at present-day research centres. A brief historical survey has been compiled by Wiedemann.[1]

Dipole magnets

Modern storage rings consist of a ring of dipolar (or dipole) magnets[2] as shown in Fig. 8.1. The arrangement of a typical dipole magnet is shown in Fig. 8.2. Ideally the magnet generates a uniform field, shown as vertical field lines in the figure. For

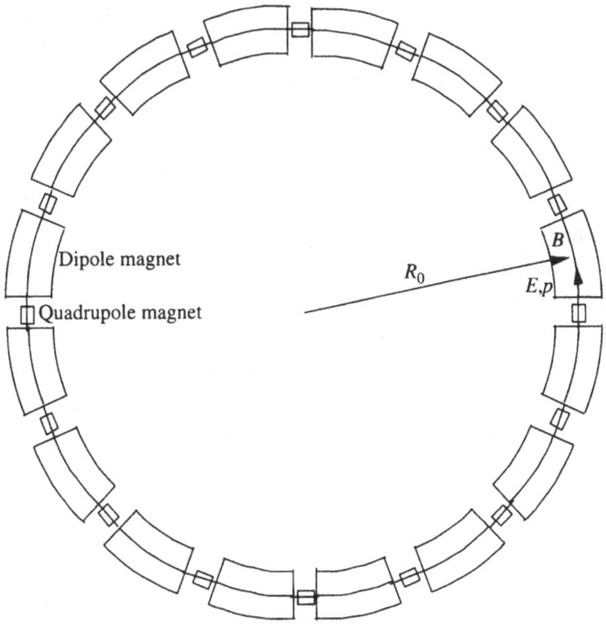

Fig. 8.1 Layout of dipolar magnets in an electron storage ring.

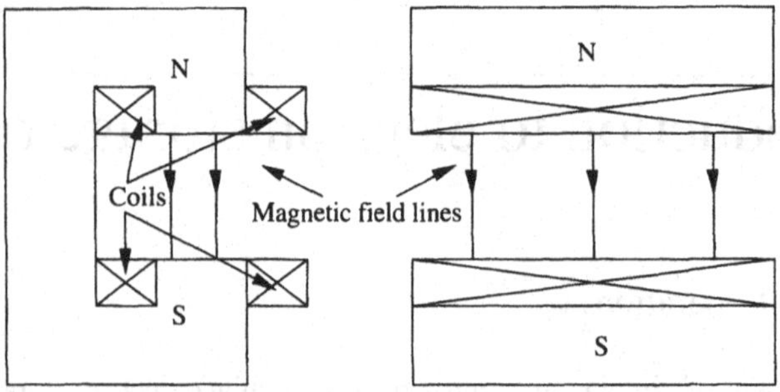

Fig. 8.2 Typical dipole magnet layout.

that reason, the force acting on an electron is always in the horizontal plane. The magnets are made from ferromagnetic material with high magnetic permeability (μ, see below) and the field strength (magnetic flux density or magnetic flux/area) in the gap is usually limited to between 1 and 1.2 T so that there is no danger of magnet saturation (non-linearity of the magnetic permeability) spoiling the uniformity of the field. In practice the magnet must be designed carefully in order to maximize the volume of uniform field within the magnet gap. Computer programs are used for this purpose. Special attention must be paid to the edges of the magnet poles where saturation occurs more easily. It is obvious that at the ends of the magnet, where the electron beam enters and leaves, there is always a field gradient. The extent of the gradient region is approximately equal to the distance between the magnet poles.

The permeability of a magnetic material differs from that of the vacuum by a numerical factor μ which allows for the magnetization of the material generated by electric currents at the atomic level. We write $\boldsymbol{B} = \mu\mu_0\boldsymbol{H}$ to allow for this effect and call $\boldsymbol{H}$ the magnetic field strength within the material. In the air gap $\mu_{\text{air}} \simeq 1$ but in the magnet steel $\mu_{\text{iron}} \simeq 1000$. By Maxwell's equations curl $\boldsymbol{B} = \mu_0\boldsymbol{j}$ so that curl $\boldsymbol{H} = \boldsymbol{j}$, whatever the value of μ.

The magnetic flux density $\boldsymbol{B}_{\text{air}} = \mu_{\text{air}}\mu_0\boldsymbol{H}$ in the air gap is generated by the current i flowing in the excitation coils which contain N turns of copper conductor producing a current density $\boldsymbol{j}$ in Amps/m^2. The yoke and the air gap form a magnetic circuit in which lines of magnetic force are continuous around the circuit so we may integrate around the closed loop (see Fig. 8.3). We assume, for this approximate calculation of the field in the gap, that the areas of the magnet yoke and the gap are equal and there is no fringe field effect.

We write

$$\int \nabla \times \boldsymbol{H} \cdot \mathrm{d}\boldsymbol{S} = \int \boldsymbol{j} \cdot \mathrm{d}\boldsymbol{S}, \tag{8.1}$$

where $\mathrm{d}\boldsymbol{S}$ is an element of area enclosed by the loop. The right-hand side of eqn (8.1) is equal to the excitation current i multiplied by the number of turns N in the coil, which is Ni ampere turns. The left-hand side of eqn (8.1) reduces, by Stokes theorem

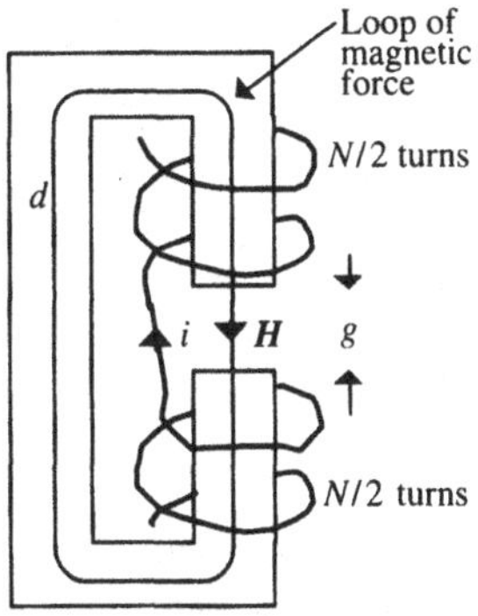

Fig. 8.3 Excitation of dipole magnet.

(Sir George Gabriel Stokes, 1819–1903), to an integration of the field strength around the loop so that, if ds is an element of length around the loop, then

$$\int \boldsymbol{H} \cdot \mathrm{d}\boldsymbol{s} = Ni. \tag{8.2}$$

Now we may evaluate the integral on the left-hand side of eqn (8.2). If g is the width of the gap between the poles and d is the length of the loop within the magnet yoke then

$$H_{\mathrm{air}} g + H_{\mathrm{iron}} d = Ni,$$

and we can use the relations between $\boldsymbol{B}$ and $\boldsymbol{H}$ to show that the excitation current required to produce a given field strength in the gap is given by eqn (8.3), which is the equivalent of the Ohm's law for a magnetic circuit (Georg Simon Ohm, 1798–1854). The magnetic flux density is given by the magnetomotive force (MMF) divided by the magnetic reluctance:

$$B = \frac{\mu_0 Ni}{(g + d/\mu_{\mathrm{iron}})}. \tag{8.3}$$

In practice, $\mu_{\mathrm{iron}} \gg 1$ so that the field strength is determined by the number of ampere turns and the width of the gap. To generate a field of 1.2 T in a 10 cm magnet gap, $\approx 10^5$ ampere turns would be required. An excitation coil with 100 turns would need to carry a current of 1000 A. Stable DC power supplies capable of delivering currents of this magnitude are required to operate the magnets and a large fraction of the running costs of the storage ring go into driving the dipole magnets.

If the dipole magnets have a uniform magnetic field B, and the energy and momentum of the electrons are E and p, respectively, then the radius of curvature, R_0, of the electron orbit in the dipole magnets is given by $p = BeR$ [eqn (5.6)], or in terms of electron energy, $E = BeR_0c$ [eqn (5.7)], where c is the velocity of light. In practical units

$$R_0 = \frac{3.33E}{B}. \tag{8.4}$$

The orbit whose radius is R_0 is called the ideal orbit.

Quadrupole magnets

The electron beam circulates in an evacuated tube and is prevented from dispersing to the edges of the tube by quadrupolar magnets (quadrupoles) whose field gradients serve to focus the electron beam. These magnets are inserted in the drift spaces between the dipole magnets. The direction of the magnetic field lines in these focusing magnets is shown in Fig. 8.4.

Electrons moving through the magnet gap close to the horizontal plane and away from the observer, as shown, are deflected towards the centre of the magnet. Because the magnetic field increases with distance from the centre, a beam of electrons is focused in this plane. On the other hand, electrons moving close to the vertical plane are deflected away from the centre and the quadrupole magnet acts as a defocusing lens in this plane. In consequence, two quadrupole magnets must be used to focus the beam in both planes simultaneously, with the second magnet rotated through 90° relative to the first. That such a symmetrical arrangement can be arranged to transport the electron beam from one focus to another is shown in Fig. 8.5, which shows the conventional ray diagram for a combination of a diverging and a converging lens.

For rays travelling from left to right, the diverging lens, one of whose principal foci is at F_1, forms a virtual image of the object O at O′. This virtual image becomes the

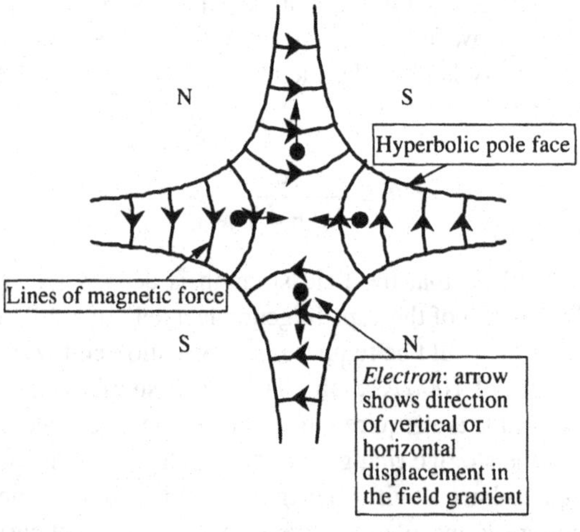

Fig. 8.4 Magnetic field in a quadrupole magnet.

object for the converging lens (focus at F_2) to form an image at I. The lens pair acts as though it were a single converging lens (an equivalent lens) with its focus at F. In the case shown in Fig. 8.5 the lens pair combines to give unit magnification. The size of the image is equal to the size of the object so, by symmetry, the combination has the same effect, independent of the order of the two lenses. In other words, such a pair transports an electron beam from O to I, to the same point on the optic axis, with unit magnification in both planes.

The focal lengths f_1 and f_2 of the lens pair are related to the focal length f of the combination by the well-known optical formula

$$\frac{1}{f} = \frac{1}{f_1} + \frac{1}{f_2} - \frac{d}{f_1 f_2},$$

where d is the distance between the two lenses. In the case shown, $f_2 = -f_1$ so that $f = f_1 f_2/d$.

How strong does this lens have to be? A magnetic field B_y acting over a length L of trajectory, perpendicular to the plane of the diagram (see Fig. 8.6) deflects an electron trajectory through an angle α proportional to the distance x (the impact parameter) of the trajectory from the optic axis which is the trajectory of zero deflection so that $\alpha = L/\rho = x/f$ and, in this case, $x = \rho$, the radius of curvature of the trajectory in the field B_y.

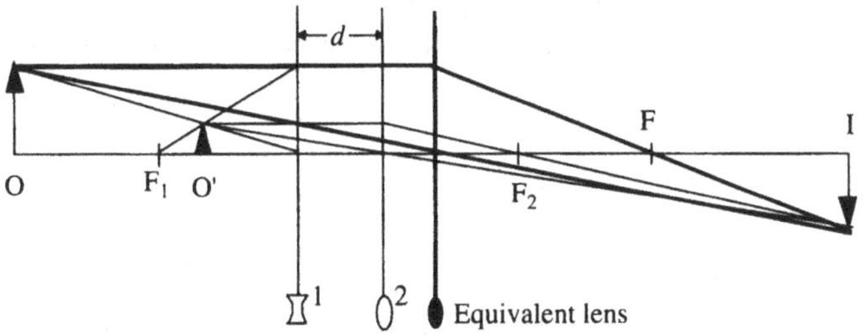

Fig. 8.5 Ray diagram for a focusing lens constructed from a diverging and a converging pair.

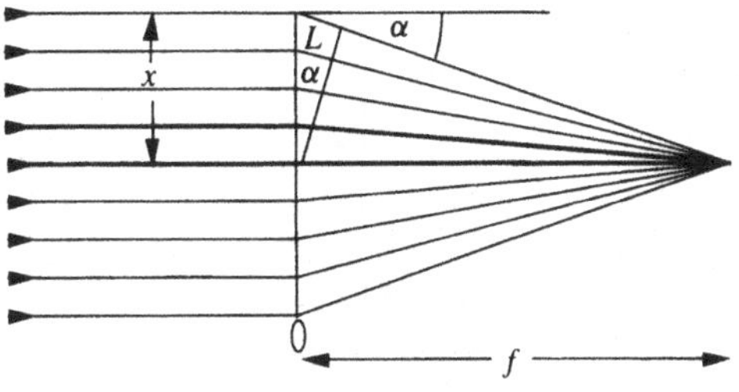

Fig. 8.6 Principle of lens action.

As usual, from eqn (8.4), for a relativistic electron with energy E, $\rho = E/B_y ec$, so that $\alpha = L(B_y ec/E)$ and

$$\frac{1}{f} = \frac{ec}{E}\frac{B_y}{x}L = \frac{gL}{3.33E}, \tag{8.5}$$

where L is now the total length of the magnet. Lens action presupposes a constant focal length, independent of the impact parameter so that $B_y = gx$, where g is the magnetic gradient, measured in T/m. In the y-direction a similar argument applies and $B_x = gy$.

As an order of magnitude calculation, consider a quadrupole pair, each of length 300 mm, separated by 3 m. If the combined focal length is 1.5 m then the focal length of each quadrupole is about 2 m. By eqn (8.5) the gradient in the quadrupole is about 10 T/m. If the bore of the quadrupole is 70 mm in diameter then the field at the surface of the pole is about 0.35 T. This low value of the magnetic field prevents saturation of the magnetic steel not only at the tip of the pole but also at the edges of the pole near the coils (e.g. near P in Fig. 8.7) where the curvature at the pole surface changes rapidly.

How can magnetic fields with constant gradient in both the x- and y-directions be generated? Figure 8.7 shows a cross section of a typical quadrupole magnet. The magnet is assumed to be long compared with the distance D between the poles. The magnet steel yoke (shown in outline), energized by an electric current in the coils shown, produces a magnetic field gradient in the region between the four poles of the magnet.

In general a magnetic field must be represented by the gradient of a vector potential (B = curl $\boldsymbol{A}$), but in this instance, in the region between the pole faces, there is

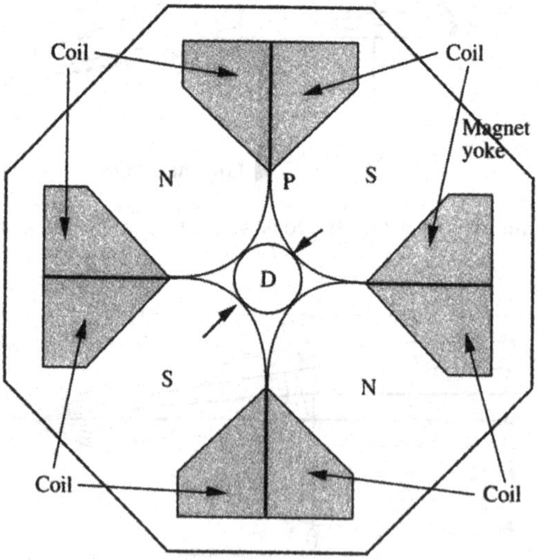

Fig. 8.7 Quadrupole magnet showing yoke, coils, and pole faces.

no electric current so the field can be described by a scalar potential function V, where $\boldsymbol{B} = \operatorname{grad} V$. In this region, a magnetic potential $V = gxy$ generates the required field $B_x = \partial V/\partial x = gy$ and $B_y = \partial V/\partial y = gx$. The equipotential lines (which extend to equipotential surfaces in the z-direction, parallel to the electron beam) are hyperbolae ($xy =$ constant), linking points with equal values of V. How do we generate this hyperbolic potential function? Consider eqn (8.3), which is the magnetic analogue of Ohm's law and states that magnetic flux = (magnetomotive force)/(magnetic reluctance). The magnetic permeability μ_{iron} is the analogue of electrical conductivity. In an electrical circuit there can be no potential difference between two points which are separated by a region of infinite conductivity (e.g. a conducting wire). If such a potential difference exists, an electric current flows to reduce the potential difference (and the electric field in the same direction, which is the gradient of the potential) to zero. The conducting wire (whatever its shape) lies along an equipotential line and all electric fields must be at right angles to it. In the same way, a pole face composed of high permeability magnetic material must form an equipotential surface and all magnetic fields must emerge at right angles to it. In the case of the dipole magnet, this theorem leads to the conclusion that to generate a uniform magnet field we require planar, parallel pole faces. In the quadrupole magnet, to form the required hyperbolic lines of equal potential we must give the appropriate hyperbolic section to the four pole faces. This ensures the linear gradient of the magnetic field throughout the region provided there is no magnetic material between the pole faces and no significant current flowing. The hyperbolic shape of the pole faces is defined by a single constant. What is its value? Figure 8.7 indicates the bore of the quadrupole with diameter D as an inscribed circle touching the tips of the pole faces. At the contact points

$$x^2 + y^2 = \frac{D^2}{4} \quad \text{and} \quad x = \pm y,$$

so that $xy = \pm D^2/8$ and it is clear that the pole face profile is defined by the bore diameter. The gradient is then determined by the excitation current in the coil windings using the same method as for the dipole magnet. Figure 8.8 shows the integration path to which we apply eqn (8.2). The section of integration path ABC within the pole and the yoke makes a negligible contribution to the integral because $\mu_{\text{iron}} \gg 1$. Because $\boldsymbol{B}$ is at right angles to the integration path along the section CO (Fig. 8.8), the scalar product $\boldsymbol{B} \cdot \mathrm{d}s = 0$ and we are left with just the section OA so that

$$\int_0^{D/2} B_r \, \mathrm{d}r = \mu_0 N i,$$

where $B_r = \sqrt{B_x^2 + B_y^2}$. The integration path has been chosen such that along OA, $B_x = B_y = gx = gy$ so that $B_r = gr$ and, therefore,

$$Ni = g\frac{D^2}{8\mu_0} = \frac{B_0 D}{4\mu_0},$$

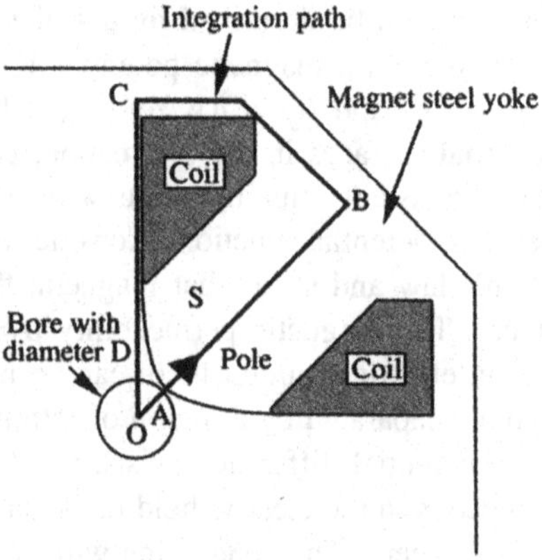

Fig. 8.8 Integration path for quadrupole magnet.

where B_0 is the magnetic field at the tip of the pole face. In practice, a current of about 5000 ampere turns in the magnet excitation coil delivers a field gradient of about 10 T/ m when the magnet bore is 70 mm as before.

Multipole magnets

Although dipoles and quadrupoles form the basic set of magnetic devices to confine the electron beam in the synchrotron radiation source, magnets with a larger number of poles are used to provide increased precision of beam control. Sextupole (six-pole) magnets are needed to reduce the chromatic aberration inherent in the quadrupolar magnetic lenses and higher order multipoles such as eight-pole (octupole) and even higher orders can be used to handle beam instabilities. The calculation of the field distribution in space and the production of this distribution by means of a suitable configuration of magnetic poles is beyond the scope of this book. An account can be found in ref. 2 and also in Wiedemann.[1]

Total energy radiated

Electrons circulating with energy E and orbit radius R_0 radiate energy in the form of synchrotron radiation at a rate given by eqn (5.20) which gives the rate of energy loss by a single electron (in J/s or W). Suppose that there are N electrons circulating in the storage ring. At any given moment only those electrons which are passing through the dipole magnets will radiate. Those traversing a drift space between the magnets will not radiate (we neglect the radiation in the weak field of the quadrupole magnets).

If L_0 is the total circumference of the electron orbit and N_{rad} is the number of radiating electrons then on average we may write

$$N_{rad} = N\frac{2\pi R_0}{L_0} \tag{8.6}$$

and, from eqn (5.20),

$$\frac{dE}{dt} = \frac{2}{3}\frac{q^2 c}{4\pi\varepsilon_0}\left(\frac{E}{m_0c^2}\right)^4 \frac{1}{R_0^2}N_{rad}. \tag{8.7}$$

We can express N_{rad} in terms of the circulating electron current I in the storage ring. We note that the current is the rate at which electric charge q passes a stationary observer located at some arbitrary point on the ring and write

$$I = \frac{dq}{dt} = \frac{dq}{ds}\frac{ds}{dt} = \frac{qN}{L_0}\beta c,$$

so that, since $\beta \approx 1$, from eqns (8.6) and (8.7) we have

$$\frac{dE}{dt} = \frac{1}{3}\frac{q}{\varepsilon_0}\left(\frac{E}{m_0c^2}\right)^4 \frac{I}{R_0}. \tag{8.8}$$

If we insert the numerical value of the electron charge $q = 1.6 \times 10^{-19}$ C and $\varepsilon_0 = 8.85 \times 10^{-12}$ F/m then we obtain

$$\frac{dE}{dt} = 88.5\frac{E^4}{R}I = 26.6E^3BI,$$

where E is measured in GeV, R_0 in m, B in T, I in A, and dU/dt in kW.

We can look at this expression in another way by expressing dE/dT in terms of the energy lost in unit distance around the ring so that

$$\frac{dE}{dt} = \frac{dE}{ds}\frac{ds}{dt}$$

and, if ΔE is the energy lost by one electron as it completes one orbit around the ring then,

$$\frac{dE}{ds} = \frac{\Delta E}{L}\frac{L}{2\pi R} = \frac{\Delta E}{2\pi R_0},$$

so that

$$\frac{dE}{dt} = \frac{\Delta E}{2\pi R_0}\beta c$$

and, from eqn (5.20), as before

$$\Delta E = \frac{2}{3}\frac{q^2 c}{4\pi\varepsilon_0}\left(\frac{E}{m_0 c^2}\right)^4 \frac{1}{R_0^2}\frac{2\pi R_0}{\beta c}$$
$$= \frac{1}{3}\frac{q}{\varepsilon_0}\left(\frac{E}{m_0 c^2}\right)^4 \frac{1}{R_0},$$

which is identical to eqn (8.8) for the power radiated for 1 A of stored current. In other words, the total power radiated (kW) is equal to the energy lost by one electron in one turn (keV) times the stored current (A).

Table 8.1 shows the power radiated in kW for 1 A of stored electron current for a number of synchrotron radiation sources.

Radio-frequency cavities

In order to maintain the average energy of the electron beam, one or more radio-frequency cavities (not shown in Fig. 8.1) must be inserted into selected drift spaces, between the ends of the quadrupole and the dipole magnet. These transfer energy to the electron beam to compensate for the energy lost by the electrons in the form of synchrotron radiation.

How do these devices operate? The force acting on a charged particle in an electromagnetic field is given by the Lorentz equation (2.6):

$$\boldsymbol{F} = e(\boldsymbol{E} + \boldsymbol{v} \times \boldsymbol{B}),$$

and, as we have shown before, the magnetic field $\boldsymbol{B}$ cannot change the energy of the particle because it produces a force at right angles to the direction of motion of the electron. On the other hand, the electric field $\boldsymbol{E}$ generates a force which is independent of the motion in a direction parallel to the field itself. It follows that we can write the energy change of the particle in the field as the integral of the scalar product of the field with the distance d$\boldsymbol{s}$ over which the field operates:

$$\Delta E = e\int \boldsymbol{E}\cdot \mathrm{d}\boldsymbol{s}. \qquad (8.9)$$

The simplest way to provide such an electric field would be as shown in Fig. 8.9(a) in which the field is generated between the plates of a capacitor C connected to the poles of a constant voltage source such as a battery or an electrostatic generator. Electrons entering the gap between the plates of the capacitor would experience a force repelling them from the negatively charged plate and attracting them to the opposite plate which carries a positive charge. They would emerge from the capacitor with an additional energy as given by eqn (8.9). This arrangement is used in electrostatic accelerators such as van de Graaff machines named after Robert Jemison van de Graaff, 1901–67. Unfortunately, it will not work in a circular accelerator because the integral in (8.9)

TABLE 8.1 Parameters of selected storage rings

Source	Electron beam energy E (GeV)	Characteristic energy ε_c (keV)	Characteristic wavelength λ_c (nm)	Magnetic field B (T)	Bending radius R_0 (m)	Radiated power (kW/A)
ALS, Berkeley, USA	1.50	1.50	0.83	1.00	4.97	88.88
SRS, Daresbury, UK	2.00	3.20	0.39	1.20	5.53	252.80
NSLS (VUV), Brookhaven, USA	0.75	0.49	2.55	1.32	1.88	14.30
NSLS (X-ray), Brookhaven, USA	2.50	5.00	0.25	1.20	6.91	493.75
ESRF, Grenoble, France*	6.00	9.60	0.13	0.40	49.73	2275.20
ESRF, Grenoble, France*	6.00	19.20	0.06	0.80	24.86	4550.40
APS, Argonne, USA	7.00	19.00	0.07	0.58	39.90	5253.50
Spring 8, Nishi-Harima, Japan	8.00	28.30	0.04	0.66	39.98	8942.80

*The field in each dipole magnet of the ESRF is split between 0.4 and 0.8 T

is zero when taken around the entire electron trajectory. The only way to make the integral non-zero is to make $\boldsymbol{E}$ time varying so that $\boldsymbol{E} = \boldsymbol{E}(s, t)$, shown schematically in Fig. 8.9(b). The capacitor (C farads) is connected in series with an inductance of L henries to form an oscillatory circuit with oscillation frequency

$$\nu = \frac{1}{2\pi\sqrt{LC}}.$$

In the circuit of Fig. 8.9(b), during the first quarter of the cycle, the field between the plates falls to zero as the energy stored in the capacitor flows into the inductance. During the next quarter, the field is restored, but in the opposite direction, as the energy flow towards the inductance continues. During the second half of the cycle, the process is reversed until the field returns to the same direction as at the start. It follows that the beam of electrons experiences only an accelerating field during half of the oscillation cycle and the circulating beam is separated into bursts of electrons, known as pulses, moving in phase with the oscillations of the electric field and the frequency of oscillation of the electric field is equal to, or a sub-multiple of, the circulation frequency of the electrons around the dipole magnet ring.

In practice, the circuit of Fig. 8.9(b) would radiate energy to its surroundings and the electrical resistance would lead to further energy loss, radiated as heat proportional to the product of the circuit resistance R and the square of the rms current (Ri^2 losses). Because of these sources of energy loss, the field would fall rapidly to zero. Table 8.1 makes it clear that prodigious amounts of power are required to replace the energy lost by the electrons as synchrotron radiation. It follows that unnecessary energy losses from the radio-frequency system cannot be tolerated. For this reason, it is necessary to develop the oscillatory circuit into a radiation cavity such as that shown in Fig. 8.9(c). The capacity and inductance are distributed over the whole structure and the cavity is totally enclosed, except for ports through which the beam enters and leaves, to minimize the radiation loss. The Ri^2 losses are minimized by making the surface of

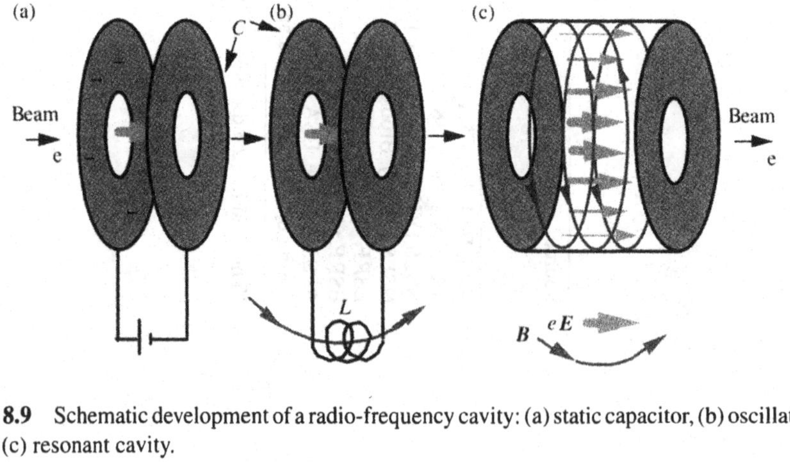

Fig. 8.9 Schematic development of a radio-frequency cavity: (a) static capacitor, (b) oscillator, and (c) resonant cavity.

the cavity of low resistive (or even superconducting) material such as electroplated copper.

The fields within the cavity can oscillate in a variety of modes which correspond to different standing wave configurations of the electric and magnetic fields. We wish to emphasize those modes which produce electric fields in the direction shown in the figure, the so-called TM modes, in which the magnetic field is transverse and the electric field is longitudinal. The mode shown in Fig. 8.9(c) is the $TM_{0,1,0}$, where the subscripts refer to the number of nodes in the azimuthal, radial, and longitudinal directions, respectively. In the $TM_{0,1,0}$, the fields are uniform except in the radial direction where there is one node, at the surface of the cavity. The electric field must be zero throughout the cavity surface (the boundary condition) if it is a perfect conductor. In order that the electric wave fits into the cavity and satisfies the boundary condition, ω, the frequency of the electric (and magnetic) waves, must be directly related to the radius of the cavity. In other words, the waveform of the electric field must satisfy the boundary condition by fitting into the cavity. The numerical relationship depends on the cavity geometry because this determines the shape of the waveform.

In the case of the simple cylindrical cavity, Fig. 8.9(c), when the radiation field is vibrating in its lowest order $TM_{0,1,0}$ mode, the electric field at time t and distance r from the centre follows a Bessel function and the longitudinal electric field can be written as

$$E(r,t) = E_0 J_0(kr) \mathrm{e}^{\mathrm{i}\omega t}$$

in which E_0 is the maximum electric field (at $r = 0$), J_0 is the zero order Bessel function, plotted in Fig. 8.10, and the wavenumber $k = \nu/c = \omega/(2\pi c)$.

In order to satisfy the boundary condition, J_0 must be zero when $r = R_{\text{cav}}$, the inside radius of the cavity. The first zero value of the Bessel function occurs when the argument $kr = 2.405$ (see Fig. 8.10) so that the frequency of the cavity is given by

$$\nu = \frac{2.405}{R_{\text{cav}}} c. \tag{8.10}$$

What determines the cavity frequency? Imagine electrons circulating in the storage ring. Each time the electrons pass through the cavity they must be in phase with

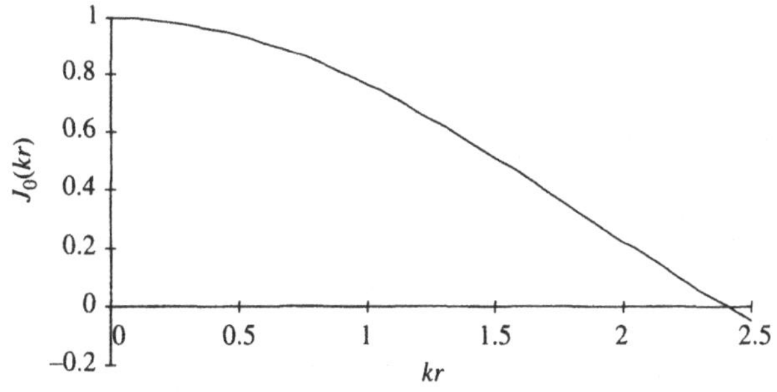

Fig. 8.10 Bessel function $J_0(kr)$.

the oscillating electric field in order to experience an accelerating force and receive energy from the radio-frequency field. It follows that the time taken for the electrons to make one turn around the ring must be equal to the time taken for the electric field in the cavity to make at least one cycle of oscillation. In other words, if electrons traverse the cavity when the electric field corresponds to a phase angle ϕ they must make their next traverse when the phase angle is $\phi + 2\pi$. In other words, the cavity must be constructed so that the oscillation frequency is equal to $n\beta c/L_0$ where L_0 is the circumference of the ring. The harmonic number n determines the number of stored bunches of electrons so that the time between individual bunches is the same as the time for one cycle of the radio frequency, $1/\nu$. The length of an individual electron bunch is some fraction of $1/\nu$.

Take the Daresbury SRS as an example: $L_0 = 96$ m so the revolution time $T_0 = L_0/c$ ($\beta = 1$) is 320 ns and the revolution frequency is 3.125 MHz. A value of 160 is chosen for the harmonic number and the radio-frequency cavity is tuned to 500 MHz so that 160 bunches of electrons are stored. These circulate around the ring with a time separation of 2 ns. The bunch length is about 120 ps. The radius of a cavity resonating at 500 MHz would be nearly 1.5 m [eqn (8.10)] so the usual design is to make the cavity doughnut-shaped, as shown in Fig. 8.11.

The resonant frequency of such a cavity can be calculated using computer programs designed for this purpose. Several of these cavities are disposed around the ring, in regions free from the magnetic field. They are fed from a radio-frequency klystron power source. Each cavity itself must have an internal surface which is smooth, clean, UHV compatible, and of high electrical (and thermal) conductivity. The cavity must be water-cooled in order to carry away the heat generated in the cavity. For example, an electron in the SRS loses about 250 keV in one turn (see Table 8.1). If the electron beam current is 300 mA then 75 kW radio-frequency power is needed (on average) just to restore the beam energy. A further 50 kW is lost in the cavities themselves. Thus the minimum total power requirement will be 125 kW at 250 kV, implying a total radio-frequency current of 0.5 A at 500 MHz. This power requirement must be split between four radio-frequency cavities in order to reduce the power loading on each cavity.

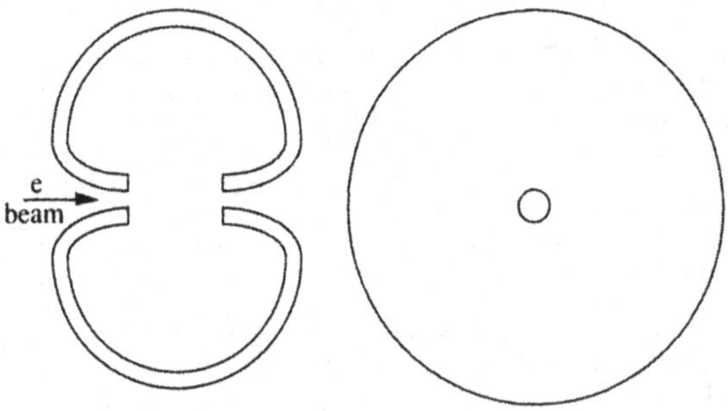

Fig. 8.11 Doughnut-shaped radio-frequency cavity.

Electron beam dimensions

Even if an individual electron is moving along the ideal orbit with the correct energy E, it does not stay there for very long. It radiates energy and immediately begins to move to an orbit of smaller radius which takes it on a new direction along a new trajectory. The radiation process is not continuous so the electron follows the new orbit for an indeterminate length of time. During that time the electron may pass through one or more radio-frequency cavities and acquire an amount of energy which may be more or less than that needed to restore the electron energy to its former value. Furthermore, the amount of energy lost at each radiation process is unpredictable, conditioned only by the synchrotron radiation spectrum. An electron may emit a photon in the infrared region of the electromagnetic spectrum and so lose a relatively small amount of energy. At the other extreme, a gamma ray (high energy X-ray photon) may be emitted. The latter causes the electron to change its orbit quite significantly. The magnetic fields in the storage rings and the parameters of the radio-frequency cavities must be chosen by the storage ring designer to limit the inevitable beam excursions and confine the electron beam to a finite size. These design considerations will be treated in Chapters 10–12.

Beam lifetime and beam movement

No storage ring is perfect in its operation. There is always a steady loss of the electron beam resulting in a gradual decrease of synchrotron radiation intensity with time. From the point of view of the synchrotron radiation user, this time–dependent intensity loss is an important factor in the design and execution of an experimental measurement such as the collection of a diffraction pattern from a single crystal or the absorption spectrum of an amorphous sample. Slow and steady beam reduction is not difficult for the experimenter to deal with. Careful monitoring of the beam intensity incident upon the sample suffices.

Rapid beam loss during data collection is a serious problem. The beam may become unstable, i.e. resonant oscillations of the electron beam trajectory in the storage ring build up rapidly to a point at which the edges of the transverse beam distribution begin to impinge on the sides of the vacuum vessel. It is the function of the storage ring designers and operators to eliminate such instabilities which tend to occur at high electron beam current. When instability develops the electron beam is lost extremely rapidly until the stored current is low enough that the instability dies away. The experimenter sees a discontinuity in the beam current or the photon flux at the experiment. An unstable beam is useless as a source of synchrotron radiation. Sudden and rapid beam loss, such as is produced by the onset of resonant instability, often forces the rejection of all the data collected during and just before that time.

Under normal operating conditions it is essential that only a negligible proportion of the beam comes into contact with the walls of the vacuum vessel. The vessel must have sufficiently large transverse dimensions so that the tails of the beam distribution do not brush against it. If the transverse beam distribution is Gaussian with standard deviations σ_x and σ_y in the horizontal and vertical directions, respectively, then the minimum width of the vacuum vessel is defined, somewhat arbitrarily, as 10σ. This restriction was not a problem in early storage rings but modern synchrotron radiation storage rings, on the other hand, require a narrow vacuum vessel in the vertical direction to accommodate insertion devices (especially undulators—see Chapter 14) which rely for their operation on magnetic fields which change sign rapidly along the electron beam path. The combination of high field and rapid longitudinal field variation means that there is correspondingly less tolerance of departures from design in both the vertical beam dimension and the permitted movement of the beam centre above or below the horizontal plane. Departure of either of these parameters from their design value can generate unacceptable beam loss and the possibility of damage to the vacuum vessel.

The factors contributing to beam size will be discussed in Chapters 10–12. Beam movement will not be discussed further except to say that the presence of beam movement obviously increases the time-averaged beam size. The reduction of beam movement to an acceptable level demands the provision of highly stable magnet power supplies and temperature regulation of all the components of the storage ring. The latter is difficult to achieve in a large storage ring, many metres in circumference. As a consequence, in modern storage rings, the position and direction of the highly collimated radiation is used to provide real-time feedback to the magnet power supplies in order to stabilize the beam position, e.g. to better than 10% of the beam diameter in the vertical direction.

Even with all of these precautions, beam is still lost even under normal, stable operation by electron scattering on gas molecules in the vacuum vessel. We can define the lifetime of the electron beam in the storage ring as the time taken for the measured beam current to fall to 1/e (e = 2.718) of its initial value. Ideally this beam lifetime must be long compared with the duration of a typical data collection period, e.g. in a single crystal diffraction experiment or for the measurement of an absorption spectrum. In order to achieve lifetimes of many hours (30 h or more in some storage rings), the residual gas pressure must be reduced to less than 10^{-7} Pa ($7.5x10^{-10}$ torr) by a variety of vacuum pumping techniques. Even then the beam lifetime is current dependent, higher current leading to shorter lifetimes because the synchrotron radiation heats the walls of the vacuum vessel and drives off the adsorbed gases which increase the residual gas pressure in the vacuum vessel and reduce the beam lifetime. Also, at high beam current, electrons in the bunch can scatter on each other, which provides another possible electron loss mechanism (known as the Touschek effect[3]). This effect reduces the beam lifetime in proportion to the electron beam density and so is most important at high beam current and small bunch size. The relative importance of the Touschek effect decreases with the fourth power of the beam energy.

Ring injection

In order to fill the storage ring with electrons (or positrons), an injection system must be provided. At the Daresbury SRS this comprises a 15 MeV linear accelerator followed by a 600 MeV electron synchrotron. Electrons are transported through a beam line of quadrupole and dipole magnets into the storage ring at an energy of 600 MeV. Once injected, the electrons are accelerated to 2 GeV energy. This process takes a few minutes. Many storage rings have adopted a similar system but others have provided for injection at full energy so that the acceleration and storage functions have been completely separated. This has the possible advantage of continuous injection so that beam loss is replaced as it occurs. In practice continuous injection has proved problematic, mainly because the beam movements induced by the injection process, when it occurs, has a negative effect on the provision of user photon beam.

Radiation shielding

The realization that high energy electrons are continuously lost from the beam under normal operation and may be lost catastrophically in the event of a serious malfunction leads one to ask the question what happens to these electrons? The answer is that these electrons and their secondary products—such as high energy gamma rays, X-rays, protons, and neutrons (from gamma-induced nuclear reactions in the vacuum vessel walls and elsewhere) must be absorbed in radiation shielding surrounding the storage ring. This shielding is designed to reduce the radiation levels outside the storage ring (i.e. in the experimental area) to levels below those permitted by current radiation safety legislation as applied in the country where the source is being operated. It is essential, in applying the legislation, to make the experimental area as open as possible, consistent with safety, so that the users of synchrotron radiation do not have to carry personal radiation monitors and can remain in the experimental area during the storage ring injection process as well as during normal operation of the storage ring. These criteria determine the thickness, density, and composition of the shield wall which separates the storage ring from the experimental area. The shield wall is, in fact, a tunnel, which completely encloses the magnet ring except for ports where synchrotron radiation can emerge, through shielded beam lines, and be directed towards the experiments (which must also be shielded). The thickness of the shield wall tunnel needs to be 1–2 m of concrete, depending on the energy of the electron beam. Heavy concrete (concrete with barium additive) is often used on and close to the median plane, with ordinary concrete above and below that plane. This concrete must be sufficiently thick to absorb neutrons which are the most serious final byproduct of the high energy electrons.

References

1. H. Wiedemann, *Particle accelerator physics I.* Springer Verlag, Berlin and Heidelberg (1993).
2. N. Marks, Conventional magnets I. *Proc. 5th General Accelerator Physics Course*, Jyväskylä, Finland, September (1992). CERN, Geneva, Switzerland. CERN 94-01 Vol. II. Also Daresbury Laboratory, Warrington, UK. DL/SCI/P861A.
3. *Synchrotron radiation. A primer* (ed. H. Winick). World Scientific, Singapore (1994); H. Wiedemann, *Particle accelerator physics II.* Springer Verlag, Berlin and Heidelberg (1995).

9

Synchrotron radiation from electron storage rings

Description of the electron beam

At any point in the storage ring, an electron can be described in terms of its position, its direction of motion, and its energy. It is convenient to measure these quantities relative to the ideal electron orbit. If we imagine a coordinate system as shown in Fig. 9.1 then at any given instant, any individual electron will be located at a point (x, y) in an x–y plane which intersects the ideal orbit at a point P, located at a distance s measured along the ideal orbit from some arbitrary starting point s_0.

This coordinate frame must be imagined as moving around the ideal orbit, so that, relative to an observer in the laboratory, the coordinate axes s and x (which with y, form a set of mutually perpendicular coordinate axes) change their direction in space as the electron moves around the storage ring. The advantage of this coordinate system is that it enables us to describe the movement of the electron in position and time relative to the ideal orbit as though the ideal orbit were a straight line.

Our imaginary electron, as well as being located at a point (x, y) in the coordinate frame, must also be travelling in some direction which will, in general, make a small angle with the s-axis. The direction of the electron can be specified by the two derivatives with respect to the s-direction, namely $\mathrm{d}x/\mathrm{d}s$ and $\mathrm{d}y/\mathrm{d}s$, which are usually denoted by x' and y'. Finally, we specify the energy of this electron by saying that it deviates from the ideal energy by an amount $\Delta E = E - E_0 = \varepsilon$. For the time being we assume that $\varepsilon = 0$.

Now imagine a bunch of electrons moving around the storage ring. It is reasonable to suppose that the bunches will be densest along the ideal orbit ($x = 0, y = 0$) and the bunch density will decrease as the values of $|x|$ and $|y|$ increase. A similar hypothesis can be made for the behaviour of the electron density in the bunch as a

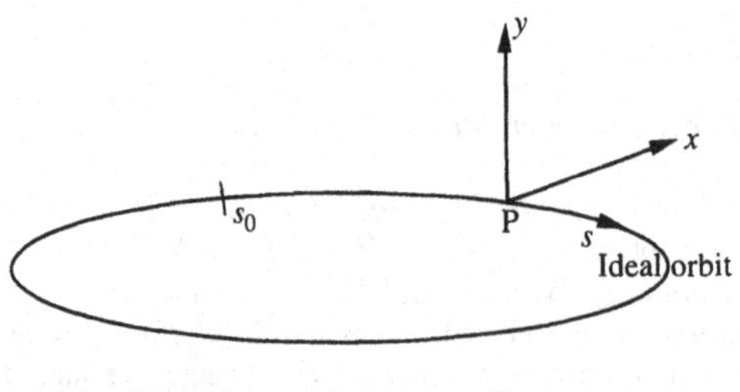

Fig. 9.1 Electron beam coordinate system.

function of x' and y'. We do not know *a priori* what the distribution of the electron density will be as a function of these variables. We will assume that the distribution function will be a Gaussian or normal distribution and that the width of the distribution in each of these variables can be specified by four standard deviations, σ_x, σ'_x and σ_y, σ'_y, so that the distribution of the electrons along the x-axis in the x–y plane at the point P can be written as

$$P(x)\,\mathrm{d}x = \frac{1}{\sigma_x\sqrt{2\pi}}\exp\left(-\frac{x^2}{2\sigma_x^2}\right)\mathrm{d}x, \tag{9.1}$$

where $P(x)$ is the probability of finding an electron at distance x from the ideal orbit when the standard deviation of the probability distribution is σ_x. The result of integrating this expression over the complete range of values of x is unity because

$$\int_{-\infty}^{\infty}\exp\left(-\frac{x^2}{2\sigma_x^2}\right)\mathrm{d}x = \sqrt{2\pi}\sigma_x.$$

If N_{tot} is the total number of electrons (or photons) in the distribution, then the number $N(x, y)$, between x and $x + \mathrm{d}x$ and $y + \mathrm{d}y$ is

$$\begin{aligned} N(x, y) &= \frac{N_{\mathrm{tot}}}{\sigma_x\sigma_y}\frac{1}{2\pi}\exp\left(-\frac{x^2}{2\sigma_x^2}\right)\mathrm{d}x\exp\left(-\frac{y^2}{2\sigma_y^2}\right)\mathrm{d}y \\ &= \frac{N_{\mathrm{tot}}}{\sigma_x\sigma_y}\frac{1}{2\pi}\exp\left(-\frac{1}{2}\left(\frac{x^2}{2\sigma_x^2}+\frac{y^2}{2\sigma_y^2}\right)\right)\mathrm{d}x\,\mathrm{d}y \end{aligned}$$

and we can use $N_{\mathrm{tot}}/(\sigma_x\sigma_y)$ as a measure of the number per mm^2 in the beam (Fig. 9.2).

The projection of the beam distribution on the x–y plane will be an ellipse with major and minor axes σ_x and σ_y. In general, $\sigma_x \gg \sigma_y$, so that the beam distribution in space will be somewhat as shown in Fig. 9.3. An observer sees a broad spread of electrons in the x-direction, in the plane of the electron orbit, and (as a consequence of

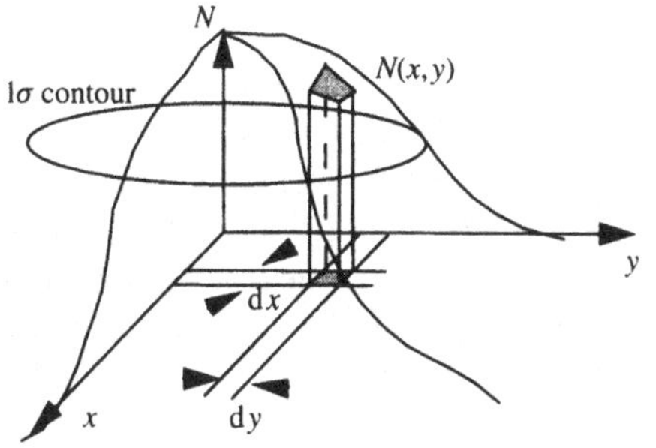

Fig. 9.2 Electron beam distribution.

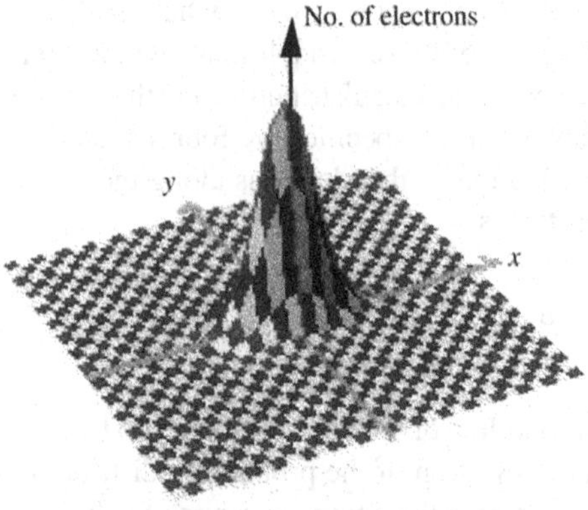

Fig. 9.3 Distribution of electron beam in x and y.

considerations which will be described later) a narrow distribution in the y-direction, perpendicular to the orbit plane.

If all of these electrons were moving parallel to the ideal orbit, and producing their photons strictly in the forward direction, then the number of observed photons in a small area $\mathrm{d}x\,\mathrm{d}y$ at the plane of detection would be equal to the expression for the synchrotron radiation spectrum, derived in an earlier chapter, multiplied by two terms of the form shown in eqn (9.1), with the insertion of the appropriate values of σ_x and σ_y. In other words, the photon distribution would be exactly that of the electron distribution, projected forward along the tangent to the ideal orbit at the point P in Fig. 9.1. This is not a bad approximation, particularly when we are interested in the X-ray end of the synchrotron radiation spectrum. However, because the electron beam has divergence as well as spatial extent, and because the photons themselves are produced with a distribution in angle relative to the direction of the individual electron which produced the photon in the first place, we must take these effects into account.

Definition of synchrotron radiation brightness

We are required to calculate the number of photons emitted from an area $\mathrm{d}x\,\mathrm{d}y$ into a cone which subtends a solid angle $\mathrm{d}\Omega$ at the source of the photons and whose axis makes an angle α with the tangent to the ideal orbit at P (the s-axis) and an angle ψ with the horizontal plane which contains the ideal electron orbit (Fig. 9.4).

Now we define a function which we call the spectral radiance or spectral brightness, $B(\omega)$ (following Born and Wolf; Mandel and Wolf[1]), then the required number of photons, which we call the photon flux, F, is

$$F = \cos\alpha \int B\left(\omega, x, y, x', y'\right) \mathrm{d}x\,\mathrm{d}y\,\mathrm{d}x'\,\mathrm{d}y'. \tag{9.2}$$

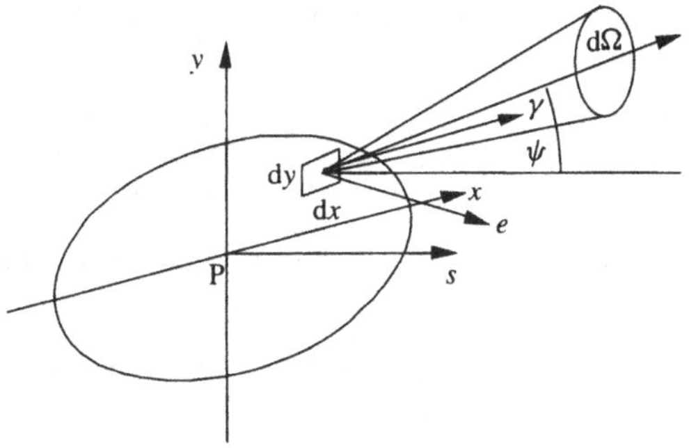

Fig. 9.4 Synchrotron radiation brightness.

Equation (9.2) is a precise definition, which states that the spectral brightness of the source is the number of photons of frequency ω emitted from unit source area into unit solid angle. The number of photons emitted by I/e electrons into unit solid angle and fractional frequency interval (sometimes called the source brilliance) is given by eqn (6.12) but this ignores the characteristics of the source. The brightness, as defined in eqn (9.2), takes into account the distribution of the electrons in both energy and angle. It must also make allowance for the distribution in angle of the radiation itself, from a single electron.

Further, we must make some approximations in order to obtain a practical expression for $B(\omega)$ which takes into account the behaviour of the angular distribution of the radiation given by eqn (6.12). In general, because α is a small angle, $\cos\alpha = 1$, this term can be dropped from the equation. The infinitesimal quantities $\mathrm{d}x$ and $\mathrm{d}y$ in eqn (9.2) simply define the source area over which the expression must be integrated to obtain the photon flux so we may calculate $B(\omega)$ from eqn (6.12) by writing

$$B\left(\omega, x, y, x', y'\right) = \frac{1}{\sigma_x \sigma_y} \frac{\mathrm{d}^2}{\mathrm{d}x'\mathrm{d}y'} \iint \left(\frac{\mathrm{d}^2 N}{\mathrm{d}\theta\, \mathrm{d}\psi}\right) \mathrm{d}\theta\, \mathrm{d}\psi .$$

Now let us examine the angular quantities $\mathrm{d}x'$ and $\mathrm{d}y'$. These must be the infinitesimal intervals of x' and y' which can project photons into the solid angle $\mathrm{d}\Omega = \mathrm{d}x'\mathrm{d}y'$. We already know that the photons themselves are projected forward from the electrons with an angular distribution given by eqn (6.12) so that in the vertical direction we may characterize the source angular distribution by approximating the shape of the angular distribution of the electrons by a Gaussian distribution with standard deviation $\sigma'_\psi(\omega)$ so that the total angular distribution is the convolution of this with the angular distribution of the electrons to give σ_γ given by

$$\sigma_\gamma'^2 = \sigma_y'^2 + \sigma_\psi'^2 . \tag{9.3}$$

In the horizontal plane, because the source is extended along the ideal orbit, not along the s-axis, the angular interval in this plane is defined by the angular distance

$d\theta$ along the orbit, and $d\Omega = d\theta\, d\psi$, not by σ_γ as above. The final expression for the spectral brightness is then

$$B(\omega, x, y, x', y') = \frac{1}{\sigma_x \sigma_y} \frac{1}{\sigma'_\gamma} \int \left(\frac{d^2 N}{d\theta\, d\psi} \right) d\psi.$$

The expression under the integral is obtained from eqn (7.41) so that

$$\frac{dN}{d\theta} = \frac{\sqrt{3}}{2\pi} \gamma \alpha \frac{d\omega}{\omega} \left(\frac{\omega}{\omega_c} \right) \int_{\omega/\omega_c}^{\infty} K_{5/3}(y)\, dy \tag{9.4}$$

and

$$B(\omega, x, y, x', y') = \frac{1}{\sigma_x \sigma_y} \frac{1}{\sigma'_\gamma} \frac{dN}{d\theta}.$$

The units of brightness are photons/(mm^2 mrad^2) into a given bandwidth interval.

Brightness is important because the number of photons reaching a sample placed at some distance for the source depends not only on the flux of photons from the source but also in their concentration. A source with a compact electron beam delivers more photons to a small sample than a source whose beam is spread over a wider area and a broader angular interval. Furthermore, as a consequence of Liouville's theorem,[2] which states that the passage of a photon beam along an optical system must be a conformal transformation, in which a change in the spatial extent of the beam must be matched by an equal and opposite change in the angular extent, the brightness of the beam reaching the target can be no greater than that entering the optical transport system, or beam line, from the source. Discussion and description of the design of beam transport systems can be found in West and Padmore.[3]

Calculation of synchrotron radiation brightness

Once the electron beam energy and the radius of the ideal orbit have been specified, the term under the integral sign can be calculated exactly. The expression for the brightness depends on the spatial and angular dimensions of the electron beam and therefore is machine dependent in a detailed way and because the electron beam parameters are dependent on the position around the ideal orbit, the brightness of the source depends on the source point of the radiation around the ring. The opening angle of the radiation also enters into the expression, thus introducing a wavelength dependence additional to that contained in the expression under the integral for the photon flux.

Use of the synchrotron radiation universal function

The expression for $dN/d\theta$ is given by eqn (9.4) in which we note that $dN/d\theta$ is the number of photons produced over the entire vertical angular range and radiated into an electron deflection angle $d\theta$ and a fractional frequency band width $d\omega/\omega$:

$$\frac{dN}{d\theta} = \frac{\sqrt{3}}{2\pi}\gamma\alpha\frac{I}{e}\frac{d\omega}{\omega}G_1(\omega/\omega_c).$$

The quantity

$$G_1(\omega/\omega_c) = \left(\frac{\omega}{\omega_c}\right)\int_{\omega/\omega_c}^{\infty} K_{5/3}(y)\,dy$$

is the synchrotron radiation universal function, plotted in Fig. 9.5.

The value of $dN/d\theta$ required for the calculation of brightness can be obtained, once the energy (E, in GeV), the value of the magnetic field (B, in T) [or the radius (R, in m) of the equilibrium orbit], and the stored current (I, in A) are known, using the following procedure to obtain G_1 from Fig 9.5.

Calculate ω_c or note that ω/ω_c can be replaced everywhere by either $\varepsilon/\varepsilon_c$ or λ_c/λ, where ε is the photon energy and λ the photon wavelength and make use of the following formulae as appropriate:

$$E = BeRc = 0.3BR,$$
$$\gamma = \frac{E}{m_e c^2} = \frac{E}{0.511 \cdot 10^{-3}},$$
$$\omega_c = \frac{3}{2}\frac{\gamma^3 c}{R} = 4.5 \cdot 10^2 \frac{\gamma^3}{R} \text{ MHz},$$
$$\varepsilon_c = \hbar\omega_c = 0.665BE^2,$$
$$\lambda_c = 5.6\frac{R}{E^3} = \frac{18.6}{BE^2},$$

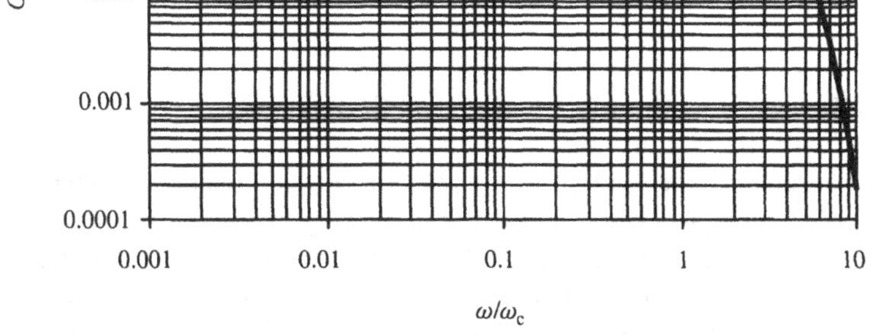

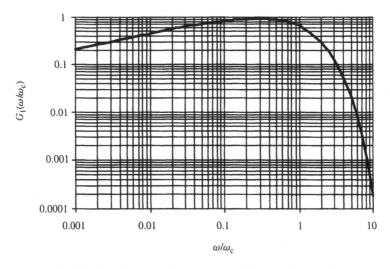

Fig. 9.5 Synchrotron radiation universal function $G_1(\omega/\omega_c)$.

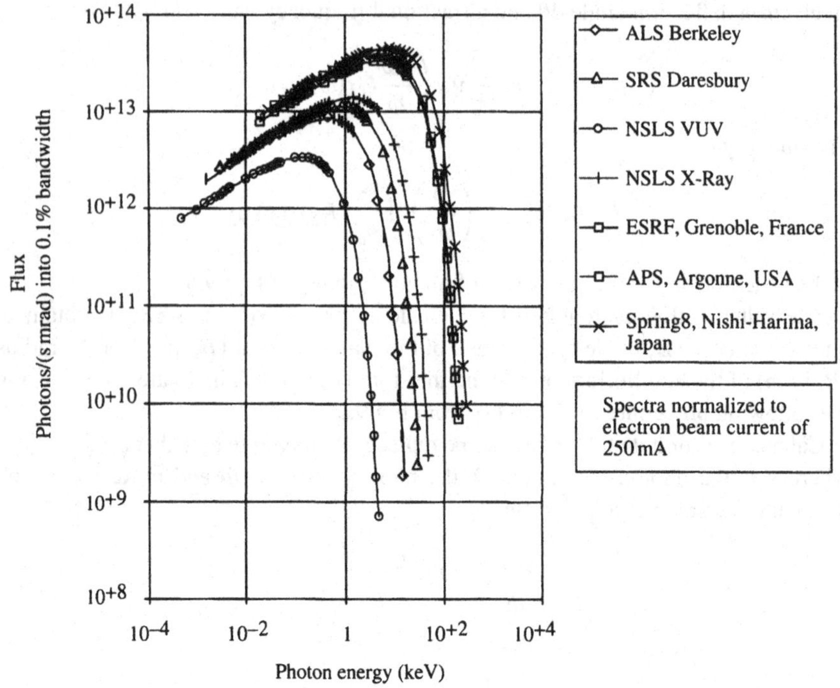

Fig. 9.6 Synchrotron radiation flux from selected sources.

and

$$\frac{\mathrm{d}N}{\mathrm{d}\theta} = 2.46 \cdot 10^{13} I E G_1(\omega/\omega_c)$$

gives the number of photons into 1 mrad and 0.1% bandwidth when the stored current is I amperes. Figure 9.6 shows the photon flux $\mathrm{d}N/\mathrm{d}\theta$ as a function of the photon energy for the selected synchroton radiation sources listed in Table 9.2. Each spectrum is normalized to an electron beam current of 250 mA.

Approximation to the photon angular distribution

In order to calculate the brightness of a synchrotron radiation beam from the value of $\mathrm{d}N/\mathrm{d}\theta$, we require a value for $\sigma'_\psi(\omega)$ to calculate σ'_γ from eqn (9.3). The expression for the angular dependence of the synchrotron radiation flux on the angle ψ does not lend itself to a simple definition of σ' [eqn (7.1)]. It is usual to proceed approximately by matching either the parallel component (the principal component) or the sum of the parallel and perpendicular components of the synchrotron radiation to a Gaussian distribution. We write the number of photons radiated into the solid angle $\mathrm{d}\Omega = \mathrm{d}\theta\,\mathrm{d}\psi$ with standard deviation σ'_ψ as

$$\frac{\mathrm{d}^2N}{\mathrm{d}\theta\,\mathrm{d}\psi} = \left.\frac{\mathrm{d}^2N}{\mathrm{d}\theta\,\mathrm{d}\psi}\right|_{\psi=0} \exp\left(-\frac{\psi^2}{2\sigma'^2_\psi}\right)\frac{\mathrm{d}\omega}{\omega}$$

into a fractional frequency bandwidth $\mathrm{d}\omega/\omega$. Integration of this expression over all vertical angles gives the total amount of radiation, which, in this approximation, is put equal to $\mathrm{d}N/\mathrm{d}\theta$ so that

$$\frac{\mathrm{d}N}{\mathrm{d}\theta} = \left.\frac{\mathrm{d}^2N}{\mathrm{d}\theta\,\mathrm{d}\psi}\right|_{\psi=0} \sigma'_\psi\sqrt{2\pi}\frac{\mathrm{d}\omega}{\omega}.$$

As before,

$$\frac{\mathrm{d}N}{\mathrm{d}\theta} = \frac{\sqrt{3}}{2\pi}\frac{\gamma\alpha}{\hbar}\frac{I}{e}\frac{\omega}{\omega_\mathrm{c}}\int_{\omega/\omega_\mathrm{c}}^{\infty} K_{5/3}(y)\,\mathrm{d}y\frac{\mathrm{d}\omega}{\omega}$$

and

$$\left.\frac{\mathrm{d}^2N(\omega)}{\mathrm{d}\theta\mathrm{d}\psi}\right|_{\psi=0} = \frac{3}{4}\frac{\alpha}{\pi^2}\gamma^2\frac{I}{e}\frac{\Delta\omega}{\omega}\left(\frac{\omega}{\omega_\mathrm{c}}\right)^2 K^2_{2/3}\left(\frac{\omega}{2\omega_\mathrm{c}}\right)$$

from eqn (6.12) with $\xi = \omega/(2\omega_c)$ so that, calling this as approximation 1, we have

$$\gamma\sigma'_{\psi 1} = \sqrt{\frac{2\pi}{3}\frac{\omega_\mathrm{c}}{\omega}\frac{\int_{\omega/\omega_\mathrm{c}}^{\infty} K_{5/3}(y)\,\mathrm{d}y}{K^2_{2/3}(\omega/2\omega_\mathrm{c})}}. \tag{9.5}$$

The alternative approximation, which ignores the relatively small contribution of the perpendicularly polarized component of the radiation to the angular distribution can be obtained very simply by multiplying eqn (9.5) by f, the fraction of the radiation polarized parallel to the horizontal plane. This is obtained from eqn (7.40) and

$$f = \frac{1}{2}\frac{K_{2/3}(\omega/\omega_\mathrm{c}) + \int_{\omega/\omega_\mathrm{c}}^{\infty} K_{5/3}(y)\mathrm{d}y}{\int_{\omega/\omega_\mathrm{c}}^{\infty} K_{5/3}(y)\mathrm{d}y},$$

so that in this alternative approximation (approximation 2),

$$\begin{aligned}\gamma\sigma'_{\psi 2} &= \sqrt{\frac{2\pi}{3}\frac{\omega_\mathrm{c}}{\omega}f\frac{\int_{\omega/\omega_\mathrm{c}}^{\infty} K_{5/3}(y)\mathrm{d}y}{K^2_{2/3}(\omega/2\omega_\mathrm{c})}}\\ &= \sqrt{\frac{2\pi}{3}\frac{\omega_\mathrm{c}}{\omega}\frac{K_{2/3}(\omega/\omega_\mathrm{c}) + \int_{\omega/\omega_\mathrm{c}}^{\infty} K_{5/3}(y)\mathrm{d}y}{2K^2_{2/3}(\omega/2\omega_\mathrm{c})}}\\ &= f\gamma\sigma'_{\psi 1}.\end{aligned}$$

In both of these expressions, $\gamma\sigma'_\psi$ is a function of ω/ω_c only. If we make use of the values of the Bessel functions tabulated previously (Tables 6.1 and 7.1) we can compare the full width at half maximum (fwhm) ($2.36 \times \gamma\sigma'_{\psi 1}$ and $2.36 \times \gamma\sigma'_{\psi 2}$), e.g. for $\omega/\omega_c = 1$. As we would expect, the approximation tends to underestimate the fwhm (Table 9.1).

Figure 9.7 shows the values of f. Figure 9.8 shows $\gamma\sigma'_{\psi 1}$ and $\gamma\sigma'_{\psi 2}$ as a function of ω/ω_c.

Figure 9.9 shows a comparison of the brightness of selected synchrotron radiation sources. In practice, the brightness of a particular beam line must be calculated using the specific parameters for that line. This is because the source parameters (horizontal beam width, etc.) are a function of the beam line and are specific to the tangent point of the line. The figure does not pretend to give detailed information but only to make a general comparison. Table 9.2 lists the electron beam parameters used in the calculation.

Values of $\sigma_{\psi 1}$ have been obtained using approximation 1 as described in eqn (9.5). The brightness also depends on the electron beam current. For the purposes of this comparison the brightness for each source has been normalized to 250 mA.

TABLE 9.1 Comparison of angular width approximation

ω/ω_c	$\int_{\omega/\omega_c}^{\infty} K_{5/3}(y)dy$	$K^2_{2/3}(\omega/2\omega_c)$	$K_{2/3}(\omega/\omega_c)$	$\gamma\sigma'_{\psi 1}$	$2.36\times$ $\gamma\sigma'_{\psi 1}$	f	$\gamma\sigma'_{\psi 2}$	$2.36\times$ $\gamma\sigma'_{\psi 2}$
1	0.651	1.454	0.494	0.65	1.53	0.88	0.57	1.35
Values of $\gamma \times$ fwhm for $\omega/\omega_c = 1$ from Fig. 6.10					1.61			1.40

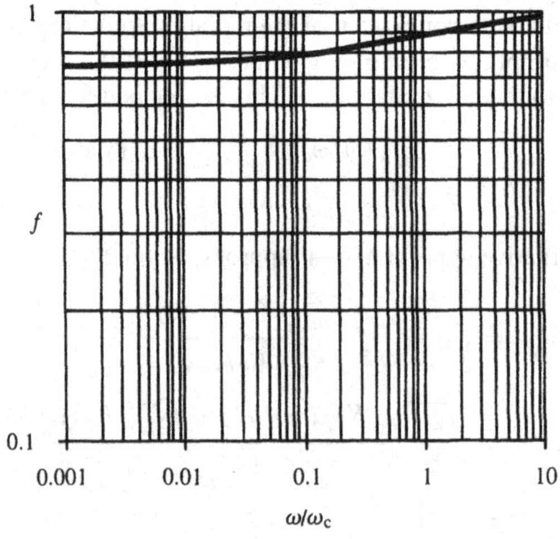

Fig. 9.7 Fraction of horizontally polarized radiation as a function of ω/ω_c.

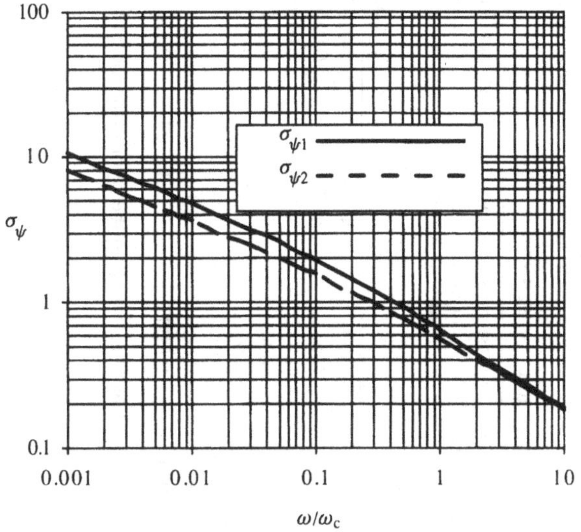

Fig. 9.8 Values of $\gamma\sigma'_{\psi 1}$ and $\gamma\sigma'_{\psi 2}$ as a function of ω/ω_c.

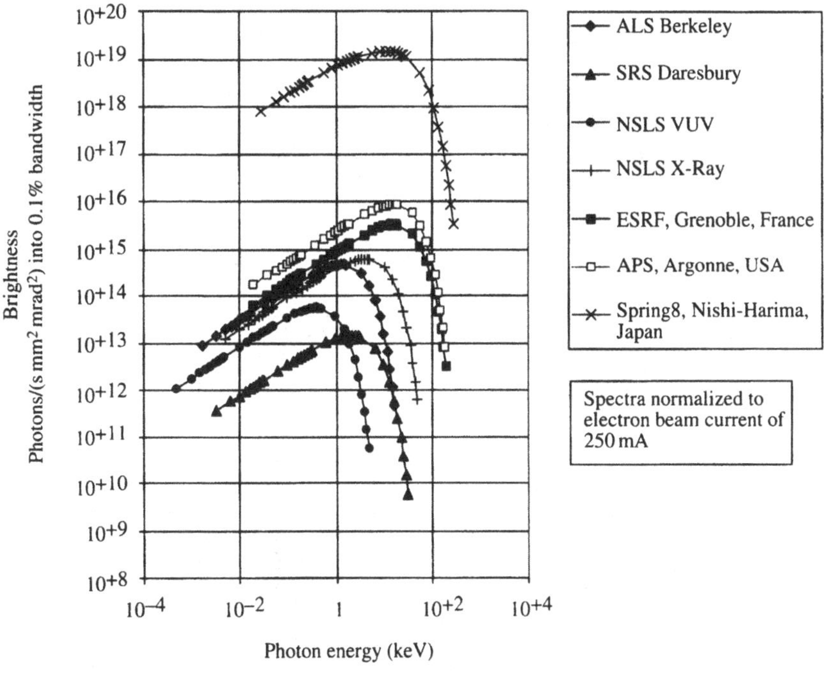

Fig. 9.9 Comparison of source brightness for selected sources.

TABLE 9.2 Electron beam parameters for selected synchrotron radiation sources

Source	Electron beam energy E (GeV)	Characteristic energy ε_c (keV)	Horizontal beam emittance ε_h (mm mrad)	Horizontal beam size (fwhm) mm	Vertical beam emittance ε_v (mm mrad)	Vertical beam size (fwhm) mm	Vertical beam divergence (fwhm) estimated (mrad)
ALS, Berkeley, USA	1.50	1.56	0.01	0.28	0.001	0.09	0.0111
SRS, Daresbury, UK	2.00	3.20	0.11	2	0.011	0.2	0.0550
NSLS (VUV), Brookhaven, USA	0.75	0.49	0.15	0.59	0.001	0.07	0.0143
NSLS (X-ray), Brookhaven, USA	2.50	5.00	0.1	0.35	0.001	0.15	0.0067
ESRF, Grenoble, France	6.00	9.60	0.007	0.266	0.0006	0.215	0.0028
ESRF, Grenoble, France	6.00	19.20					
APS, Argonne, USA	7.00	19.00	0.001	0.3	0.0001	0.1	0.0010
Spring8, Nishi-Harima, Japan	8.00	28.30	0.007	0.004	0.0007	2.00E−03	0.3500

Note: The field in each dipole magnet of the ESRF is split between 0.4 and 0.8 T. The 0.4 T field has been used in the figures for the flux and brightness comparison.

References

1. M. Born and E. Wolf, *Principles of optics* (6th edn), Chapter 4.8, p. 181, Pergamon Press, Oxford, England (1980); L. Mandel and E. Wolf, *Optical coherence and quantum optics*, Chapter 5.7, p. 293, Cambridge University Press, Cambridge, England (1995).
2. M. Born and E. Wolf, *Principles of optics* (6th edn), Chapter 4.2, p. 146, Pergamon Press, Oxford, England (1980).
3. J. B. West and H. A. Padmore, *Handbook on synchrotron radiation*, Vol. 2, North Holland, Amsterdam, The Netherlands (1987).

10

Behaviour of the electron beam in a synchrotron radiation storage ring. The concept of phase space

Introduction

In this chapter, we develop a formal description of the electron beam in a storage ring. We begin by defining the magnetic field function of an elementary storage ring which includes only the dipole and quadrupole fields. This will be sufficient for a basic understanding of the subject. Next, we develop an equation which describes the motion of the electron in this field and in Chapter 11 we shall show how the solutions of this equation of motion describe the oscillatory behaviour of the electron beam in the horizontal and vertical planes.

The magnetic field function

The coordinate system is the same as in Fig. 9.1.

Throughout the region of interest the magnetic field $\boldsymbol{B}(x, y, s)$ must satisfy the Maxwell equation curl $\boldsymbol{B} = \nabla \times \boldsymbol{B} = 0$ which means, in particular, that

$$\frac{\partial B_y}{\partial x} - \frac{\partial B_x}{\partial y} = 0. \tag{10.1}$$

Equation (10.1) is quite general. To proceed further we must make several assumptions:

(i) All magnetic field lines lie in a plane perpendicular to the horizontal plane which contains the ideal orbit. In other words, $B_s = 0$. There is no field component parallel to the direction of the ideal orbit at the point P.

(ii) In addition, along the ideal orbit itself, there is no field component parallel to the x-axis so that, at P, $\boldsymbol{B}(s) = B_{y,0}(s) = B_0(s)$. Assumptions (i) and (ii) imply that the magnets have no fringe fields.

(iii) There are no multipole fields of higher order than quadrupole so that the only magnetic field terms are the field itself and its gradient. All second derivatives such as $(\partial/\partial x)\left(\partial B_y/\partial x\right)$ and $(\partial/\partial y)\left(\partial B_y/\partial x\right)$ and, of course, third and higher derivatives are all zero. This means that the fields $B_x(x, y, s)$ and $B_y(x, y, s)$ can be described in terms of the field at the ideal orbit, $B_0(s)$, and the components of the gradients measured at the ideal orbit. The latter are $(\partial B/\partial x)_{0,s}$ and $(\partial B/\partial y)_{0,s}$ which must be equal to each other from eqn (10.1).

(iv) These magnetic field gradients are symmetric above and below the plane of the ideal orbit.

Putting these assumptions together we can write, for the three components of the magnetic field at any arbitrary point P,

$$\begin{aligned} B_x(x, y, s) &= y\left(\frac{\partial B}{\partial x}\right)_{0,s}, \\ B_y(x, y, s) &= x\left(\frac{\partial B}{\partial x}\right)_{0,s} + B_0(x, y, s), \\ B_s(x, y, s) &= 0. \end{aligned} \tag{10.2}$$

Suppose now that a particle with charge $+e$ is moving along the s-axis in Fig. 10.1 in the direction of the arrow so that the vector describing this motion has components $(0, 0, v_s)$. The components of the magnetic field vector are $(B_x, B_y, 0)$ so that the force acting on this particle is given in the usual way by the Lorentz equation

$$\boldsymbol{F} = e(\boldsymbol{v} \times \boldsymbol{B}) = e\begin{vmatrix} \varepsilon_x & \varepsilon_y & \varepsilon_s \\ 0 & 0 & v_s \\ B_x & B_y & 0 \end{vmatrix} = e(-\varepsilon_x v_s B_y + \varepsilon_y v_s B_x), \tag{10.3}$$

where ε_x, ε_y, and ε_s are unit vectors in the appropriate directions. Equation (10.3) establishes the sign convention and serves to emphasize that account must be taken of the sign of the electronic charge of the particle in the storage ring.

The magnetic fields described by eqns (10.2) are a function of E_0, the operating energy of the storage ring. A change to the operating energy requires a corresponding change to the magnetic fields everywhere around the ring. What is constant around the ring is the ratio of field to energy (at least for relativistic electrons), so it is convenient to multiply the field functions of eqns (10.2) by ec/E_0 and to define two new functions which are independent of energy and functions of x, y, and s only:

$$\begin{aligned} G(s) &= \frac{ec}{E_0} B_0(s), \\ K_x(s) = K_y(s) &= \frac{ec}{E_0}\left(\frac{\partial B}{\partial x}\right)_{0,s}. \end{aligned} \tag{10.4}$$

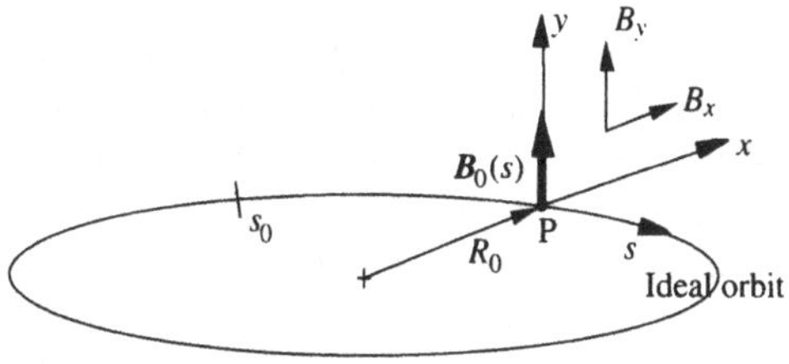

Fig. 10.1 Magnetic field definition.

From eqn (5.7), $G(s) = 1/R_0$, where R_0 is the radius of the ideal orbit and has dimensions m^{-1}; $K_x(s)$ and $K_y(s)$ have dimensions m^{-2}. In a modern storage ring, $G(s)$ is produced by the dipole magnets, where $K_x(s)$ is zero. In the quadrupole magnets, where $G(s) = 0$, $K_x(s)$ and $K_y(s)$ alternate in sign around the ring to provide the required focusing action. In drift spaces (or straight sections) between magnets, G, K_x, and K_y are zero. A storage ring with these properties is said to be isomagnetic.

The object in the design of a storage ring is to provide an electron beam which has the properties needed (starting with the beam emittance) to produce synchrotron radiation with the required specification. By means of a suitable choice of magnet geometry, the designer of the storage ring must produce a variation of the values of $G(s)$, $K_x(s)$, and $K_y(s)$ around the ring to do this. Within this framework, the functions are subject to some general constraints:

(i) They must be cyclic functions, returning to their starting values after a complete turn around the ring, whatever the starting point s_0. In practice, a further constraint is applied in that the values of the functions are repeated a number of times around the ring to generate what is often called a magnetic lattice so that the entire ring is generated by repeating a (usually even) number of identical unit cells. In other words, $G(s)$ and $K(s)$ are periodic functions of s.

(ii) The function $G(s)$, often called the curvature function, must generate a closed orbit. Fig. 10.1 shows the ideal orbit as a circle with a constant value of R_0 but in a real storage ring this is not the case. At the very least, the circle is broken by the insertion of focusing magnets and drift spaces between them. Some rings (of which the ESRF is an example) (see Table 8.1 or 9.2) have two field values [and therefore two $G(s)$ values] within each dipole magnet so that R_0 is a function of s. However, as one goes round the ring, a change ds in the azimuthal coordinate s generates a corresponding change $d\theta$ in the rotation angle: $d\theta = ds/R_0(s) = G(s)\,ds$. The total change in θ around the ring, whatever its shape, must be equal to 2π so, if L is the total length of the circumference, then

$$\int_0^L G(s)\,ds = 2\pi.$$

The electron equations of motion—motion in the horizontal plane

In this section we must relate the rate of change of the coordinates of an electron as it travels around the storage ring to the magnetic field components which it encounters along its trajectory. This is not a simple exercise because the fields are themselves functions of the coordinates through eqns (10.2) and (10.4). We consider first the radial motion, which in the coordinate system of Fig. 10.1 is also the motion transverse to the ideal electron orbit in the horizontal plane. At first sight, the problem is complicated further because the linear z-coordinate with which we are familiar has been replaced by the curvilinear coordinate s. A moment's consideration shows that if we were to use a rectilinear coordinate system, the stored electron, which spends its life close to the ideal orbit would move further and further away from the z-axis.

There are several ways of treating this problem. We follow the treatment given by Sands[1] and also by Duke[2] which attempt to clarify the principles rather than present a rigorous mathematical treatment. The latter can be found, for example, in Krinsky.[3]

Figure 10.2 illustrates these principles. The origin of the s-axis is the arbitrary point s_0 where the radius vector from the point O to the point N intersects with the ideal orbit. An electron which has reached the point P has coordinate s along the s-axis, measured around the ideal orbit from s_0. The corresponding x-coordinate is $x(s)$, measured along the extension of the radius vector $\mathrm{OP}' = R_0$ from P′ to the electron trajectory at P. The line OP′P makes an angle $\theta(s)$ with the line ON from which the rotation angle of the radius vector around the ideal orbit is measured. At P the tangent to the electron trajectory makes an angle $x'(s) = \mathrm{d}x/\mathrm{d}s$ with a line parallel to the tangent to the ideal orbit which is the direction of the s-axis at that point. In general, and as shown in the figure, the electron trajectory is not parallel to the ideal orbit. The radius vector of the trajectory at P is the line O′P of length $R(s)$ and makes an angle $\alpha(l)$ with the line O′N. In other words, as the radius vector joining the electron to O rotates through the angle $\theta(s)$, the radius vector joining the electron to O′ rotates through an angle $\alpha(l)$. At the same time, the electron travels a distance l along its trajectory which projects to a distance s along the ideal orbit which is the s-axis.

Suppose now that the electron moves a further distance Δl along its trajectory, projecting to an increment Δs along the ideal orbit, so that the new s-coordinate is $s + \Delta s$ and the new x-coordinate is $x(s + \Delta s)$. The electron has reached the point Q and the radius vector OQ to the ideal orbit now makes an angle $\theta(s + \Delta s)$ with the

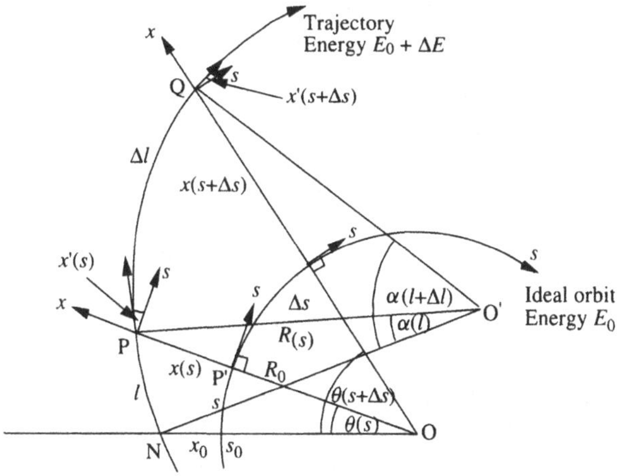

Fig. 10.2 Radial electron motion. Note that, for clarity, here and in Fig. 10.3 Δl and Δs have been drawn much larger than l and s respectively. In reality they are much smaller.

line ON. Likewise, the radius vector O′Q makes an angle $\alpha(l + \Delta l)$ with O′N. The new direction of motion is $x'(s + \Delta s)$ and the change in the direction of the electron, measured in the curvilinear coordinate system, is $(x'(s + \Delta s) - x'(s))$ and simple geometry shows that

$$x'(s + \Delta s) - x'(s) = \{\theta(s + \Delta s) - \theta(s)\} - \{\alpha(l + \Delta l) - \alpha(l)\}.$$

The rate of change of x' with respect to distance s along the ideal orbit is

$$\frac{x'(s + \Delta s) - x'(s)}{\Delta s}$$

and we can obtain $\mathrm{d}x'/\mathrm{d}s = \mathrm{d}^2x/\mathrm{d}s^2$, which we write as x'' by taking the limit of this expression as $\Delta s \to 0$ (and $\Delta l \to 0$). In other words:

$$\begin{aligned}\frac{\mathrm{d}^2x}{\mathrm{d}s^2} &= \lim \left\{\frac{x'(s + \Delta s) - x'(s)}{\Delta s}\right\}_{\Delta s \to 0} \\ &= \lim \left\{\frac{\theta(s + \Delta s) - \theta(s)}{\Delta s} - \frac{\alpha(l + \Delta l) - \alpha(l)}{\Delta s}\right\}_{\Delta s \to 0},\end{aligned}$$

so that

$$x'' = \frac{\mathrm{d}^2x}{\mathrm{d}s^2} = \frac{\mathrm{d}\theta}{\mathrm{d}s} - \frac{\mathrm{d}\alpha}{\mathrm{d}l}\frac{\mathrm{d}l}{\mathrm{d}s} \qquad (10.5)$$

and we can now relate the right-hand side of eqn (10.5) to the magnetic field and the electron energy to give the required equation of motion in the radial direction.

Consider the two terms on the right-hand side of eqn (10.5). The second term contains two factors. The first factor $\mathrm{d}\alpha/\mathrm{d}l$ is equal to $1/R$, R being the radius of curvature of the electron orbit at the point P which is given, as usual by $R = E/ecB$, where B is the magnitude of the magnetic field at P. In the approximation we are using, $B = B_y$ as given by eqn (10.2), so that

$$\frac{\mathrm{d}\alpha}{\mathrm{d}l} = \frac{ec}{E_0 + \Delta E}\left(B_0 + x\frac{\partial B_x}{\partial x}\right) \qquad (10.6)$$

and we have allowed for the possibility that E, the energy of the electron at P may differ from the ideal E_0 by an amount ΔE. We can now rearrange eqn (10.6) and insert the magnetic field functions from eqn (10.4) to give

$$\frac{\mathrm{d}\alpha}{\mathrm{d}l} = \left(1 - \frac{\Delta E}{E_0}\right)(G(s) + xK_x(s)), \qquad (10.7)$$

where we have neglected all second order terms.

We require next the factor $\mathrm{d}l/\mathrm{d}s$. Because the angle of deviation of the electron orbit and the energy deviation from the ideal are both small we may write

$$\frac{\mathrm{d}l}{\mathrm{d}s} = \frac{R_0 + x}{R_0} = 1 + \frac{x}{R_0} = (1 + G(s)x). \qquad (10.8)$$

We now combine eqns (10.7) and (10.8), keeping only the first order terms, to give

$$\frac{d\alpha}{ds} = \left(1 - \frac{\Delta E}{E_0}\right)(G(s) + xK_x(s))(1 + xG(s)). \tag{10.9}$$

The first term in eqn (10.5), $d\theta/ds$, is equal to $1/R_0$ which is $G(s)$, so combining this with eqn (10.9) and sorting out the algebra gives

$$x'' = G(s)\frac{\Delta E}{E_0} - \left[G(s)^2 + K_x(s)\right]x, \tag{10.10}$$

which is the required equation of motion in the radial direction describing the motion of the electron projected onto the horizontal plane.

The electron equations of motion—motion in the vertical plane

The corresponding equation for the vertical direction is easier to derive because the s-axis does not curve in the vertical plane. In the equivalent of eqn (10.5), $d\theta/ds = 0$ and

$$y'' = \frac{d^2y}{ds^2} = -\frac{d\alpha}{dl}\frac{dl}{ds}. \tag{10.11}$$

This can be seen by employing the same argument as was used in the radial direction so that $(y' + \Delta y) - y' = -\Delta\alpha$. The negative sign arises in eqn (10.11) [and in eqn (10.5)] because the trajectory turns in the negative direction (y' decreases) as the angle α increases (Fig. 10.3).

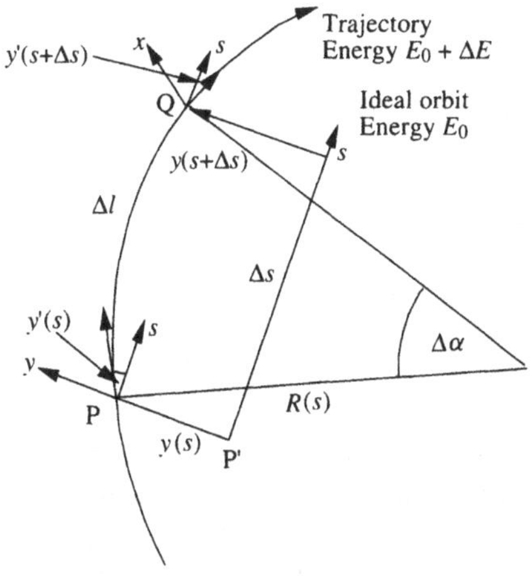

Fig. 10.3 Vertical electron motion.

This time $d\alpha/dl = -ecB_x/E$, from the sign convention defined in eqn (10.3) so that, from eqn (10.2)

$$\begin{aligned}\frac{d\alpha}{dl} &= -\frac{ec}{E_0 + \Delta E}\left(y\frac{\partial B_x}{\partial x}\right)\\ &= -\left(1 - \frac{\Delta E}{E_0}\right)K_y(s)y.\end{aligned}$$

The increment Δs is equal to the increment Δl along the orbit times the cosine of y' which is $\sqrt{1-y'^2}$ which can be set equal to 1 because y'^2 is a second order quantity so that $dl/ds \approx 1$ and, in the same approximation,

$$y'' = \frac{d^2 y}{ds^2} = K_y(s)y. \tag{10.12}$$

As we would expect, the focusing introduced by the gradient function is opposite in sign in the x- and y-directions. What is, perhaps, unexpected is that the dipole magnetic field generates a focusing force, proportional to $G(s)^2$ in the x–s plane as seen in eqn (10.10). In general this force is smaller than that produced by the quadrupole magnets.

Electron beam trajectories

We are now in a position to trace the progress of a particle through the magnetic fields of the storage ring. Equations (10.10) and (10.12) can both be written in the form

$$u'' + K(s)u = 0, \tag{10.13}$$

where, in the x–s plane,

$$u'' = x'' - G(s)\frac{\Delta E}{E_0} \tag{10.14}$$

and

$$K(s) = -[G(s)^2 + K_x(s)] \tag{10.15}$$

and in the y–s plane

$$u'' = y'' \quad \text{and} \quad K(s) = -K_y(s).$$

In any particular element of the storage ring, whether it be a quadrupole magnet, a dipole magnet or a drift space, $K(s)$ is constant and eqn (10.13) has the form of the harmonic oscillator equation in which $\sqrt{K}$ plays the role of the frequency of the oscillation. For a focusing quadrupole, for which $K > 0$, two particular solutions for u are of the form $A_0 \cos\sqrt{K}s$ and $B_0 \sin\sqrt{K}s$. In these solutions, A_0 and B_0 are constants which depend on the position and direction of the particle as it enters the region of magnetic field defined by $K(s)$. We would like to obtain a general solution which would apply to any particle in the region, so we combine these two independent solutions and write, for the generalized position of the particle at s,

$$u = A_0 \cos \sqrt{K}s + B_0 \sin \sqrt{K}s.$$

Clearly the generalized direction of the particle at s will be

$$u' = -A_0\sqrt{K} \sin \sqrt{K}s + B_0\sqrt{K} \cos \sqrt{K}s.$$

At the entrance to the field region $s = 0$, the boundary conditions $u = u_0$ and $u' = u'_0$ serve to fix the values of A_0 and B_0 so that

$$\begin{aligned} u(s) &= u_0 \cos \sqrt{K}s + \frac{u'_0}{\sqrt{K}} \sin \sqrt{K}s, \\ u'(s) &= -u_0\sqrt{K} \sin \sqrt{K}s + u'_0 \cos \sqrt{K}s. \end{aligned} \tag{10.16}$$

When $K < 0$, the quadrupole is defocusing and $\sqrt{K}$ must be replaced by $\mathrm{i}\sqrt{|K|}$ and the corresponding solutions are

$$\begin{aligned} u(s) &= u_0 \cosh \sqrt{|K|}s + \frac{u'_0}{\sqrt{|K|}} \sinh \sqrt{|K|}s, \\ u'(s) &= u_0\sqrt{|K|} \sinh \sqrt{|K|}s + u'_0 \cosh \sqrt{|K|}s. \end{aligned} \tag{10.17}$$

If we denote the position and direction of the particle by the matrix $\begin{bmatrix} u(s) \\ u'(s) \end{bmatrix}$ then we can write the solutions (10.16) and (10.17) in the form of a matrix equation

$$\begin{bmatrix} u(s) \\ u'(s) \end{bmatrix} = \boldsymbol{M} \begin{bmatrix} u_0(s) \\ u'_0(s) \end{bmatrix} \tag{10.18}$$

and $\boldsymbol{M}$ is given by

$$\boldsymbol{M} = \begin{bmatrix} \cos \sqrt{K}s & \dfrac{1}{\sqrt{K}} \sin \sqrt{K}s \\ -\sqrt{K} \sin \sqrt{K}s & \cos \sqrt{K}s \end{bmatrix} \tag{10.19}$$

for a focusing quadrupole and

$$\boldsymbol{M} = \begin{bmatrix} \cosh \sqrt{|K|}s & \dfrac{1}{\sqrt{K}} \sinh \sqrt{|K|}s \\ \sqrt{|K|} \sinh \sqrt{|K|}s & \cosh \sqrt{|K|}s \end{bmatrix}$$

for a defocusing quadrupole.

In the simple case where the energy of the particle is equal to the ideal energy E_0, $\Delta E = 0$ and u and u' can be replaced by x and x' or y and y' [see eqn (10.14)] and K by K_x or K_y to obtain the expressions for the horizontal or the vertical motion.

The problem of tracing an individual particle through the magnetic lattice of the storage ring has been reduced to the multiplication of the matrix for a given initial position and direction of the particle by a succession of matrices describing the magnetic fields and drift spaces between them. In such a drift space between magnets the appropriate transformation matrix is

$$\boldsymbol{M} = \begin{bmatrix} 1 & s \\ 0 & 1 \end{bmatrix}. \tag{10.20}$$

To illustrate the effect of this matrix on the particle trajectory, we expand the equation

$$\begin{bmatrix} u(s) \\ u'(s) \end{bmatrix} = \begin{bmatrix} 1 & s \\ 0 & 1 \end{bmatrix} \begin{bmatrix} u_0(s) \\ u_0'(s) \end{bmatrix}$$

to give

$$u(s) = u_0(s) + u_0'(s)s \quad \text{and} \quad u'(s) = u_0'(s)$$

showing, as we would expect, that the effect of a drift space is to leave the gradient unchanged and to shift the position of the particle laterally by an amount depending on the product of the gradient of the particle multiplied by the length of the drift space.

The transformation matrix for a dipole magnet can be obtained by substitution of eqn (10.15) with $K_x = 0$ into the matrix of eqn (10.19) to give

$$\boldsymbol{M} = \begin{bmatrix} \cos Gs & \dfrac{1}{G}\sin Gs \\ -G\sin Gs & \cos Gs \end{bmatrix}$$

in the horizontal plane. In the vertical plane the matrix for the dipole magnet is the same as that for a drift space.

In general, bearing in mind eqns (10.16) and (10.17), the transformation matrices are often written as

$$\boldsymbol{M} = \begin{bmatrix} C(s) & S(s) \\ C'(s) & S'(s) \end{bmatrix} \tag{10.21}$$

where $C(s)$ and $S(s)$ are cosine-like and sine-like functions and C' and S' are their first derivatives. These functions correspond, in the case of a focusing quadrupole, to the standard ray-tracing trajectories shown in Fig. 8.5.

An important property of all these matrices is that their determinant is equal to unity. In other words

$$|\boldsymbol{M}| = 1. \tag{10.22}$$

This is a consequence of the fact that the forces acting on the particle in the storage ring are non-dissipative, i.e. they do not lead to any loss of energy by the particle.

Phase space

The matrices which transform the particle trajectory from one point in the storage ring to another operate on the position and direction of the particle, denoted by u and u'. It is useful to describe these operations graphically using the concept of phase space. We are all familiar with depiction of a particle as a straight line starting at the position (u, s) of the particle and whose direction is (u', s). This pictorial representation is easy to visualize but does not take into account an important property of the particle motion as we shall see.

When we make use of the phase space description, the particle is represented by a point whose coordinates are (u, u'). The advantage of this representation is that we can use it to describe the motion of an assemblage of particles at a particular position around the ring. If we consider a plane which intersects the ideal orbit at any arbitrary point around the ring then we can provide a picture of the particle distribution by plotting the u and u' coordinates of each particle at that point.

Figure 10.4, on the left, shows an extended source of particles. These can be photons, electrons, or any kind of radiation. Their nature does not matter. The source has a width $2w$ and emits radiation over all forward angles. The radiation emission is limited by an aperture, also $2w$ wide (this is a convenient simplification, any width would do), placed downstream from the source, symmetrically with respect to the centre of the source at C. Again, the symmetry does not matter, it just makes the description simpler. The right-hand side of Fig. 10.4 shows the phase space representation of this source, comprising the radiating surface and the aperture. How is this representation constructed? Particles emitted from the point A of the surface must follow a trajectory which lies between AA′ and AA″ if they are to pass through the aperture. Consider the extreme ray AA′ which starts out from the top edge of the source and skirts the edge of the aperture. This ray has the position coordinate $u = +w$ and direction $u' = 0$. In the phase diagram this ray is plotted as the point P $(+w, 0)$ at $+w$ on the

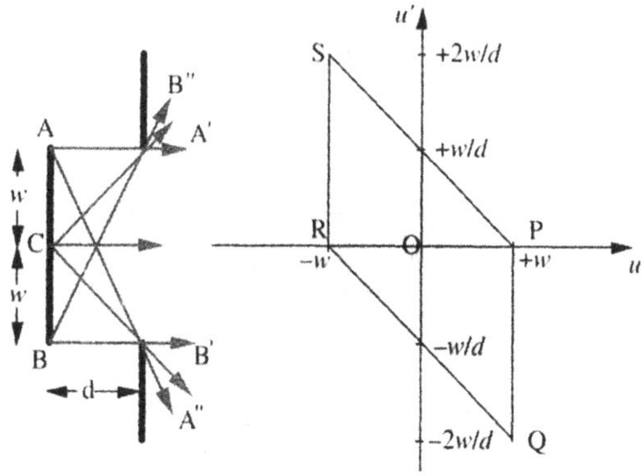

Fig. 10.4 Phase space description of an extended source.

u-axis. The other extreme ray starting from A is the ray AA″ which has $u = +w$ as before and $u' = -2w/d$. This translates into the point $Q\ (+w, -2w/d)$ when plotted in the phase diagram. The line PQ contains all rays which emerge from the point A at the source and pass through the aperture. Similarly, radiation from the point B at the other end of the source has $u = -w$ and u' extends over the range 0 to $+2w/d$ and is plotted as the line RS on the phase diagram. Particles from the point C within the source are limited to the angular range $\pm w/d$ by the aperture and have u equal to zero. These particles map onto a line along the u'-axis between the points $(0, +w/d)$ and $(0, -w/d)$. In fact, all particles emitted from the source are contained within the parallelogram PQRS.

The particles satisfying this description could be considered as a possible distribution, in one plane, of the particles in the beam of an electron storage ring. The portrayal of the beam in the phase description on the right-hand side of Fig. 10.4 is often called the phase portrait of the beam. The full utility of the phase portrait becomes clearer when we consider what happens to it as a function of time, i.e. as the beam moves around the storage ring.

Suppose we are required to determine the distribution of the beam at a distance s along the ideal orbit from that imagined in Fig. 10.4. If we take the simplest case, in which the beam is supposed to be moving through a drift space, the coordinates u and u' of each particle in the beam are transformed by the matrix of eqn (10.20). In particular particles at the points S and Q in the phase portrait will move through a distance $2sw/d$ in the direction shown by the arrows in Fig. 10.5 to points S′ and Q′. Particles at the points P and R do not move at all in phase space because u' is equal to zero at these points. The transformed space portrait of the beam is thus a new parallelogram PQ′RS′ different in shape from the first but covering the same area.

That the new phase portrait has the same area as the old one can be shown by a simple geometrical argument. If we denote the area of the initial portrait by Æ PQRST

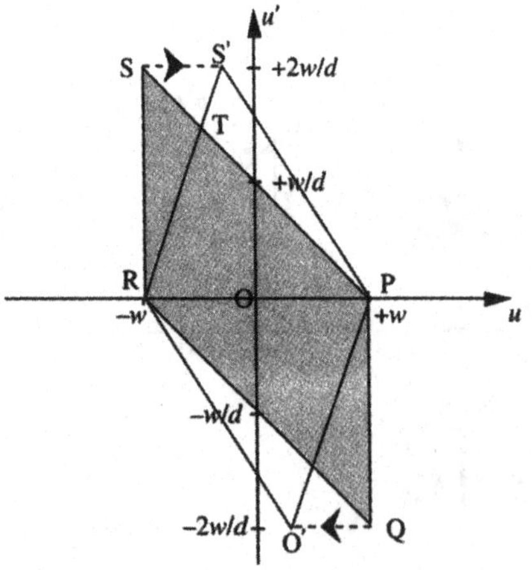

Fig. 10.5 Development of the phase portrait with time.

then, referring to Fig. 10.5 and by symmetry,

$$\text{Æ}\,\mathrm{PQ'RS'T} = \text{Æ}\,\mathrm{PQRST} + 2\{[\text{Æ}\,\mathrm{PSS'} - \text{Æ}\,\mathrm{TSS'}] - [\text{Æ}\,\mathrm{RSS'} - \text{Æ}\,\mathrm{TSS'}]\}$$

so that

$$\text{Æ}\,\mathrm{PQ'RS'T} = \text{Æ}\,\mathrm{PQRST} + 2\{\text{Æ}\,\mathrm{PSS'} - \text{Æ}\,\mathrm{RSS'}\}$$

but Æ PSS′ and Æ RSS′ are both equal to $(1/2)(2sw/d)(2w/d)$ so that the area under the phase portrait remains unchanged under the transformation defined by (10.20).

The preservation of the area under the phase portrait, described in detail for the particular case above, is a particular example of a very general theorem known as Liouville's theorem (Joseph Liouville 1809–1882), which states that the number of particles in unit volume of phase space, the phase space density, remains constant, for all time, provided that the forces acting on the particles do not lead to loss of energy. In phase space the particle beam behaves like an incompressible, frictionless, fluid. Constriction of the fluid in one direction leads to expansion in another direction and/or a change in the velocity of the fluid. In general the phase space is a six-dimensional volume, defined by the components of the position and momentum vectors $(x, y, z; p_x, p_y, p_z)$. In the representation used here z has been replaced by the coordinate s and the momentum components by x', y', and s'. This replacement is possible because, for example, p_x is equal to $m\,\mathrm{d}x/\mathrm{d}t = mx'\,\mathrm{d}s/\mathrm{d}t$ and m and $\mathrm{d}s/\mathrm{d}t$ are constant ($\mathrm{d}s/\mathrm{d}t = \beta c$ and $\beta \approx 1$). This procedure reduces the phase space to a four-dimensional volume bounded by x, x' and y, y' which are each a function of s, and s' is equal to 1.

In the approximation which we are considering, the motions of the particles in the horizontal and vertical planes of the storage ring are independent. There is no coupling between them. This supposes an ideal magnetic field in which vertical fields (in the dipole and quadrupoles) are truly vertical and affect only the horizontal motion. Likewise, horizontal magnetic fields affect only the vertical motion. As a consequence, Liouville's theorem can be applied separately to each plane using u and u' as generalized coordinates. The theorem can be proved very simply as follows.

Figure 10.6 shows an element of phase space defined by the vectors $\boldsymbol{U}_\mathrm{a}$ and $\boldsymbol{U}_\mathrm{b}$ in the u–u' plane. The area of this phase space element is the vector product $\boldsymbol{U}_\mathrm{a} \times \boldsymbol{U}_\mathrm{b}$ whose magnitude is the determinant of the components of the two vectors, namely,

$$\begin{vmatrix} u_\mathrm{a} & u_\mathrm{b} \\ u'_\mathrm{a} & u'_\mathrm{b} \end{vmatrix} = u_\mathrm{a}u'_\mathrm{b} - u_\mathrm{b}u'_\mathrm{a}.$$

After a time t which is equal to $s/\beta c$, the phase space element has been transformed by a matrix $\boldsymbol{M}$ into a new element defined by $\boldsymbol{V}_\mathrm{a}$ and $\boldsymbol{V}_\mathrm{b}$ whose area is

$$\begin{vmatrix} v_\mathrm{a} & v_\mathrm{b} \\ v'_\mathrm{a} & v'_\mathrm{b} \end{vmatrix} = v_\mathrm{a}v'_\mathrm{b} - v_\mathrm{b}v'_\mathrm{a}.$$

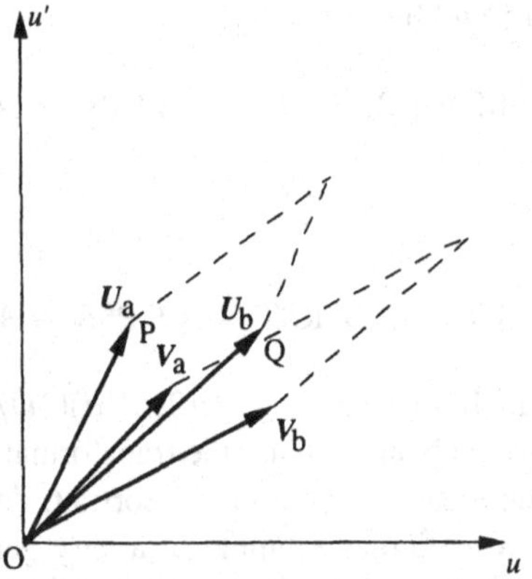

Fig. 10.6 Liouville's theorem.

The components of $\boldsymbol{U}$ and $\boldsymbol{V}$ are related by

$$\begin{bmatrix} v_a(s) \\ v'_a(s) \end{bmatrix} = \boldsymbol{M} \begin{bmatrix} u_a(s) \\ u'_a(s) \end{bmatrix}$$

and

$$\begin{bmatrix} v_b(s) \\ v'_b(s) \end{bmatrix} = \boldsymbol{M} \begin{bmatrix} u_b(s) \\ u'_b(s) \end{bmatrix}$$

so that

$$\begin{vmatrix} v_a & v_b \\ v'_a & v'_b \end{vmatrix} = |\boldsymbol{M}| \begin{vmatrix} u_a & u_b \\ u'_a & u'_b \end{vmatrix}$$

as can be shown by carrying out the matrix operation explicitly. But, from eqn (10.22), $|\boldsymbol{M}|$, the determinant of the transformation matrix, is always equal to unity so long as there is no gain or loss of energy. Furthermore, this is true whether $\boldsymbol{M}$ is one of the transformation matrices for a single element of the storage ring or the product of a succession of such transformations. It follows that as the particle beam travels around the ring, the phase space area in both the horizontal and the vertical planes remains constant and the number of particles within the phase space area remains constant as well. The shape of the phase space element changes but the enclosed area of phase space is invariant. This proof of Liouville's theorem can be extended to cover the six-dimensional phase space invariance as shown in Wiedemann.[4]

The phase space ellipse

It is clear from the previous paragraph that we can select any shape we like to describe the behaviour in phase space of an assembly of particles. Once we have made our selection, Liouville's theorem ensures that the application of the transform equations preserves the density of particles within the area of the phase space. Naturally, it is most convenient to choose a shape which bears an obvious relationship to the problem in hand. For example, the parallelogram used to represent the distribution in phase space of the rays from an aperture (Fig. 10.4) arose naturally from the geometry of the physical situation.

The geometry here is similar to that of Figs 9.2 and 9.3 where we described the distribution of the beam particles in the x–y plane as the product of two Gaussian distributions whose standard deviations were a measure of the beamwidth in the x- and y-directions. The elliptical 1σ contour of the distribution gave us a snapshot of the beam at each point in the storage ring. It is reasonable to employ a similar description to characterize the beam in the x–x' or y–y' planes. The width of the beam distribution in x and x' or y and y' is related to the major and minor axes of the elliptical 1σ contour of the two-dimensional distribution of the beam in phase space. The coordinates of any point on this contour transform according to the transformation equations as the beam circulates around the storage ring. As the beam of particles circulates, the ellipse changes in shape and orientation but its area remains unchanged.

The simplest phase ellipse occurs at a point in the storage ring where either the beam width or the divergence is at a minimum (or a maximum). In that case the ellipse is vertical or horizontal as shown in Fig. 10.7.

The ellipse B shows the beam contour at what is often called a focus but is better described as a point where the beam width is a minimum. We use the generalized coordinates u and u' to remind us that there are separate phase ellipses for both the x–x' and y–y' planes. Such an ellipse can be described by

$$au^2 + bu'^2 = \varepsilon. \tag{10.23}$$

Comparison with the standard equation for an ellipse with the origin at the centre and major and minor axes a and b, respectively, i.e.

$$\frac{u^2}{a^2} + \frac{u'^2}{b^2} = 1,$$

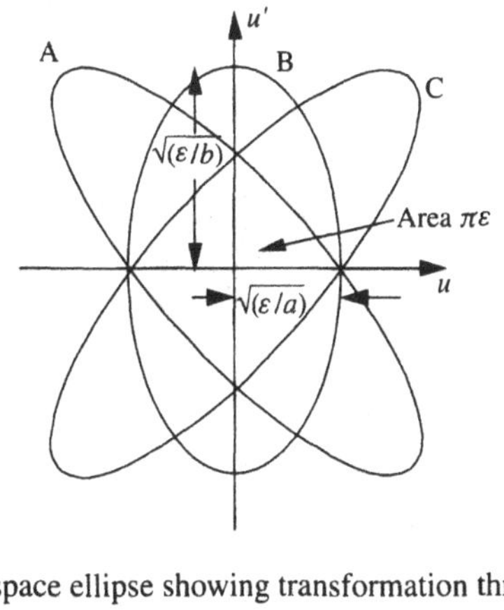

Fig. 10.7 Phase space ellipse showing transformation through a drift space.

shows that the ellipse of eqn (10.23) has axes with lengths $2\sqrt{\varepsilon/a}$ and $2\sqrt{\varepsilon/b}$ and area $\pi\varepsilon$ provided the constants a and b are restricted by the condition $ab = 1$. The area $\pi\varepsilon$ is called the beam emittance. It is usually measured in metre · radians (m · rad) and can be expressed with or without the factor π (absence of this factor can cause confusion when the properties of a storage ring beam are being described).

As the beam travels through a drift space, the behaviour of the phase space contour is similar to that described in detail for the parallelogram in Fig. 10.5. An observer moving along the s-axis sees the ellipse gradually rotate from position A through B to something like position C. Only at position B does the length of the axis of the ellipse in the u-direction corresponds to the width of the beam. At B the projection of the major axis (in this case, the axis in the u'-direction) is zero and the half width of the beam (which corresponds to the projection of the 1σ contour on the u-axis) is $\sqrt{\varepsilon/a}$. Away from B, the rotation of the ellipse causes the major axis to generate a non-zero projection on the u-axis and the beam size measured along that axis is consequently larger. How much larger is determined by the length of the drift space which is the only parameter in the transformation matrix [eqn (10.20)].

In order to be precise about the relation between the phase ellipse and the beam dimensions, we need the equation of the tilted ellipse. If we refer to Fig. 10.8 we can see that the equation of an ellipse tilted through an angle ϕ relative to the u-axis can be obtained from the symmetrical case by a rotation of the coordinate axes through an angle ϕ in the opposite direction to the angle of tilt, i.e. through $-\phi$.

Suppose the symmetrical ellipse [eqn (10.23)] to be described by

$$av^2 + bv'^2 = \varepsilon$$

in the (v, v') frame shown in Fig. 10.8.

The (v, v') frame is related to the (u, u') frame by the matrix equation

$$\begin{vmatrix} u \\ u' \end{vmatrix} = \begin{vmatrix} \cos\phi & -\sin\phi \\ \sin\phi & \cos\phi \end{vmatrix} \begin{vmatrix} v \\ v' \end{vmatrix}. \qquad (10.24)$$

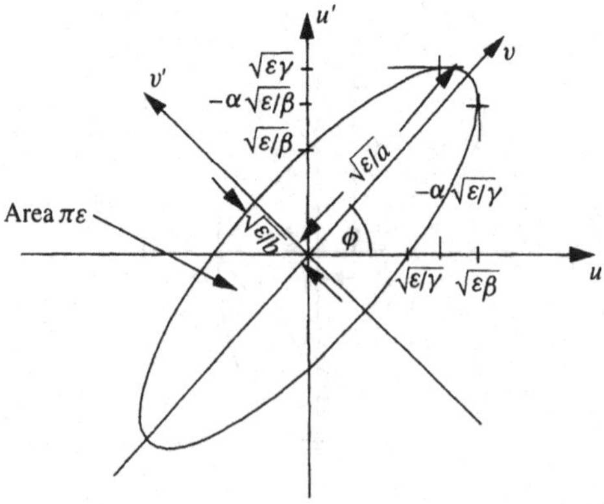

Fig. 10.8 Parameters of the phase space ellipse.

Unlike the beam line transform matrices, this rotation matrix preserves not only the area but also the shape of the ellipse so that the major and minor axes have the same length before and after the transformation. The inverse transformation is obtained by inverting the transformation matrix so that

$$\begin{vmatrix} v \\ v' \end{vmatrix} = \begin{vmatrix} \cos\phi & \sin\phi \\ -\sin\phi & \cos\phi \end{vmatrix} \begin{vmatrix} u \\ u' \end{vmatrix} \tag{10.25}$$

and

$$v = u\cos\phi + u'\sin\phi, \qquad v' = -u\sin\phi + u'\cos\phi$$

which describes a rotation of the coordinate axes through the angle $-\phi$.

Insertion of eqn (10.25) into (10.24) leads to the general equation for the phase space ellipse:

$$\begin{aligned} \gamma u^2 + 2\alpha u u' + \beta u'^2 &= \varepsilon \\ \gamma &= a\cos^2\phi + b\sin^2\phi, \\ \beta &= a\sin^2\phi + b\cos^2\phi, \end{aligned} \tag{10.26}$$

where

$$2\alpha = (a-b)\sin 2\phi \quad \text{and} \quad \tan 2\phi = \frac{-2\alpha}{\beta - \gamma} \tag{10.27}$$

with the condition

$$\beta\gamma = 1 + \alpha^2 \tag{10.28}$$

because $ab = 1$. The parameters α, β, and γ are called the Twiss parameters.[5]

The usual algebraic procedure can be worked through to obtain the extreme points of the ellipse (the maximum and minimum values of u and u') as shown in Fig. 10.8 which are related to the beam dimensions. If the ellipse outlines the 1σ contour of the beam distribution so that the full width at half height of the beam in the storage ring is $2.36\sigma_u$ and the full width at half height of the angular distribution of the beam at the same point is $2.36\sigma_{u'}$. The quantities $2\sigma_u$ and $2\sigma_{u'}$ are the projections of the phase ellipse on the u and u' axes and σ_u and $\sigma_{u'}$ are given by

$$\sigma_u = \sqrt{\varepsilon\beta}, \qquad \sigma_{u'} = \sqrt{\varepsilon\gamma}. \tag{10.29}$$

When $\phi = \pi/2$ the ellipse is upright (position B in Fig. 10.7), $\alpha = 0$, $\beta\gamma = 1$, and the product of σ_u and $\sigma_{u'}$ is equal to ε, the beam emittance.

It should now be clear that the Twiss parameters are functions of position and they determine the shape and orientation of the phase ellipse around the storage ring. The coordinates of each point in the phase diagram transform according to the transformation matrices of eqns (10.18) and (10.21). In particular, u_0 and u'_0 are the phase space coordinates of a particle at the position s_0 and are related to u and u' at position s by

$$\begin{bmatrix} u(s) \\ u'(s) \end{bmatrix} = \begin{bmatrix} C(s-s_0) & S(s-s_0) \\ C'(s-s_0) & S'(s-s_0) \end{bmatrix} \begin{bmatrix} u_0(s_0) \\ u_0'(s_0) \end{bmatrix}. \tag{10.30}$$

The Twiss parameters at s_0 and s are given by

$$\gamma_0 u_0{}^2 + 2\alpha_0 u_0 u_0' + \beta_0 u_0'^2 = \varepsilon \tag{10.31}$$

so we can find the equation of the phase ellipse at s by solving eqn (10.30) for u_0 and u_0' in terms of u and u' and inserting the result in eqn (10.31). The solution for u_0 and u_0' in matrix form is

$$\begin{bmatrix} u_0(s_0) \\ u_0'(s_0) \end{bmatrix} = \begin{bmatrix} C(s-s_0) & S(s-s_0) \\ C'(s-s_0) & S'(s-s_0) \end{bmatrix}^{-1} \begin{bmatrix} u(s) \\ u'(s) \end{bmatrix}. \tag{10.32}$$

The inverse of the transformation matrix is found in the usual way, remembering that the determinant of the matrix is unity [eqn (10.22)], so that

$$\begin{bmatrix} C(s-s_0) & S(s-s_0) \\ C'(s-s_0) & S'(s-s_0) \end{bmatrix}^{-1} = \begin{bmatrix} S'(s-s_0) & -C'(s-s_0) \\ -S(s-s_0) & C(s-s_0) \end{bmatrix}$$

and eqn (10.32) becomes

$$\begin{bmatrix} u_0(s_0) \\ u_0'(s_0) \end{bmatrix} = \begin{bmatrix} S'(s-s_0) & -C'(s-s_0) \\ -S(s-s_0) & C(s-s_0) \end{bmatrix} \begin{bmatrix} u(s) \\ u'(s) \end{bmatrix}.$$

We can now expand eqn (10.32) to give the required expressions for u_0 and u_0', and insert these into eqn (10.31) to give

$$\gamma u^2 + 2\alpha u u' + \beta u'^2 = \varepsilon,$$

with

$$\begin{aligned} \gamma &= S'^2\gamma_0 - 2S'C'\alpha_0 + C'^2\beta_0, \\ \alpha &= -SS'\gamma_0 + (S'C + SC')\alpha_0 - CC'\beta_0, \\ \beta &= S^2\gamma_0 - 2SC\alpha_0 + C^2\beta_0, \end{aligned} \tag{10.33}$$

which we can write in matrix form as

$$\begin{bmatrix} \gamma \\ \alpha \\ \beta \end{bmatrix} = \begin{bmatrix} S'^2 & -2S'C' & C'^2 \\ -SS' & (S'C + SC') & -CC' \\ S^2 & -2SC & C^2 \end{bmatrix} \begin{bmatrix} \gamma_0 \\ \alpha_0 \\ \beta_0 \end{bmatrix}. \tag{10.34}$$

Equation (10.34) relates the Twiss parameters at each point around the storage ring. The transformation matrix can be obtained from the known values of the magnetic field and once the parameters of the phase ellipse have been given at one position, the shape and orientation of the phase ellipse can be found everywhere else.

In the next chapter we shall show how the motion of the electrons in the storage ring is related to the equations of motion and the phase space parameters.

References

1. M. Sands, The physics of electron storage rings. SLAC 121. US Atomic Energy Commission (1970).
2. P. J. Duke, In *X-ray science and technology*, Chapter 3 (eds A. G. Michette and C. J. Buckley).
3. S. Krinsky, M. L. Perlman, and R. E. Watson, In *Handbook of synchrotron radiation*, Vol. 1a. Chapter 3 (ed. E.-E. Koch).
4. H. Wiedemann, *Particle accelerator physics*, Chapter 5, p. 151. Springer Verlag.
5. R. Q. Twiss and N. H. Frank, *Review of Scientific Instruments*, **20**, 1 (1949).

11

Behaviour of the electron beam in a synchrotron radiation storage ring. Betatron oscillations

Introduction

In Chapter 10 we obtained the equation of motion of a particle in the magnetic fields of a storage ring and we showed how the motion of an individual particle could be traced throughout the storage ring using the transformation matrices. In order to proceed further, we must now describe the individual behaviour of a particle as it travels around the storage ring and relate this motion to the concept of the phase space ellipse which we developed in the last chapter.

Betatron oscillations

If we imagine the particle starting from some arbitrary position and direction relative to the ideal orbit, the system of focusing magnets exerts a force on the particle which causes it sooner or later to move towards the ideal orbit. Eventually the particle crosses the ideal orbit (we assume motion in the plane of the ideal orbit for the time being) where it again experiences a restoring force back towards the ideal orbit. As we would expect, the particle executes oscillations about the ideal orbit in the horizontal plane. A similar argument applies in the vertical plane. These oscillations are called betatron oscillations because they were first observed in a type of accelerator called a betatron.

We speak of a particle with some arbitrary position and direction (i.e. arbitrary values of u and u') because the loss of energy in synchrotron radiation is a statistical process. The energy loss can be of any amount, constrained only by the properties of the synchrotron radiation loss process described previously. Furthermore, this energy loss can take place anywhere around the storage ring where the orbit has non-zero curvature. It follows that even if the particles began their oscillations all with the same phase and the same amplitude and direction, this variable energy loss would serve to randomize the betatron oscillations after a few turns around the ring. Even if all the particles were injected into the storage ring so that they began their motion by moving along the ideal orbit, the energy loss process (and the replacement of energy in the radio-frequency cavities) leads to a random oscillatory motion of the particles after a few turns around the ring. In any case, the particles are injected with a finite distribution of positions, directions, and energies though these distributions are sufficiently narrow so that most of the particles are captured into stable orbits in the storage ring.

The concept of a beam of particles circulating around the storage ring gives only an overall, macroscopic view of what is going on. Within the beam any individual particle is executing its own betatron oscillation about the ideal orbit. The description of the beam as a phase ellipse gives an average view of the beam at each point around the ring but it does not tell us how an individual particle is behaving. The general theorem of Liouville tells us that the phase ellipse maintains its area around the ring (though even this can only be an average because of the loss and gain of energy of the particles referred above) which means that the values of u and u' for each and every particle must satisfy an ellipse equation similar to eqn (10.26):

$$\gamma u^2 + 2\alpha u u' + \beta u'^2 = \varepsilon. \tag{11.1}$$

However, the meaning of this equation is not the same as in Chapter 10, where the perimeter of the ellipse was chosen to coincide with the 1σ contour of the beam distribution in phase space and the area ε of the ellipse was therefore equal to the beam emittance. In this case the equation describes the motion of an individual particle and tells us that its motion in phase space is an ellipse whose area is ε and whose parameters α, β, and γ describe its orientation and shape. As before α, β, and γ are related to each other through eqns (10.27) and from one position round the ring to the next by the transformation matrices. Equation (11.1), in this context, is called the Courant–Snyder invariant[1] and tells us that an individual particle always moves around the phase ellipse. That ellipse changes its shape around the ring, as we have seen before, but (neglecting energy loss or gain) the particle remains on the elliptical contour even though that contour is changing as a function of position around the ring.

Betatron tune values

Because the values of β and the other phase ellipse parameters (known collectively as the betatron functions) depend on the magnetic fields in the ring magnets, they repeat themselves after a single turn round the storage ring. In practice they usually repeat themselves more quickly than that because the ring is made up as an integral number of identical magnetic units, the magnetic lattice. At the start of each unit, the betatron functions have the same value and the ellipse has the same shape and orientation.

The oscillations of the individual electrons do not have this property. Each electron moves around the circumference of its own phase ellipse but in general it does not return to the same point on the contour at the start of each unit magnetic cell. Indeed, it is important that it does not do so because such orbits would not be stable. Imperfections in the magnetic fields which perturb the particle orbits would combine together in phase to drive the particles further and further away from the ideal orbit until eventually the particles would be lost from the ring.

An important quantity is the number of betatron oscillations in one turn around the ring or, in other words, the number of times the betatron oscillation wavelength fits into the total length of one turn around the ideal orbit. In the simple linear theory which we are considering, this quantity is independent of the area ε of the phase ellipse, just as the period of a simple pendulum is independent of the amplitude of the oscillation. This number, the number of betatron oscillations in one turn, is called Q or sometimes ν.

Figure 11.1 (identical to Fig. 10.8) shows the parameters for the phase ellipse of an electron executing betatron oscillations. As the electron moves around the circumference of the ellipse, it reaches a maximum distance from the ideal orbit $u_{\max} = \pm\sqrt{\varepsilon\beta}$ so we can write the equation of motion and its solution as

$$u'' + K(s)u = 0, \tag{11.2}$$

$$u(s) = \sqrt{\varepsilon\beta(s)}\cos(\psi(s) - \psi_0), \tag{11.3}$$

respectively, where the dependence of u, β, and the phase angle ψ on the position variable s has been shown explicitly. As before, ε is constant and ψ_0 is the phase angle at some arbitrary starting point s_0. The corresponding solution for $u'(s)$ is obtained as the first derivative of eqn (11.3):

$$u'(s) = \sqrt{\frac{\varepsilon}{\beta(s)}}\frac{1}{2}\beta'(s)\cos(\psi(s) - \psi_0) - \sqrt{\varepsilon\beta(s)}\psi'(s)\sin(\psi(s) - \psi_0) \tag{11.4}$$

which implies that $\alpha(s) = -\beta'(s)/2$ and $\psi'(s) = 1/\beta(s)$ as can be seen from Fig. 11.1. If we eliminate $(\psi(s) - \psi_0)$ between eqns (11.3) and (11.4) we obtain the expression [eqn (11.1)] for the Courant–Snyder invariant.

The identity $\psi'(s) = 1/\beta(s)$ is very important because it relates the rate of change of phase of the betatron oscillations to the parameter $\beta(s)$ which is related in turn to the magnetic field values around the ring [eqns (10.33) and (10.34)]. The phase at any point is given by

$$\psi(s) = \int_{s_0}^{s} \frac{1}{\beta(s)}\,\mathrm{d}s.$$

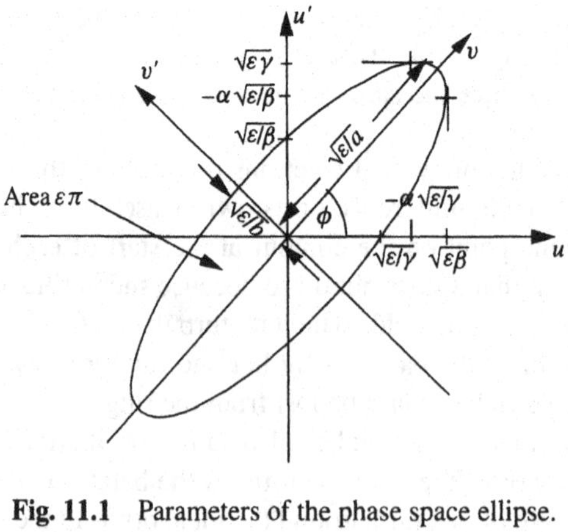

Fig. 11.1 Parameters of the phase space ellipse.

After one whole turn around the ring, the phase of the oscillation must advance by $2\pi Q$, where Q is the number of betatron oscillations defined above. If L is the length of the ideal orbit then

$$2\pi Q = \int_{s_0}^{s_0+L} \frac{1}{\beta(s)}\,\mathrm{d}s = \oint \frac{1}{\beta(s)}\,\mathrm{d}s \tag{11.5}$$

which is independent of the starting point s_0 because of the circular symmetry of the ring. The symbol $\oint$ denotes an integration, from an arbitrary starting point, around one turn of the ring. It is also clear that the form of the betatron oscillation is a function of the whole ring. Any change in the field at one point changes the value of Q and the shape of the oscillation at every point along the ideal orbit.

The values of the magnetic field parameters for the SRS at Daresbury are shown in Table 11.1 for one section of the magnetic lattice. The whole ring is composed of 16 sections, each with the same properties.

The beta functions corresponding to this lattice are shown in Fig. 11.2, calculated using the program package Agile230.[2] In the horizontal plane Q1 is a diverging quadrupole with a positive K-value in Table 11.1. The electron beam width, in the horizontal plane, as a function of position is determined using the expression

TABLE 11.1 SRS lattice parameters

No.	Name	Type	Length (m)	H-bend (rad)	K-value (m^{-2})
1	D1	Drift space	0.4725	0	0
2	Q1	Quadrupole	0.3	0	1.55
3	D2	Drift space	2.167	0	0
4	Q2	Quadrupole	0.5	0	−1.27
5	D3	Drift space	0.3725	0	0
6	BD	Dipole	2.188	0.3927	0
Total length of section			6.0000		

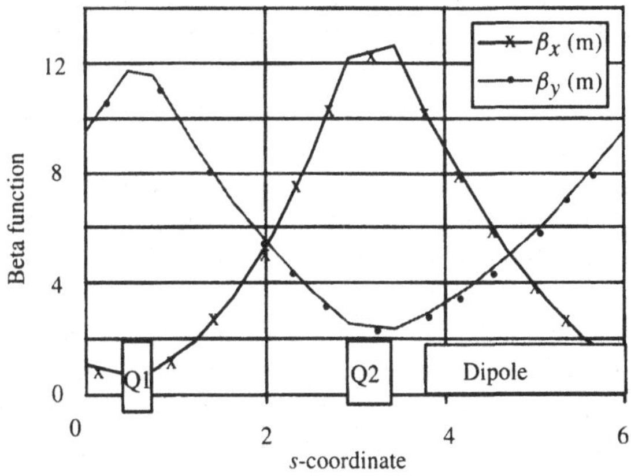

Fig. 11.2 Beta functions for the SRS lattice.

$\sigma_{\beta h} = \sqrt{\varepsilon\beta}$ as shown in Fig. 11.1. In the case of the SRS the beam emittance $\varepsilon = 0.11 \times 10^{-3}$ m rad so that the beam size determined by the betatron oscillations near the centre of the dipole magnet is 0.55 mm. The phase advance of the betatron oscillation of an individual electron can now be obtained by inserting the calculated values of the beta function into eqn (11.5) and the behaviour of an electron trajectory with a given starting point is obtained from the solutions of the equations of motion [eqns (11.3) and (11.4)]. The most important quantity is the Q-value, obtained by integrating the beta function around the whole ring. For the SRS these Q-values turn out to be 6.184 in the horizontal plane (a betatron oscillation wavelength of 15.53 m) and 3.201 (a betatron oscillation wavelength of 30 m) in the vertical plane.

It is very important that the actual Q-values are chosen so that the electron trajectories do not repeat themselves after a small number of revolutions around the ring. For example, integral values of Q ensure that an individual electron orbit repeats itself identically in each revolution. In this case, the effect of any small imperfections in the magnet lattice is amplified after each turn around the ring and the loss of the entire electron beam follows quickly. Such resonances can occur not only at integral Q values but also at half integral or any simple fractional value. Furthermore, magnet imperfection, in general, causes coupling between the horizontal and vertical electron beam oscillations. This is because a magnetic imperfection, e.g. a misalignment of a dipole magnet, causes the electron trajectory to receive a small vertical deflection as well as the designed horizontal deflection. A useful rule is that, in order to avoid resonant beam losses, the tune values must satisfy eqn (11.6), in which L, M, and N are integers:

$$L \neq MQ_x + NQ_y. \tag{11.6}$$

A typical orbit of a single electron, starting at an arbitrary point 1 mm from the central orbit and executing three successive turns around the ring are shown in Fig. 11.3. It can be seen that the rapid variation of fields and gradients leads to trajectories which are far from being simply sinusoidal, as well as being non-repetitive (at least on the scale shown). Nevertheless, the general pattern follows the form of the beta functions as expected from eqn (11.3) and as shown in Fig. 11.4. Maximum excursions of the electron trajectories occur at values of the s-coordinate where the beta function is a maximum.

In order to calculate the other beam parameters of the phase space ellipse shown in Fig. 11.1, we require the α and γ functions, which are shown in Figs 11.5 and 11.6. By definition [eqn (10.28)], the function γ remains constant in a drift space.

Finally we need to know how the tilt angle of the phase space ellipse [given by eqn (10.27)] changes through the magnetic lattice. The behaviour for SRS2 is shown in Fig. 11.7.

At points around the ring, where the tilt angle of the phase ellipse is zero, the axes of the ellipse coincide with the coordinate axes, so that $\alpha = 0$ and $\beta\gamma = 1$. At these points, the beam emittance, which is constant around the ring, is equal to the product of the standard deviations of the beam width and beam divergence, i.e.

$$\varepsilon_x = \sigma_x {\sigma'}_x \quad \text{and} \quad \varepsilon_y = \sigma_y \sigma'_y.$$

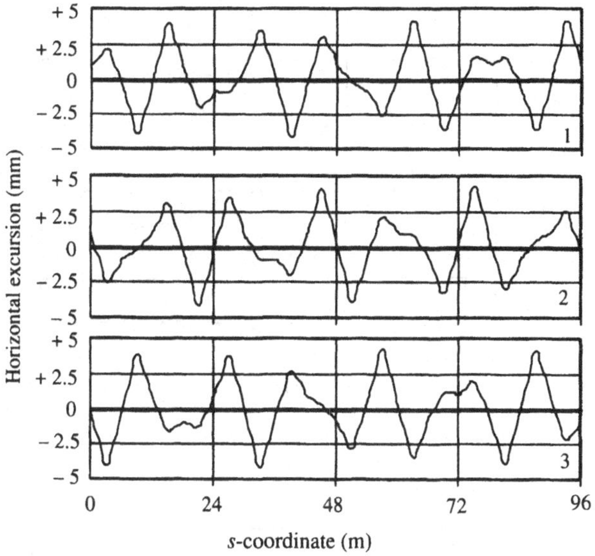

Fig. 11.3 Electron orbits for three successive turns.

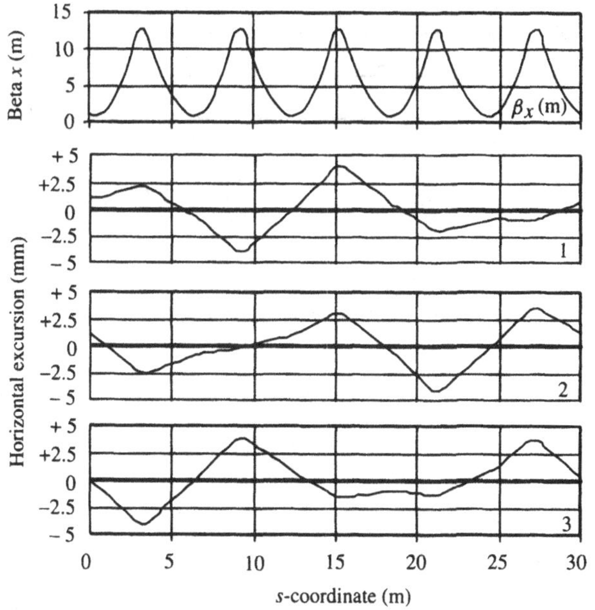

Fig. 11.4 Comparison of trajectories and beta function values over five unit cells of the SRS2 magnetic lattice.

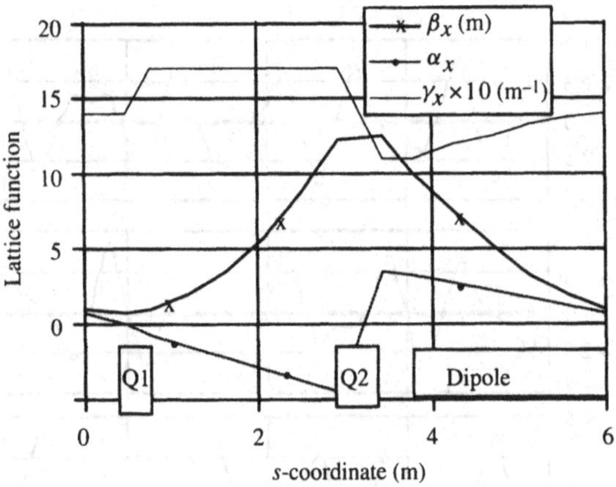

Fig. 11.5 Horizontal lattice functions for SRS2.

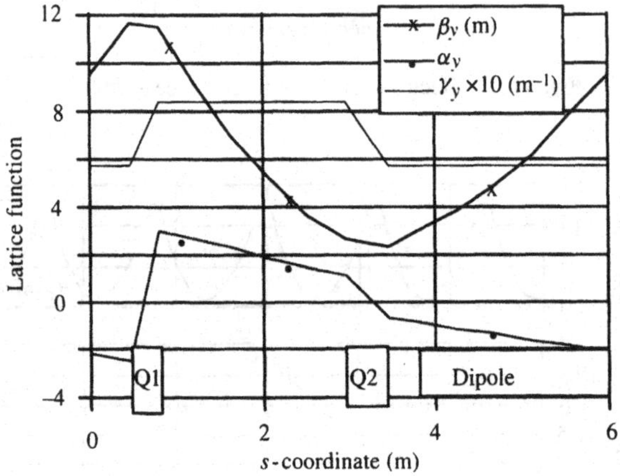

Fig. 11.6 Vertical lattice functions for SRS2.

Elsewhere, the product $\sigma\sigma' > \varepsilon$ in both dimensions, so that ε_x and ε_y are given by

$$\varepsilon_x = \frac{\sigma_x \sigma_x'}{\sqrt{1+\alpha_x^2}} \tag{11.7}$$

and

$$\varepsilon_y = \frac{\sigma_y \sigma_y'}{\sqrt{1+\alpha_y^2}}. \tag{11.8}$$

It follows that the synchrotron radiation beam itself will have a minimum emittance where $\alpha = 0$ and the phase ellipse is upright.

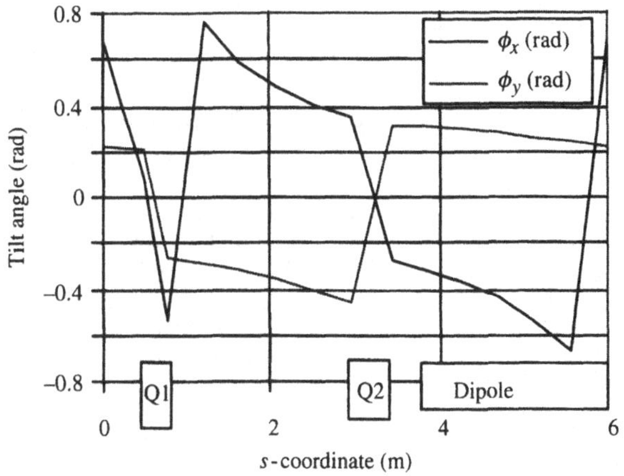

Fig. 11.7 Variation of the angle of tilt of the phase ellipse in SRS2.

Energy dispersion

We have considered what happens to particles whose trajectory takes them away from the ideal orbit. We must now enquire about particles whose energy is different from the design energy of the storage ring. Clearly all electrons fall into this category for most of the time as they lose energy by the emission of synchrotron radiation and gain energy from the radio-frequency system to compensate for this loss.

Suppose an electron has an energy E which is higher than the design energy E_0, then $E - E_0 = \Delta E$. This electron executes betatron oscillations with an amplitude x_β about a new 'ideal' orbit appropriate for an electron with energy $E = E_0 + \Delta E$. This orbit is at a distance x_Δ from the reference orbit for the electron with energy E_0. The x-coordinate of this off-energy electron (see Fig. 11.8) is given by the sum of x_β and x_Δ:

$$x(s) = x_\beta(s) + x_\Delta(s).$$

An off-energy electron can be generated in many ways, the most obvious being a loss of energy to synchrotron radiation in a dipole magnet. Such an electron does not change its x-coordinate when its energy changes but its ideal orbit, from which x_β is measured, jumps instantaneously through a distance x_Δ. The equation of motion of this electron is given by eqn (10.10), repeated here for convenience,

$$\begin{aligned} x'' &= G(s)\frac{\Delta E}{E_0} - [G(s)^2 + K_x(s)]x \\ &= G(s)\frac{\Delta E}{E_0} + K(s)x, \end{aligned}$$

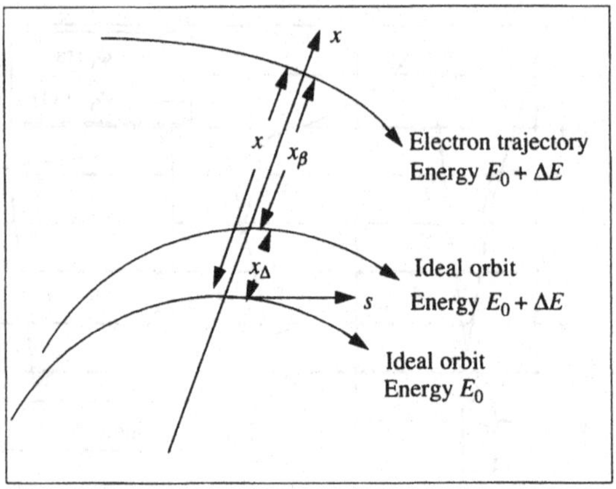

Fig. 11.8 Radial displacement from the ideal orbit.

so we can write

$$x''_\beta = K(s)x_\beta, \tag{11.9a}$$

$$x''_\Delta = K(s)x_\Delta + G(s)\frac{\Delta E}{E_0}, \tag{11.9b}$$

where eqn (11.9b) is the equation of motion of a notional off-energy electron with energy difference ΔE from the norm. Suppose we describe this orbit by the dispersion function $D(s)$ [some authors use $\eta(s)$ for the dispersion function] and write

$$x_\Delta = D(s)\frac{\Delta E}{E_0} \tag{11.10}$$

so that $D(s)$ (which has the dimensions of length) represents the deviation from the ideal orbit of a fictitious off-energy electron with $\Delta E/E_0$ equal to unity. We insert this definition into eqn (11.9b) so that

$$D''(s) = K(s)D(s) + G(s) \tag{11.11}$$

and solve for $D(s)$ to obtain an expression for the dispersion function in terms of the magnet parameters which are contained in the function $K(s)$. Because $K(s)$ is a periodic function of the magnetic lattice, $D(s)$ is periodic as well and we can impose that condition in order to solve eqn (11.11). The general solution for $D(s)$, with the periodicity condition can be written as

$$D(s) = \frac{\sqrt{\beta(s)}}{2\sin(\pi Q)}\int_0^s G(s')\sqrt{\beta(s')}\cos(\psi(s) - \psi(s') - \pi Q)\,\mathrm{d}s'.$$

An important feature of this solution is that $D(s)$ is infinite when Q, the number of betatron oscillations around the ring, is an integer. When this is so, the betatron wavelength is a multiple (or a sub-multiple) of the wavelength of the dispersion function and a resonant condition exists which renders the off-energy electron orbits unstable and forces a rapid loss of the electron beam. That this is to be expected can be seen from eqn (10.10) where the off-energy term acts as a periodic force exciting the betatron oscillatory motion defined by the other two terms in the equation. When the frequency of the exciting force is an integral multiple of the natural frequency of the betatron oscillations, resonance ensues and the beam is lost.

The behaviour of the dispersion function is calculated from the parameters of the magnetic lattice and is shown for SRS2 in the horizontal plane (in Fig. 11.9). In the vertical plane there is no dispersion, for an ideal magnetic lattice of the type we are considering here, and $D_y(s) = 0$ throughout. The shape of the dispersion function follows that of the beta function, as can be seen from Fig. 11.9.

Orbit length and energy

An off-energy particle with energy $E_0 + \Delta E$, has a shorter orbit radius in the dipole magnets than that of a particle with energy E_0. Such a particle follows a tighter orbit and we would expect the orbit time of such a particle to be less than one which follows the slightly longer ideal orbit. However, that is not necessarily so because a particle which has a higher energy also has a higher velocity which tends to reduce the orbit time. These two effects might be expected to cancel each other out to some extent but, in fact, because the particles are travelling at a speed almost equal to the speed of light, there is essentially no change in the velocity and it is the length of the orbit which determines the orbiting time.

The dependence of orbiting time on particle energy could mean that off-energy particles do not remain in step with the radio-frequency electric field. Energy changes could put the particles into unstable orbits, leading to loss of stored current. Let us calculate the change in the length of the orbit (and equivalently, the change in the

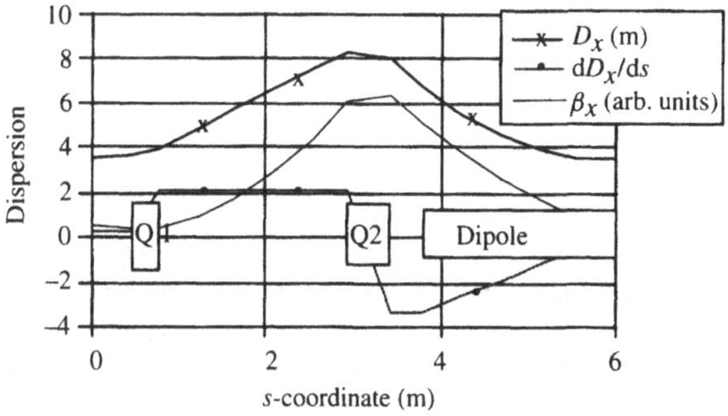

Fig. 11.9 SRS2 dispersion function ($D_x(s)$) and its derivative ($D'_x(s)$) compared with the horizontal beta function.

orbiting time) for a particle with energy $E_0 + \Delta E$. If T_0 is the time taken to complete an ideal orbit whose length is L_0 then we can write, to first order,

$$\frac{\Delta T}{T_0} = \frac{\Delta L}{L_0} = \alpha \frac{\Delta E}{E_0}. \tag{11.12}$$

The quantity α is called the momentum compaction or the dilation factor (not to be confused with the Twiss parameter α) and is a measure of the degree of compression (or expansion) of the orbit resulting from unit fractional change of particle energy. Clearly, this quantity must be related in some way to the dispersion function $D(s)$ and in particular to the integral of this function around the orbit. Let us determine this relationship.

The change of path length $\mathrm{d}l$ along the actual orbit is related to $\mathrm{d}s$, the change, along the ideal orbit by eqn (10.8)

$$\mathrm{d}l = (1 + G(s)x)\,\mathrm{d}s,$$

where x, the displacement of the particle from the ideal orbit in the horizontal plane, is related to the dispersion function through eqn (11.10).

We neglect the change in orbit length and orbit time produced by the betatron oscillations because as the particle travels around the ring the lengthening of the orbit during the excursion of the particle outside the ideal orbit is compensated by a reduction of the orbit length during inward excursions.

With this approximation, the total length of the actual orbit is

$$L = \oint (1 + G(s)x_\Delta)\,\mathrm{d}s = L_0 + \oint G(s)x_\Delta\,\mathrm{d}s$$

so that

$$\Delta L = \oint G(s)x_\Delta\,\mathrm{d}s = \frac{\Delta E}{E_0}\oint G(s)D(s)\,\mathrm{d}s$$

and

$$\alpha = \frac{1}{L_0}\oint G(s)D(s)\,\mathrm{d}s. \tag{11.13}$$

Isomagnetic guide field

In most storage rings, $G(s)$ is constant in the dipole magnets and equal to zero everywhere else. Such a magnetic lattice is said to be isomagnetic and $G(s) = G_0 = 1/R_0$. In such a lattice there is no contribution to the momentum compaction factor from $D(s)$ in the quadrupole magnets where $G(s) = 0$, so we may write

$$\alpha = \frac{G_0}{L_0}\oint_{\mathrm{d}} D(s)\,\mathrm{d}s,$$

where $\oint_{\mathrm{d}}$ indicates that the integral around the orbit is evaluated over the dipoles only. If we write for the average value of $D(s)$ in the dipole magnets

$$\langle D\rangle_{\mathrm{d}} = \frac{\oint_{\mathrm{d}} D(s)\,\mathrm{d}s}{\oint_{\mathrm{d}} \mathrm{d}s} = \frac{\oint_{\mathrm{d}} D(s)\,\mathrm{d}s}{L_{\mathrm{d}}},$$

where $L_{\mathrm{d}} = 2\pi R_0 = 2\pi/G_0$ is the length of the orbit in the dipoles so that

$$\oint_{\mathrm{d}} D(s)\,\mathrm{d}s = \frac{2\pi\langle D\rangle_{\mathrm{d}}}{G_0}$$

and

$$\alpha = \frac{2\pi}{L_0}\langle D\rangle_{\mathrm{d}}. \tag{11.14}$$

Figure 11.9 shows that $D(s)$ is always positive and, in particular, $D(s)$ is positive in the dipole magnet as is $\langle D(s)\rangle_{\mathrm{d}}$, so α must also be positive, which gives the important result that a particle whose energy is greater than the ideal energy executes an orbit which is longer than L_0 and takes a time $>T_0$ to complete the orbit. Similarly, an electron which loses energy to synchrotron radiation finds itself executing an orbit whose orbiting time is shorter than the orbiting time T_0 for a particle with the ideal energy. The numerical value of α for the SRS is found to be 0.03 from eqn (11.14) by obtaining $\langle D(s)\rangle_{\mathrm{d}}$ (from Fig. 11.9) as 0.46 m and dividing by the circumference (96 m).

In the next chapter, we will explore these energy oscillations in more detail and relate their properties to the parameters of the radio-frequency cavities.

References

1. E. Courant, M. S. Livingston, and H. Snyder, *Phys. Rev.* **88**, 1190 (1952).
2. P. J. Bryant, A brief history and review of accelerators. CERN Accelerator School Proceedings. CERN 91-04 (1991) CERN, Geneva, Switzerland.

12

Behaviour of the electron beam in a synchrotron radiation storage ring. Energy oscillations

Introduction

Just as deviations form the ideal orbit lead to betatron oscillations, so also deviations from the ideal energy lead to energy oscillations which are sometimes called synchrotron oscillations. These oscillations contribute to the size of the electron beam in the horizontal plane because energy deviations lead to deviations in position and direction of the electrons through the dispersion function [see eqn (11.9)]. Because the dispersion function depends on the value of the s-coordinate around the ring, the contribution of the energy oscillations to beam size also depends on $D(s)$. Furthermore, because the dispersion function tends to follow the contour of the beta function [see eqn (11.11)], the effects of the betatron and synchrotron oscillations tend to combine to increase the beam size instead of cancelling each other out. If the energy distribution of the electrons follows a Gaussian function with standard deviation $\sigma(E)$, then from eqn (11.9) the contribution of the energy oscillations to the total beam width is $(\sigma(E)/E_0) \times D(s)$. In any practical storage ring, Q is a long way from resonance so we can assume that the two types of oscillations are uncorrelated and add the contributions from the betatron [eqn (10.29)] and synchrotron oscillation in quadrature. This gives, for the total standard deviation of the position and direction distributions as a function of position around the ring:

$$\begin{aligned} \sigma_x^2(s) &= \varepsilon_x \beta(s) + \left(\frac{\sigma_E}{E_0}\right)^2 D(s)^2, \\ \sigma_x'^2(s) &= \varepsilon_x \gamma(s) + \left(\frac{\sigma_E}{E_0}\right)^2 D'(s)^2. \end{aligned} \tag{12.1}$$

Let us now examine the properties of the energy oscillations and explore their effects on beam behaviour in more detail. An electron moving along the ideal orbit with energy equal to the ideal energy loses energy as it radiates but gains energy as it passes through the radio-frequency cavities. On average, after several turns around the ring, these gains and losses balance so that

$$\Delta E_{\mathrm{rf}} + \Delta E_{\mathrm{sr}} = 0. \tag{12.2}$$

The energy lost to synchrotron radiation, ΔE_{sr}, is always negative and it is a statistical process. The energy gain, ΔE_{rf}, from the radio-frequency system can be positive or negative, depending on the phase angle of the electron as it passes through the cavity.

Figure 12.1 represents the electric field in the radio-frequency cavities as a simple function of time (or phase). An imaginary electron with the ideal energy is represented by a series of open circles shown, in the diagram, on the trailing edge of the curve. Such an electron remains exactly in phase with the radio frequency from one orbit to the next and its orbiting time is T_0. In order to have these ideal properties, the phase angle of this electron must be arranged so that eqn (12.2) is satisfied. It is shown in the figure as though it gained the same amount of energy from the radio-frequency field at each turn around the orbit. Such an electron is stationary in phase, with a phase angle ψ_0. Consider now an electron, represented by the black circle at P, which passes through the radio-frequency cavity ahead of the ideal particle, i.e. at an earlier time and, therefore, with a smaller phase angle. Whatever the energy of that particle, it gains more energy than the ideal particle because the field is stronger as it passes through the cavity. Because the momentum compaction factor, α, of the particle is always positive [eqn (11.14)], the particle completes its next orbit in a time $T_1 = T_0 + \Delta T_0$ which is longer than the ideal by an amount ΔT (the particle has further to travel) so that the phase angle of the particle is closer to that of the ideal particle when they next traverse the radio-frequency cavity and it is closer in time to the ideal phase angle, at the point Q. This process continues and on the next orbit we can imagine the off-energy particle overshooting so that its phase angle moves ahead of the ideal. Once this state is reached, the energy gain is less than the ideal (it may even decelerate if the electric field has changed sign by the time the particle passes through) and the particle tends once again to move closer in phase to the ideal particle. The net result is that the particle executes phase oscillations, often called synchrotron oscillations relative to the ideal, equilibrium, radio-frequency phase. The period of these synchrotron oscillations is long compared to T_0. It follows that the particles form a compact bunch in which individual particles execute stable oscillations, over a range of phase angles, and about a point of stationary phase. This important principle of phase stability was first realized by Veksler[1] at Dubna, in what was then the USSR, and, independently,

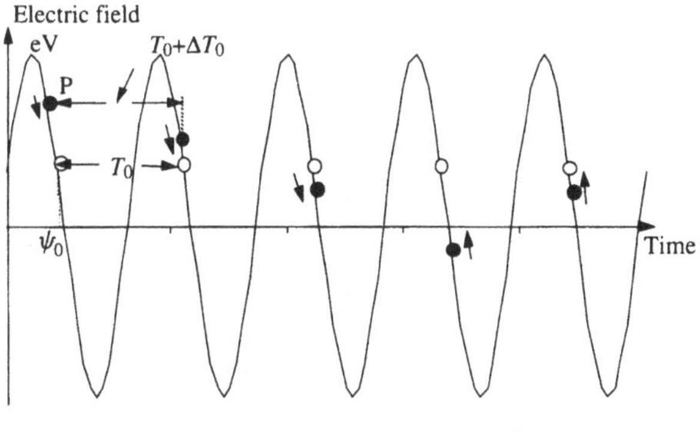

Fig. 12.1 Phase oscillations over several orbits.

by McMillan[2] at Berkeley in the USA, in connection with the design of the first synchrocyclotrons. (Edwin Mattison McMillan, 1907–91; Vladimir Iosivich Veksler, 1907–66).

We could also consider the behaviour of particles whose phase angle places them on the leading edge of the cosine curve. Can a bunch form around that phase angle, equal to $\psi_0 - \pi$? Application of the same argument shows that these particles are unstable and move away from ψ_0 instead of towards it. Actually, particles with negative momentum compaction factor would form a stable bunch at this point but negative values of the momentum compaction factor cannot occur for relativistic electrons in a storage ring.

Damping of energy oscillations

How can we describe the oscillation of a particle within a phase stable bunch? It will be helpful to use a time coordinate τ measured relative to the radio-frequency phase ψ so that $\omega\tau = \omega t - \omega t_0$, where t is the time at which a typical particle P traverses the cavity. When $\tau = 0$, $t = t_0$ and the radio-frequency phase is equal to ψ_0. The energy of the particle at P is measured in terms of its difference ε from the ideal energy E_0 as shown in Fig. 12.2. Note that ε is equal to ΔE. During the time T_0 which it takes the particle at the bunch centre to travel around the ring, a particle at point P in Figs 12.1 and 12.2 moves from P at time τ_1 to Q at time τ_2, and the time of passage of the particle through the radio-frequency cavity moves closer in time to the centre of the bunch by an amount $\Delta\tau = \tau_2 - \tau_1$ given by eqn (11.12):

$$\Delta\tau = -\alpha \frac{\Delta E}{E_0} T_0.$$

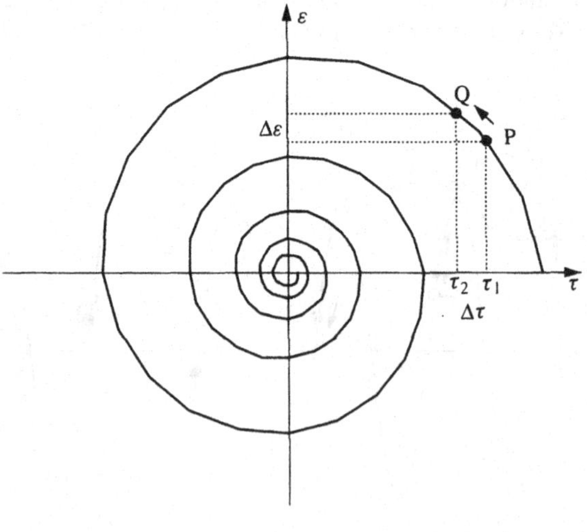

Fig. 12.2 Electron oscillation and damping within the electron bunch.

The minus sign arises because $\tau_2 < \tau_1$. The particle moves from P to Q during the time T_0 so we may write, from eqn (11.12),

$$\frac{\mathrm{d}\tau}{\mathrm{d}t} = \frac{\Delta\tau}{T_0} = -\frac{\alpha\varepsilon}{E_0}. \tag{12.3}$$

At the same time, the difference between the energy of the particle and the ideal energy increases by an amount $\Delta\varepsilon$ which is equal to $eV(\tau)$, the energy gained by the particle from the radio-frequency field in one turn around the ring and $U(\varepsilon)$, the energy lost to synchrotron radiation in one turn, so that

$$\frac{\mathrm{d}\varepsilon}{\mathrm{d}t} = \frac{eV(\tau) - U(\varepsilon)}{T_0}. \tag{12.4}$$

The radio-frequency field acting on the particle is a function of the time of passage of the particle through the cavity but for a small deviation of the particle (small oscillation amplitude) from the synchronous phase we may use Taylor's theorem to write the behaviour of the function [in this case the electric field, $eV(\tau) = eV(t - t_0)$] in the vicinity of t_0 as an expansion:

$$\begin{aligned} eV(\tau) = eV(t - t_0) &= eV(t_0) + (t - t_0)\left.\frac{\mathrm{d}V}{\mathrm{d}t}\right|_{t=t_0} + \cdots \\ &= eV(t_0) + \tau\dot{V}|_{t=t_0} \end{aligned} \tag{12.5}$$

in which all terms of higher order than linear have been dropped and $\dot{V}$ is to be evaluated at the synchronous phase.

A similar expression for the synchrotron radiation energy loss in the vicinity of E_0 is, in a linear approximation,

$$U(\varepsilon) = U(E_0) + \varepsilon\left.\frac{\mathrm{d}U}{\mathrm{d}E}\right|_{E=E_0}. \tag{12.6}$$

These approximations are valid only for oscillations with small amplitude. In practice, large deviations are to be expected, demanding a more complex treatment. This linear approximation is sufficient to describe the principal features of the motion.

If we differentiate expression (12.4) with respect to time, we obtain

$$\frac{\mathrm{d}^2\varepsilon}{\mathrm{d}t^2} = \frac{e}{T_0}\frac{\mathrm{d}V}{\mathrm{d}\tau}\frac{\mathrm{d}\tau}{\mathrm{d}t} - \frac{1}{T_0}\frac{\mathrm{d}U}{\mathrm{d}\varepsilon}\frac{\mathrm{d}\varepsilon}{\mathrm{d}t}.$$

Substitution of the expressions for $\mathrm{d}\tau/\mathrm{d}t$ and $\mathrm{d}\varepsilon/\mathrm{d}t$ from eqns (12.3) and (12.4) with the approximations (12.5) and (12.6) [remembering that $eV(t_0) = U(E_0)$] gives

$$\begin{gathered} \frac{\mathrm{d}^2\varepsilon}{\mathrm{d}t^2} + 2\mu\frac{\mathrm{d}\varepsilon}{\mathrm{d}t} + \Omega^2\varepsilon = 0 \\ \text{with} \quad \mu = \left.\frac{1}{2T_0}\frac{\mathrm{d}U}{\mathrm{d}E}\right|_{E=E_0} \quad \text{and} \quad \Omega^2 = \left.\frac{e}{T_0}\frac{\alpha}{E_0}\dot{V}\right|_{t=t_0} \end{gathered} \tag{12.7}$$

for the energy oscillations of the particle contained within the phase stable bunch. The same procedure, applied to the variable τ, gives an identical equation for the phase oscillations along the time axis of Fig. 12.2. Both equations describe damped harmonic oscillations with frequency Ω. The effect of the damping is to reduce the amplitude of the oscillations, $1/\mu$ being the damping time. If we plot the variables ε and τ in the form of a phase diagram (see Fig. 12.3) the particle moves around the diagram in a path which, in the absence of damping, is an ellipse. In the presence of damping the curve is a spiral as shown in Fig. 12.2.

The time taken to complete one revolution around this phase ellipse is $1/\Omega$. If we assume, correctly as it will appear, that the damping time is much longer than the revolution time then the damping term can be neglected, so that

$$\frac{d^2\varepsilon}{dt^2} = -\Omega^2\varepsilon \quad \text{and} \quad \frac{d^2\tau}{dt^2} = -\Omega^2\tau$$

with Ω given by eqn (12.7). The position of the electron on the phase ellipse as a function of time can be written as

$$\varepsilon = \varepsilon_0 \sin(\Omega t - \phi_n) \quad \text{and} \quad \tau = \tau_0 \cos(\Omega t - \phi_n)$$

in which ϕ_n is the arbitrary phase of the nth particle in the bunch at time $t = 0$.

The oscillations in electron energy and temporal position in the bunch are 90° out of phase. In other words, each electron executes a pendulum motion in both energy (ε) and time (τ) within the bunch, but when τ reaches its maximum excursion at τ_0, $\varepsilon = 0$ and vice versa. An electron at the head of the bunch (which therefore arrives early, relative to the design phase ψ_0 at the radio-frequency cavity) gains energy and gradually swings towards the tail of the bunch. As the phase of the oscillating electron passes through ψ_0 the electron reaches its maximum energy excursion at $\tau = 0$ and $\varepsilon = \varepsilon_0$ and continues on its way towards the tail of the bunch. There it reverses its direction and moves back towards the head of the bunch, taking a time equal to $1/\Omega$ to complete one cycle of this synchrotron oscillation. At any moment an individual electron can be anywhere in this cycle, depending on its starting phase, as indicated by the arbitrary phase ϕ_n. From eqn (12.3) the amplitudes of these coupled oscillations of phase and energy are related by

$$\tau_0 = \frac{\alpha}{\Omega}\frac{\varepsilon_0}{E_0}. \tag{12.8}$$

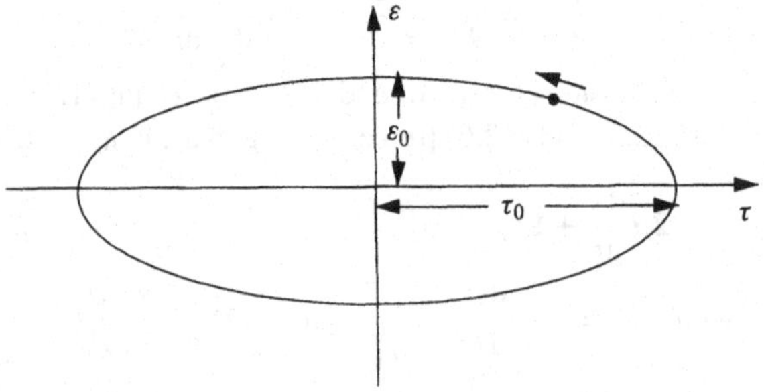

Fig. 12.3 Electron bunch as a phase ellipse.

To evaluate Ω we describe the radio-frequency waveform by the equation

$$V = V_0 \sin \omega t,$$

then

$$\dot{V} = \left.\frac{\mathrm{d}V}{\mathrm{d}t}\right|_{t=t_0} = \omega V_0 \cos \phi_0.$$

The revolution time $T_0 = 2\pi/\omega_0$ and $n\omega_0 = \omega$, where ω is the radio frequency and n the harmonic number, equal to the maximum number of stored bunches within the circumference of the storage ring, so that

$$\Omega = \omega \sqrt{\frac{\alpha e V_0 \cos \phi_0}{2\pi E_0}}.$$

If we suppose $V_0 = 250\,\mathrm{kV}$ (see Chapter 8), set $\cos \psi_0 \approx 1$, and use the value of $\alpha = 0.03$ calculated above, we obtain a value of approximately 400 kHz for the frequency of the synchrotron oscillations, which shows that these are slow compared with the revolution frequency of the particle bunches. This is what would be expected from the arguments given above.

The damping term causes the particle to lose energy and follow a spiral path, drifting in towards the centre of the diagram as shown in Fig. 12.2. There is a similarity to the dynamics of the Earth–Moon system in which the elliptical orbit of the Moon around the Earth becomes a spiral because the tidal forces induced mainly on the Earth cause energy to be radiated in the form of heat. In the Earth–Moon case, the drift is very slowly outwards and the Moon is slowing down at a rate of about 40 arcsec/century.[2]

What is the damping time for the electron beam? The quantity we require is $\mathrm{d}U/\mathrm{d}E$ evaluated at $E = E_0$ [eqn (12.7)]. We can write

$$\frac{\mathrm{d}U}{\mathrm{d}E} = \frac{\mathrm{d}}{\mathrm{d}E} \oint \frac{\mathrm{d}U}{\mathrm{d}t}\mathrm{d}t = \frac{\mathrm{d}}{\mathrm{d}E} \oint \frac{\mathrm{d}U}{\mathrm{d}t}\frac{\mathrm{d}t}{\mathrm{d}s}\mathrm{d}s$$

with the integral calculated around the storage ring, and, from eqn (10.8),

$$\frac{\mathrm{d}t}{\mathrm{d}s} = \frac{1}{c}\left(1 + \frac{x}{R_0}\right) = \frac{1}{c}\left(1 + \frac{D(s)}{E_0 R_0}\varepsilon\right).$$

We must write the expression in this way because when the energy of a particle changes by an amount $\mathrm{d}E$, the change in energy radiated, $\mathrm{d}U$, is a function of the magnetic field and the length of the orbit as well as the particle energy itself.

From eqn (5.20), we know that

$$\frac{\mathrm{d}U}{\mathrm{d}t} = \frac{2}{3}\frac{e^2 c}{4\pi\varepsilon_0}\frac{E^2 B^2}{M^4}$$

which we write as $\dot{U}$ and $\mathrm{d}\dot{U}/\mathrm{d}E$, evaluated at $E = E_0$, is

$$\frac{\mathrm{d}\dot{U}}{\mathrm{d}E} = 2\frac{\dot{U}}{E_0} + 2\frac{\dot{U}}{B_0}\frac{\mathrm{d}B}{\mathrm{d}E}.$$

Let us first determine the order of magnitude of the damping time by calculating only the zero order term, which means that we neglect terms which are first or higher order in $D(s)$. In other words, $(1/B_0) \times (\mathrm{d}B/\mathrm{d}E)$ is assumed to be small compared to $(1/E_0)$, so that

$$\frac{\mathrm{d}U}{\mathrm{d}E} = \frac{2}{E_0}\oint \frac{\mathrm{d}U}{\mathrm{d}t}\mathrm{d}t = 2\frac{U_0}{E_0}$$

because $\oint (\mathrm{d}U/\mathrm{d}t)\,\mathrm{d}t = U_0$, the total energy radiated in one turn around the ring. So, from eqn (12.7) the damping constant is

$$\mu = \frac{1}{T_0}\frac{U_0}{E_0}.$$

Since U_0/T_0 is the average rate of loss of energy, the rule of thumb is that in zero order the damping time $1/\mu$ is the time which would be taken for the electron to lose all its energy. If the electron loses 250 keV/turn then, for a 2 GeV electron, 8000 turns are required and the damping time is about 3 ms.

The contribution to the damping time from the effect of the change of orbit on $\dot{U}$ when the energy of the electron changes from E to $E + \mathrm{d}E$ is, in general, small but not negligible. It can be included by following through the above argument with care to retain all the terms which depend on the dispersion function $D(s)$. The result is that

$$\mu = \frac{1}{2T_0}\frac{U_0}{E_0}(2 + \mathfrak{D}), \tag{12.9}$$

where $\mathfrak{D}$ denotes an integral around the ring, which is a function of the magnetic lattice properties, independent of the particle energy. Following Sands,[3] we may write

$$\mathfrak{D} = \frac{\oint D(s)\left(G(s)^2 + 2K(s)\right)\mathrm{d}s}{\oint G(s)^2\,\mathrm{d}s}. \tag{12.10}$$

For an isomagnetic, separated function lattice, this integral reduces to the simple form

$$\mathfrak{D} = \alpha\frac{R}{R_0}$$

in which $\mathfrak{D}$ is formed from the product of α, the momentum compaction factor, and the ratio of the mean orbit radius $(2\pi/L_0)$ to the orbit radius in the bending magnets. For the SRS $\alpha = 0.03$, $R_0 = 5.56$ m, and $R = 15.28$ m so that $\mathfrak{D} = 0.08$, which confirms the validity of the approximate calculation of the damping time given above.

The damping of the synchrotron oscillations determines the energy spread of the electrons within the bunch and the length of the bunch along the s-direction. The latter is the longitudinal bunch width. Let us consider the energy spread first. If we imagine an oscillating pendulum which is slowly, but steadily, losing energy (long damping time) the pendulum eventually comes to rest. If we refer to Fig. 12.2, the electron loses energy to synchrotron radiation and gains energy from the radio-frequency field at a rate which is short compared to the damping time, so the electrons move towards the centre of the figure where each has energy E_0 and there is zero energy spread.

However, the loss of energy by the electrons is not a steady process. The energy is lost in finite amounts (quanta) and the amount of energy lost in one quantum can vary over a wide range. The emitted photon can have an energy corresponding to a gamma ray or to radiation in the infrared. In fact, it can be anywhere inside or outside those rather general limits. In terms of Fig. 12.2, a photon close to the origin in the upper half of the diagram can suddenly find itself shifted instantaneously to an extreme position in the lower half of the diagram from which it must work its way around the spiral back towards the origin. This kind of behaviour is called quantum excitation. It is a random statistical process, random in time and random in energy loss, although the distribution of emitted photons must approximate more and more to the synchrotron radiation spectrum over a time comparable with the damping time. There is no correlation between the energies of successive photons so we expect that the eventual distribution of electron energies within the bunch follows a Gaussian form with a mean value E_0 and a standard deviation σ_E given approximately by the average energy of the emitted photons multiplied by the square root of the number of photons emitted by each electron during the damping time $1/\mu$. In other words, we can approximate the standard deviation of the energy distribution by

$$\sigma_E^2 = \frac{\mathrm{d}N_\gamma}{\mathrm{d}t}\left\langle E_\gamma^2\right\rangle\frac{1}{\mu}. \tag{12.11}$$

We can approximate $\mathrm{d}N_\gamma/\mathrm{d}t$ by the power radiated divided by the characteristic energy, so that $\mathrm{d}N_\gamma/\mathrm{d}t \approx \dot{U}/\varepsilon_\mathrm{c}$ and $\mu \approx \dot{U}/E_0$. So, if we further approximate by putting $\langle E_\gamma^2\rangle \approx \varepsilon_\mathrm{c}^2$ then the energy spread is the geometric mean of the electron energy and the characteristic energy of the radiation spectrum. Specifically,

$$\sigma_E = \sqrt{E_0\varepsilon_\mathrm{c}} \quad \text{and} \quad \frac{\sigma_E}{E_0} = \sqrt{\frac{\varepsilon_\mathrm{c}}{E_0}}.$$

From eqn (5.30) $\varepsilon_\mathrm{c} = 3\hbar c\gamma^3/2R$ so that

$$\left(\frac{\sigma_E}{E_0}\right)^2 = \frac{3}{2}\frac{\hbar}{m_\mathrm{e}c}\frac{\gamma^2}{R}$$

which depends only on the energy of the electrons and the radius of the mean orbit in the dipole magnets.

Let us now obtain a more precise result by sharpening this argument. At a time t we can write the energy deviation ε as

$$\varepsilon = \varepsilon_0 \exp(-\mathrm{i}\Omega(t - t_0))$$

which represents an electron undergoing synchrotron oscillations (frequency Ω) which have energy amplitude ε_0 at $t = t_0$. At some later random time t_1 the electron emits a photon with energy E_γ which causes the electron to lose energy and take up a new oscillation which has an amplitude

$$\varepsilon = \varepsilon_0 \exp(-\mathrm{i}\Omega(t - t_0)) - E_\gamma \exp(-\mathrm{i}\Omega(t - t_1)) = \varepsilon_1 \exp(-\mathrm{i}\Omega(t - t_1))\,.$$

We can imagine what is happening if we represent the oscillation amplitudes in a vector diagram (energy is directly proportional to momentum for photons and relativistic electrons). It is clear that

$$\varepsilon_1^2 = \varepsilon_0^2 + E_\gamma^2 - 2\varepsilon_0 E_\gamma \cos(\Omega(t_1 - t_0)) .$$

The phase angle in Fig. 12.4 is completely random (E_γ can point in any direction) so that the average change $\langle\varepsilon^2\rangle$ in the synchrotron oscillation amplitude when a photon with energy E_γ is emitted is

$$\langle\varepsilon^2\rangle = \langle\varepsilon_1^2\rangle - \langle\varepsilon_0^2\rangle = E_\gamma^2$$

because the cosine term averages to zero over the interval 0 to 2π.

We know that the period of the synchrotron oscillations is long compared to the time taken for the electron to go around the ring so we can obtain the average rate of change of the oscillation amplitude generated by the quantum excitations, to a good approximation, as

$$\left\langle \frac{d\varepsilon^2}{dt} \right\rangle_q = \oint E_\gamma^2 \frac{dN_\gamma}{dt} dE_\gamma .$$

The damping equation is $\varepsilon^2 = \varepsilon_0^2 \exp(-2\mu t)$ so that, when equilibrium between excitation and damping is reached, the value of $\langle\varepsilon^2\rangle$ settles down to an asymptotic value reached when

$$\left\langle \frac{d\varepsilon^2}{dt} \right\rangle_q = \left\langle \frac{d\varepsilon^2}{dt} \right\rangle_d = -2\mu\langle\varepsilon^2\rangle$$

so that

$$\langle\varepsilon^2\rangle_q = \frac{1}{2\mu} \oint E_\gamma^2 \frac{dN_\gamma}{dt} dE_\gamma .$$

This is the average amplitude squared of an approximately sinusoidal synchrotron oscillation $\langle\varepsilon^2\rangle = \langle\varepsilon^2\rangle_q \sin^2 \Omega t$, so the rms amplitude σ_E^2 is obtained by taking the average value of $\langle\varepsilon^2\rangle$ over the angular interval from 0 to π so that

$$\sigma_E^2 = \frac{1}{2}\langle\varepsilon^2\rangle_q = \frac{1}{4\mu} \oint E_\gamma^2 \frac{dN_\gamma}{dt} dE_\gamma \qquad (12.12)$$

to be compared with eqn (12.11).

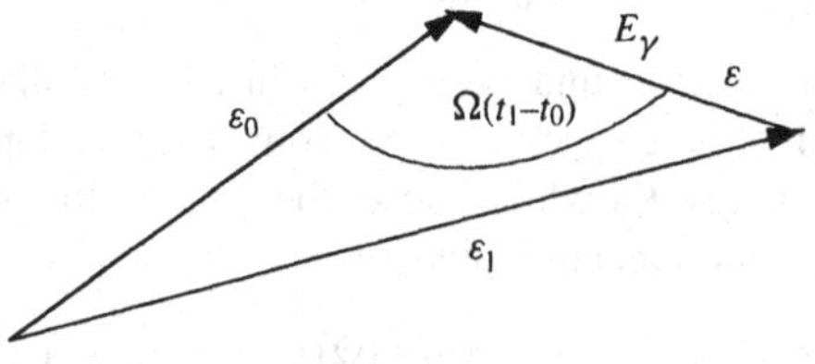

Fig. 12.4 Addition of synchrotron oscillation amplitudes.

The integral in eqn (12.12) is the same as that evaluated in eqn (7.58) where the mean square photon energy was obtained. In fact,

$$\oint E_\gamma^2 \frac{\mathrm{d}N_\gamma}{\mathrm{d}t} \mathrm{d}E_\gamma = \langle \varepsilon^2 \rangle N_\gamma \frac{c}{2\pi R}$$

and making use of eqns (7.56) and (7.58) and the definition of P_γ in eqn (7.55) we can write

$$\oint E_\gamma^2 \frac{\mathrm{d}N_\gamma}{\mathrm{d}t} \mathrm{d}E_\gamma = \frac{55}{24\sqrt{3}} \varepsilon_c P_\gamma .$$

From the definition of μ [eqn (12.9)], inserting ε_c and remembering that $P_\gamma = U_0/T_0$ we find that

$$\begin{aligned} \left(\frac{\sigma_E}{E_0}\right)^2 &= \frac{55}{32\sqrt{3}} \frac{\hbar}{m_e c} \frac{\gamma^2}{R} \frac{1}{(2+\mathcal{D})} \\ &= 3.83 \times 10^{-13} \frac{\gamma^2}{R} \frac{1}{(2+\mathcal{D})} . \end{aligned} \tag{12.13}$$

In eqn (12.13) we note that $\hbar/m_e c$ is the Compton wavelength (Arthur Holly Compton, 1892–1962) of the electron (3.86×10^{-13} m) and the numerical constant $55/32\sqrt{3}$ is very close to unity. For the SRS, $\sigma_E/E_0 = 0.7 \times 10^{-3}$ and the size of the bunch, expressed as the standard deviation of a Gaussian energy distribution, $\sigma_E = 1.4$ MeV. The standard deviation of the bunch length can be obtained from eqn (12.8):

$$\sigma_t = \frac{\alpha}{\Omega} \frac{\sigma_E}{E_0} .$$

If we use $\alpha = 0.03$ and $\Omega = 400$ kHz from above, then $\sigma_t = 50$ ps (50×10^{-12} s), which means that the full width at half height of the time distribution (assumed to be Gaussian) is 120 ps and the equivalent bunch length is 35 mm compared to 37 mm for the observed value.

In practice, the phase oscillations are not small. Electrons can make large excursions in phase and still maintain phase stability. This is shown in Fig. 12.5 in which phase contours are shown for a range of phase oscillation amplitudes.

In order to compensate for synchrotron radiation energy loss, the equilibrium phase ψ_0 given by

$$\frac{1}{\sin \psi_0} = \frac{eV_0}{U_0} = q \tag{12.14}$$

must be different from zero or π. For small oscillations about the equilibrium phase, the phase contour is the small ellipse shown centred on $\psi = \psi_0$. Because of quantum excitation the phase oscillations can increase in amplitude so that the small angle approximation is no longer valid and the corresponding phase contour becomes elongated in the direction of earlier times. For some electrons, their deviation from the equilibrium energy is so large that their phase contour is that shown as the heavy line,

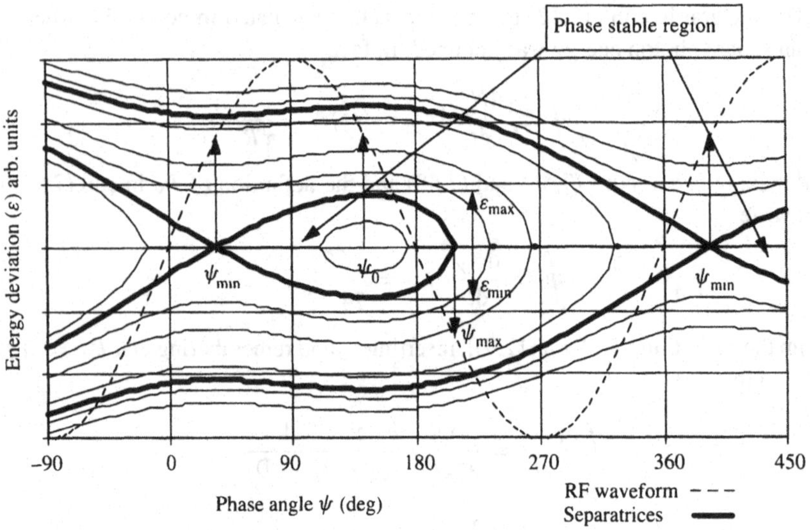

Fig. 12.5 Limits to the region of phase stable oscillations.

known as the separatrix. This limiting contour is often called the radio-frequency bucket. It is bounded by ψ_{min} and ψ_{max}, and ε_{min} and ε_{max}. Beyond this range the phase contour is no longer closed, the phase oscillations are unstable and the electron is lost from the bucket. Electrons which remain in the beam execute damped synchrotron oscillations within the bucket. Their oscillations are analogous to a damped pendulum whose resting position is at ψ_0. The pendulum receives a kick every so often which keeps it in motion but the damping makes it gradually return to ψ_0. Occasionally, the kick (or a series of kicks) causes the amplitude of the swing to reach quite large values. If the kick is large enough, the pendulum reaches the top of its swing at $\psi_{min} = \pi - \psi_0$ and (if there were no damping) it swings back towards ψ_{max}. Any further kick takes it 'over the top', outside the stable region. Figure 12.5 shows that the phase stable range for the electrons between ψ_{min} and ψ_{max} extends over a wide range of values of ψ. The stable region of phase space is no longer elliptical but elongated towards low ψ, which means towards the forward region of the electron bunch. At the same time, the energy width of the beam is reduced by a factor $F(q)/2q$, where q, the over-voltage, is defined in eqn (12.14) and $F(q)$ is given by

$$F(q) = 2\left(\sqrt{q^2 - 1} - \cos^{-1}\left(\frac{1}{q}\right)\right).$$

The behaviour of the reduction factor is shown in Fig. 12.6.

Quantum excitation, combined with energy gain from the radio-frequency cavities, not only induces damped longitudinal synchrotron oscillations but also causes the betatron oscillations to be damped as well. How does this happen? Consider an electron, in the usual reference frame, with transverse coordinate u, moving in

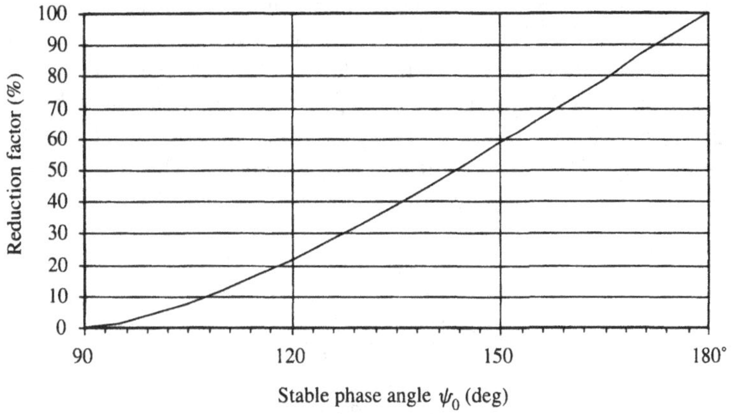

Fig. 12.6 Energy width reduction factor.

direction u'. When the emission of a synchrotron radiation photon takes place, the electron loses energy ΔE, but the coordinates (u, u') do not change because, to a first approximation, the photon is emitted in the direction of motion of the electron. The magnitude of the photon momentum changes by an amount $\Delta E/c$ but the direction of the momentum vector is unaltered by the emission of the photon. When the electron reaches the radio-frequency cavity the energy lost by the electron is replaced and the magnitude of its momentum vector is restored, on average, to its former value. However, the restoration is not like the loss. The loss affects all the components of the momentum vector but the restoration increases only the component of the electron momentum parallel to the direction of the ideal orbit (and makes it larger than it was before). This mechanism rotates the total momentum vector towards the direction of the ideal orbit, thereby decreasing the amplitude of the betatron oscillations. Eventually, the amplitude of the transverse betatron oscillations reaches a limiting value.

Alternatively, assume that the beam consists of two groups of electrons, one with energy $E_0 + \Delta E$ and one with energy $E_0 - \Delta E$. Synchrotron radiation energy loss is proportional to the fourth power of the electron energy (for constant radius) but the energy replacement process is independent of the electron energy so that electrons with positive energy deviation (i.e. off-energy electrons with energy higher than the ideal energy) lose energy on average whereas electrons with a negative energy deviation gain on average. The net result, over a time comparable with the damping time, is that the mean energy of the two groups drifts closer to the ideal energy and the energy spread of the beam decreases.

To first order, the radiation damping is shared equally between the longitudinal and the transverse degrees of freedom, i.e. between the s-direction and the x- and y-directions, taken together. The radiation damping in the transverse directions is then shared equally between the x and y degrees of freedom. As usual, this is not the whole story because at regions in the magnetic lattice where the dispersion is different from zero, the loss of energy causes the electron to move to a region of different field strength. In the longitudinal direction this effect produces the damping coefficient of eqn (12.9). In the vertical plane there is no dispersion [$\mathcal{D}(s)$ is zero throughout] so that the integral ratio $\mathcal{D}$ is also zero. The sharing of the damping between the longitudinal and transverse directions has been analysed by Robinson.[4] This leads to the results for the damping coefficients in the three directions:

$$\mu_x = \frac{1}{2T_0}\frac{U_0}{E_0}(1-\mathcal{D}) = \frac{1}{2T_0}\frac{U_0}{E_0}J_x,$$
$$\mu_y = \frac{1}{2T_0}\frac{U_0}{E_0} = \frac{1}{2T_0}\frac{U_0}{E_0}J_y,$$
$$\mu_s = \frac{1}{2T_0}\frac{U_0}{E_0}(2+\mathcal{D}) = \frac{1}{2T_0}\frac{U_0}{E_0}J_s,$$

where J_x, J_y, and J_s are the damping partition numbers in the three directions and

$$\sum_{i=x,y,s} J_i = 4.$$

Note that the coefficient μ_s is the same as that derived in eqn (12.9).

The general result, that the sum of the partition numbers is constant and independent of the magnet field configuration, is known as Robinson's theorem.

We must now determine the limiting values of the betatron oscillation amplitudes in the transverse directions. Consider the horizontal direction first. Figure 11.8 reminds us that the motion of an off-energy electron can be regarded as a betatron oscillation about a reference orbit which is offset from the ideal orbit by an amount $x_\Delta = D(s)\Delta E/E_0$ so that

$$x(s) = x_\Delta(s) + x_\beta(s) = D(s)\frac{\Delta E}{E_0} + x_\beta(s) \tag{12.15}$$

and

$$x'(s) = x'_\Delta(s) + x'_\beta(s) = D'(s)\frac{\Delta E}{E_0} + x'_\beta(s). \tag{12.16}$$

The emission of a photon with consequent loss of energy ε does not change the x-coordinate of the electron but does send it on a new orbit. The reference orbit offset changes by an amount $\delta x_\Delta = D(s)\varepsilon/E_0$ and x_β changes by an equal and opposite amount δx_β to compensate.

Figure 12.7 shows this in terms of the phase ellipse. An electron moving around the ellipse reaches the point P and emits a photon with energy ΔE. There is no immediate change in the position of the electron but its motion continues along a new ellipse whose origin has moved to the point $(D\Delta E/E_0, D'\Delta E/E_0)$. The new and the old ellipses share the same tangent at P.

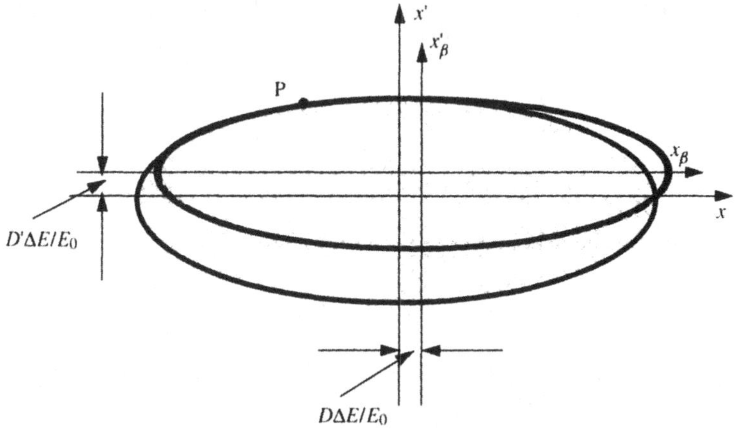

Fig. 12.7 Displacement of phase space ellipse.

Now the betatron oscillation as a function of s can be represented by equations similar to

$$x_\beta(s) = a_0\sqrt{\beta(s)}\cos(\psi(s) - \psi_0),$$
$$x'_\beta(s) = -\frac{a_0}{\sqrt{\beta(s)}}\alpha(s)\cos(\psi(s) - \psi_0) - \frac{a_0}{\sqrt{\beta(s)}}\sin(\psi(s) - \psi_0) \quad (12.17)$$

and

$$\gamma(s)x_\beta^2(s) + 2\alpha(s)x_\beta(s)x'_\beta(s) + \beta x'^2_\beta(s) = a_0^2$$

is the equation of the new phase ellipse after the emission of the photon. As before α, β, and γ are the Twiss parameters which depend only on the magnetic field geometry in the horizontal plane of the storage ring and $a_0\sqrt{\beta(s)}$ is the maximum amplitude of the betatron oscillations.

Imagine now that the electron beam consists of a very large number of such ellipses, with their centres changing all the time as photons are emitted and reabsorbed from the radio-frequency cavities. The effect of the damping of the betatron oscillations causes the phase ellipses to compress in both the x- and x'-directions until the beam width at a point on the ideal orbit defined by the coordinate s is given by

$$\sigma_x^2(s) = \beta(s)\langle a_0^2\rangle_{\text{rms,q,s}} = \tfrac{1}{2}\beta(s)\langle a_0^2\rangle_{\text{q,s}},$$

where $\langle a^2_{0,\text{rms}}\rangle$ denotes the root mean square value of a_0^2 averaged over all phase angles $\psi(s)$ between 0 and π at s and the subscripts q and s denote an integral over all photon energies and over one turn around the ring.

How do we calculate $\langle a_0^2\rangle$? When the energy of the electron changes by an amount ΔE, there is no instantaneous change to the position and direction of motion of the electron. In other words, the coordinates of P in Fig. 12.7 do not change, so we may write [from eqns (12.15) and (12.16)]

$$\delta x(s) = D(s)\frac{\Delta E}{E_0} + \delta x_\beta(s) = 0,$$
$$\delta x'(s) = D'(s)\frac{\Delta E}{E_0} + \delta x'_\beta(s) = 0,$$

so that, for the change in the betatron oscillations [from (12.17)],

$$D(s)\frac{\Delta E}{E_0} = -\delta a_0\sqrt{\beta(s)}\cos(\psi(s)-\psi_0),$$
$$D'(s)\frac{\Delta E}{E_0} = \frac{\delta a_0}{\sqrt{\beta(s)}}\alpha(s)\cos(\psi(s)-\psi_0) + \frac{\delta a_0}{\sqrt{\beta(s)}}\sin(\psi(s)-\psi_0).$$

Elimination of the sine and cosine functions of ψ between these two equations gives

$$\delta a_0^2 = \left(\frac{\Delta E}{E_0}\right)^2 \{\gamma(s)D^2(s) + 2\alpha(s)D(s)D'(s) + \beta D'^2(s)\}$$
$$= \left(\frac{\Delta E}{E_0}\right)^2 H(s),$$

where $H(s)$ is a function of the magnetic field parameters. We now follow the same procedure as before when we obtained the rms energy spread by balancing the damping rate of the synchrotron oscillations with the rate of quantum excitation to obtain

$$\frac{\sigma_x^2(s)}{\beta(s)} = \frac{55}{32\sqrt{3}}\frac{\hbar}{m_e c}\frac{\gamma^2}{R}\frac{\langle H\rangle_s}{(1-\mathfrak{D})}$$
$$= 3.83\times10^{-13}\frac{\gamma^2}{R}\frac{\langle H\rangle_s}{(1-\mathfrak{D})}, \tag{12.18}$$

which is equal to the beam emittance ε_x in the horizontal plane. The observed value of σ_x is greater than this because the energy spread of the synchrotron oscillations provides an additional contribution to the beam width as described in eqn (12.1).

Minimizing the electron beam emittance

So far we have considered only the simplest form of magnetic lattice, the FODO lattice which contains just two focusing elements. In optics this is the equivalent of a single focusing lens—a magnifying glass, for example. We know perfectly well that a typical optical instrument contains many focusing (and defocusing) elements which are arranged to optimize the particular properties (freedom from spherical and chromatic aberrations, wide acceptance angle and the like). In synchrotron radiation sources, high beam brightness is one criterion for a powerful source. This means a low electron beam emittance at least in those elements which are to be used for synchrotron radiation production. A second important criterion is low dispersion within insertion devices (see Chapters 13 and 14). This is particularly important because an insertion device produces a sudden and substantial electron beam energy loss. Zero lattice dispersion function in the region of the insertion device prevents an increase in the beam emittance when the insertion device is brought into operation.

Figure 12.8 shows a comparison of several generic types of magnetic lattices which are in use at several synchrotron radiation sources. Figure 12.9 shows the behaviour of the dispersion functions for the more advanced forms of lattices. In these lattices, the dispersion function is either low or approaches zero in the dipole magnets and in drift spaces where insertion devices can be inserted. This contrasts with the simple FODO lattice (such as that used at the SRS) where the dispersion function, shown in Fig. 11.9, is far from zero in the drift space between the quadrupole magnets where insertion devices must be located.

A good approximation for the beam emittance in any isomagnetic lattice can be obtained from eqn (12.18) by evaluating the average value of $H(s)/R$ for the dipole magnets within the unit cell. The starting point is the matrix of the Twiss parameters α_0, β_0, and γ_0 at the entrance to the magnet multiplied by the matrix which transforms the electron trajectory through the magnet. A derivation is given, for instance, by Wiedemann.[5] The integral $\mathcal{D}$ of eqn (12.10) is set equal to zero and, assuming that

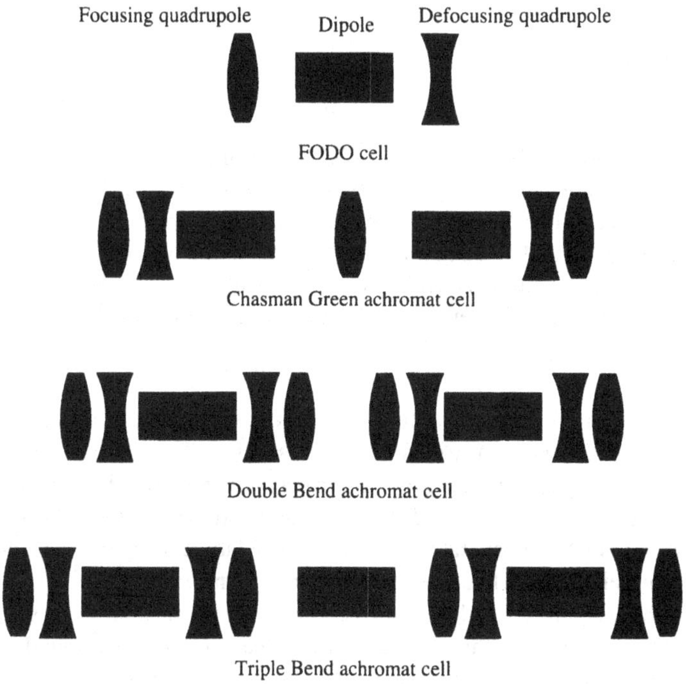

Fig. 12.8 Magnet layout for various lattice types.

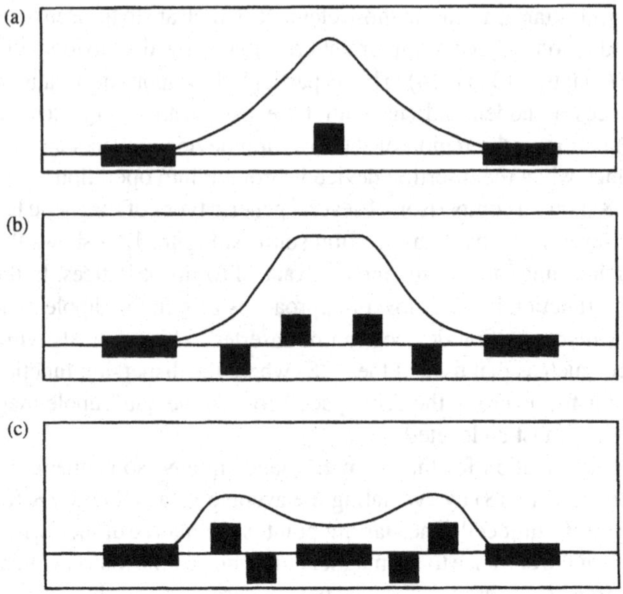

Fig. 12.9 Dispersion function for various lattice types: (a) Chasman Green achromat; (b) double bend achromat; and (c) triple bend achromat.

the bending angle in the magnet is small, the equation reduces to

$$\varepsilon_x = 3.83 \times 10^{-13} \gamma^2 \theta_\mathrm{d}^3 F\,(\alpha_0, \beta_0, \gamma_0, s_\theta)\,.$$

Evidently, the horizontal beam emittance is proportional to the cube of the deflection angle θ_d in the individual dipole magnet or to the inverse cube of the number of deflecting magnets. The function F of the lattice parameters, including s_θ, the length of the trajectory in the deflecting magnet, depends on the type of magnetic lattice. A choice of lattice parameters which leads to a minimum value for F gives a minimum value of the horizontal beam emittance for that particular lattice. If we write

$$\varepsilon_{x,\mathrm{min}} = K E^2 \theta_\mathrm{d}^3 \;\; (\mathrm{m\,rad})$$

with E in GeV and θ in radians then for the Chasman–Green lattice as used, for example, in the Brookhaven storage rings, $K \approx 10^{-7}$. The FODO lattice has $K \approx 100 \times 10^{-7}$ and the triple bend achromat is in between the two at $K \approx 20 \times 10^{-7}$. The design emittance of a wide range of synchrotron radiation sources has been compared by Suller[6] with that of a generalized low emittance lattice using the same bend angle per dipole by calculating a figure of merit [ε_x (minimum)/ε_x (design)] for each source. Values of this quantity (as calculated by Suller and for selected sources) are shown in Table 12.1. At the time this comparison was made (1992), horizontal emittances some ten times the theoretical minimum were being obtained. This indicated both the need and the possibility of improved performance from existing magnetic lattices.

TABLE 12.1 Values of Suller's figure of merit for selected radiation sources

Radiation source	Energy (GeV)	Lattice type	Horizontal emittance (nm rad)	Figure of merit = ε_x(minimum)/ ε_x(design)
NSLS VUV	0.744	CG	88	1.0×10^{-1}
ALS	1.5	TBA	3.4	1.1×10^{-1}
SRS	2.0	FODO	110	0.7×10^{-1}
NSLS X-ray	2.5	CG	80	1.5×10^{-1}
ESRF	6.0	DBA	7	1.5×10^{-1}
APS	7.0	DBA	8	0.9×10^{-1}
SPring8	8.0	DBA	7	1.1×10^{-1}

Vertical beam emittance

The dispersion function in an ideal dipole magnet in the vertical plane is zero so that the fluctuations in electron energy and the energy spread of the beam have no effect on the transverse beam dimensions in the vertical plane. The damping of the vertical betatron oscillations due to synchrotron radiation emission might be expected to reduce the transverse beam emission to a level dominated by space charge effects (mutual Coulomb repulsion of the charges within the bunch). This is not realized in practice because no magnetic lattice is perfect. The locations of the magnets themselves are subject to error so that, for example, the magnetic field directions in the dipole magnets are neither perfectly parallel nor are they located exactly in the horizontal plane. Similar errors are apparent in the location and orientation of the quadrupole magnets. Insertion devices such as single and multipole wigglers (see Chapter 13) also make a contribution to these effects. The small errors induce a coupling between the horizontal and vertical betatron oscillations. All of these effects can be lumped together into a coupling constant k defined via

$$\varepsilon_{\mathrm{v}} = \frac{k}{1+k}\varepsilon_{\mathrm{h},0}, \qquad \varepsilon_{\mathrm{h}} = \frac{1}{1+k}\varepsilon_{\mathrm{h},0},$$

where $\varepsilon_{\mathrm{h},0}$ is the horizontal beam emittance in the absence of coupling ($k = 0$). The vertical beam divergence can be defined in a similar way. The value of k is characteristic of the particular storage ring under consideration and its precise operating conditions. In general, k is usually between 1% and 10%. Coupling increases the vertical beam size and is the dominant effect in that direction. Coupling also produces a small decrease in the horizontal beam emittance compared to the expected value under ideal conditions. The practical effect is that in all synchrotron radiation sources (unless deliberate steps are taken to increase the coupling between the horizontal and vertical betatron oscillations) the vertical beam size is typically 1/3 ($1/\sqrt{10}$

Fig. 12.10 An APS storage ring sector, looking downstream.

when $k = 10\%$) of the horizontal beam size at any point around the ring. More precisely,

$$\frac{\sigma_{\mathrm{v}}^2(s)}{\beta_{\mathrm{v}}(s)} = \frac{k}{1+k}\frac{\sigma_{\mathrm{h}}^2(s)}{\beta_{\mathrm{h}}(s)}\bigg|_{k=0}.$$

Figure 12.10 shows one sector of the magnetic lattice of the Advanced Photon Source at the Argonne National Laboratory near Chicago, USA. This lattice would provide a horizontal beam emittance, in the absence of coupling, of 8.2 nm rad but 10% coupling between the horizontal and vertical betatron oscillations reduces the horizontal emittance to 7.5 nm rad and increases the vertical beam emittance to 0.75 nm rad.

References

1. V. I. Veksler, *Dokl. Akad. Nauk* (USSR), **44**, 393 (1944).
2. E. M. McMillan, *Phys. Rev.*, **68**, 144 (1945).
3. M. Sands, The physics of electron storage rings. SLAC 121. US Atomic Energy Commission (1970).

4. K. W. Robinson, *Phys. Rev.*, **111**, 373 (1958).
5. H. Weidemann, *Particle accelerator physics*, Chapter 13. Springer Verlag, Berlin and Heidelberg (1993).
6. V. P. Suller, Review of the status of synchrotron radiation storage rings. *Proc. 3rd European Particle Accelerator Conference*, Berlin, March 1992, pp. 77–81. Published by Editions Frontières, 91192 Gif-sur-Yvette, France.

13

Insertion devices—wigglers

Introduction

Insertion devices are a general term for a range of magnetic elements which can be added to a storage ring (or designed in from the beginning). They are designed to provide radiation whose characteristics are enhanced compared with that available from the dipole magnets. The insertion device is placed in a drift space between the existing magnets of the storage ring. There are two principal types of such devices—the multipole wiggler and the undulator which are illustrated and compared with the dipole source in Fig. 13.1.

Synchrotron radiation from dipole magnets has many desirable characteristics and the aim of the insertion device is to improve on these. These improvements are best described with reference to Table 13.1.

Single and multipole wigglers

Radiation from the dipole magnets of a storage ring is produced with a total horizontal angular spread of 2π radians, divided equally between each individual magnet. The power radiated is proportional to the magnetic field. The simplest form of a multipole wiggler is one with a single pole with two 'half' poles at the entrance and exit of the single dipole so that the ideal orbit is unchanged by this magnetic perturbation. The arrangement (ref. 1) is shown in Fig. 13.2.

The diagram (Fig. 13.2) shows the symmetrical arrangement in which the magnetic field integral along the electron beam orbit in the drift space ensures that the electrons return to the ideal orbit after their passage through the wiggler. In order to achieve this, the single pole, which is the principal source of radiation, must be matched by end poles whose angle of bend cancels out the bend induced by the single pole. These end poles add to the radiation from the central pole and enhance the power output. This may be an advantage but because the source now appears elongated in the horizontal plane (or even as two separate sources), the gain in flux from the end poles is not necessarily accompanied by a gain in brightness. The radiation from the poles is emitted in a broad horizontal fan but equipment receiving radiation emitted to the right or the left of the central axis sees even more source structure. However, the source structure is a function of the radiation energy and the characteristic energy of the radiation from the end poles can be reduced (by reducing the field and lengthening the pole) so long as the total field integral of the endpoles is equal and opposite to that of the central pole so that the total deflection angle is zero.

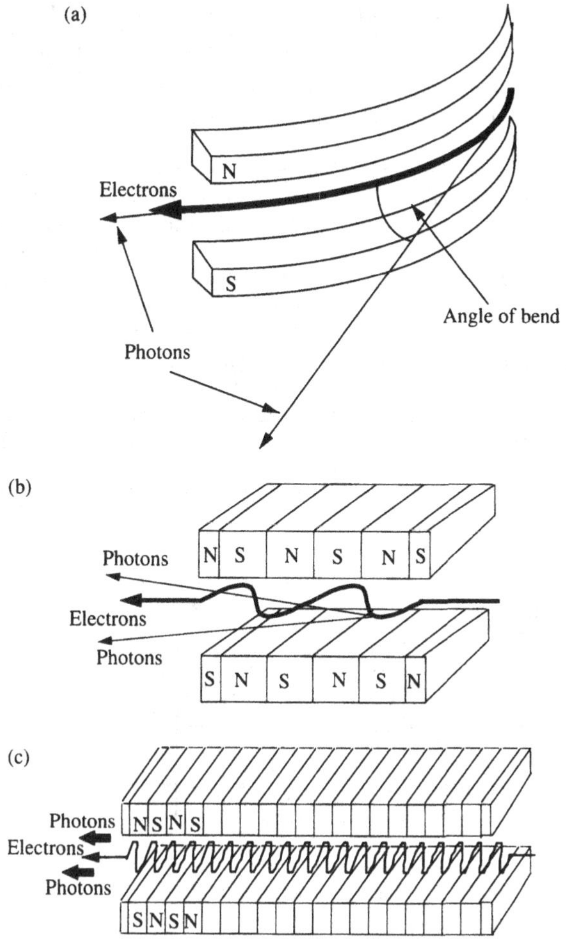

Fig. 13.1 Types of synchrotron radiation sources: (a) dipole magnet; (b) multipole wiggler; and (c) undulator.

The principal source of radiation is the central pole which in the two magnets of this type installed in the SRS at Daresbury have field strengths of 5 and 6 T, respectively. The maximum synchrotron radiation power output from these magnets is 170 and 200 mW/(mA mrad), respectively. This is to be compared with 40 mW/(mA mrad) emitted by the 1.2 T dipole magnets. Superconducting magnetic materials (cooled by liquid helium) are used to provide these high magnetic fields.

Since the principal reason for installing single pole wigglers is to provide radiation harder than that from the dipole magnets, the use of extended end poles (as in the 6 T wiggler at the wiggler laboratory) provides a useful method to clean up the source appearance for experimenters working at short wavelengths.

TABLE 13.1 Comparison of synchrotron radiation source types

	Dipoles	Multipole wigglers	Undulators
Horizontal angular distribution	Equal to the bending angle of the magnet	Collimated according to the deflection angle	Approaching the diffraction limit
Source appearance	Source size determined by magnet lattice	Multiple sources may be seen. Off-axis viewing leads to elongated source	Source size determined by magnet lattice. Some expansion in the undulating plane
Radiation spectrum	Continuous	Continuous with adjustable λ_c depending on the magnetic field	Narrow peak with harmonics
Energy radiated	As from a single pole	Proportional to the number of poles, N	Proportional to N^2
Polarization	Horizontal polarization 100%	Polarization zero when N even, as for dipole when N odd	Variable, depending on magnet geometry

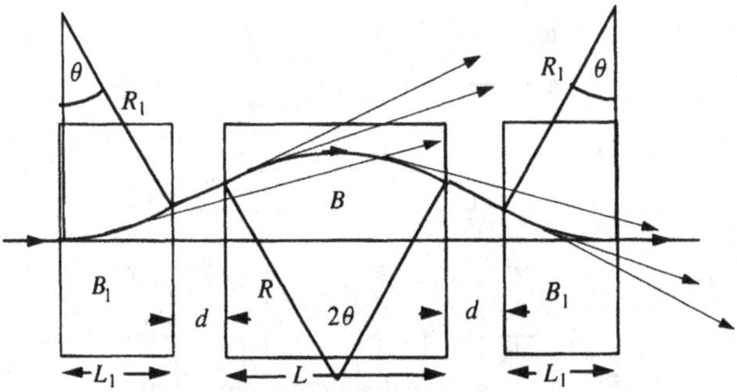

Fig. 13.2 Magnet layout in a single pole wiggler.

The actual displacement of the beam from the ideal orbit is an important parameter which can be easily calculated once the field profile of the magnets has been determined. The high magnetic field leads to a large displacement, which in the case of the 5 T wiggler magnet at Daresbury is about 9 mm (refs 2 and 3).

The radiation output from these insertion devices can be multiplied further by increasing the number of magnetic poles. In general, the length of drift space available for these magnets is fixed by the design of the storage ring so that an increase in the number of poles can only be achieved by decreasing the magnetic field in each pole. The problem is illustrated at the SRS where the drift space available to each magnet is just over 1 m. The single pole wigglers (with their endpoles) each fit into a space of this length and represent the maximum field volume that can be compressed into the space available. This device acts as wavelength shifter so that the synchrotron radiation spectrum is extended (for the stations associated with the device) to shorter wavelengths (higher energy).

To obtain more synchrotron radiation output, one must get more poles into the same drift space, which inevitably means that the maximum field must be reduced. At the SRS, this has been done by replacing the section of the vacuum vessel in the drift space with a specially designed narrow aperture unit (ref. 4) so that the gap between the pole faces can be reduced to just under 20 mm. The drift space can then be filled with nine permanent magnets, each producing a field of 2 T along the beam axis. Each permanent magnet is surrounded by magnetic steel blocks which provide a return loop for the lines of force generated by the 2 T magnet blocks.

The magnetic field generated along the electron beam axis by this arrangement follows an approximately sinusoidal pattern, with $B_y = B_0 \sin(2\pi z/\lambda_0)$ and λ_0, the magnetic wavelength, is 200 mm. The arrangement is shown in Fig. 13.3. It makes it possible to insert nine 2 T poles into the 1 m drift space. The resulting synchrotron radiation spectrum is the sum of the radiation emitted from each pole. A spectral comparison in terms of radiation flux for the SRS dipoles, the 6 T single pole wiggler, and the 2 T wavelength shifter is shown in Fig. 13.4.

Electron trajectory in a multipole wiggler

Let us consider the motion of an electron in the field of a magnet such as that shown in Fig. 13.3. As remarked above, the magnetic field variation along the axis of the

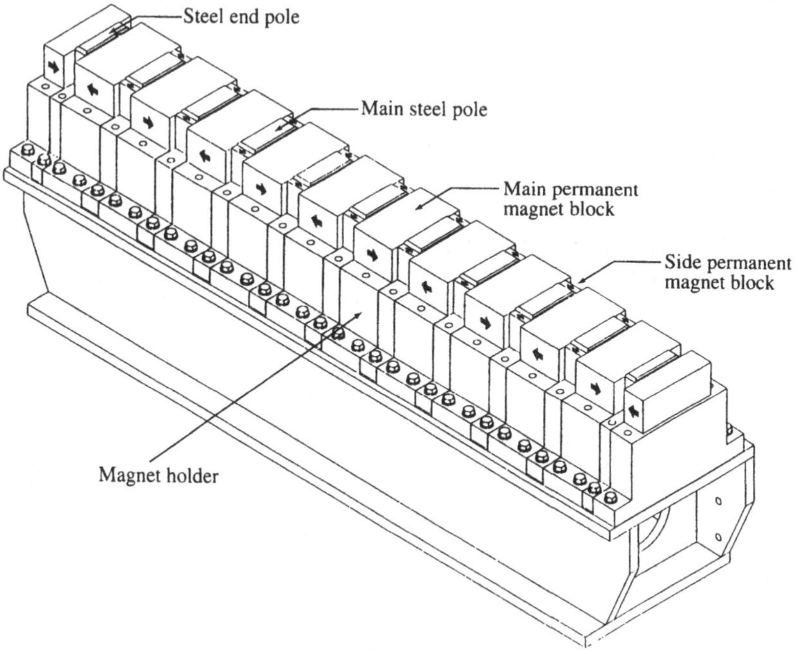

Fig. 13.3 Layout of magnetic poles showing the direction of magnetization.

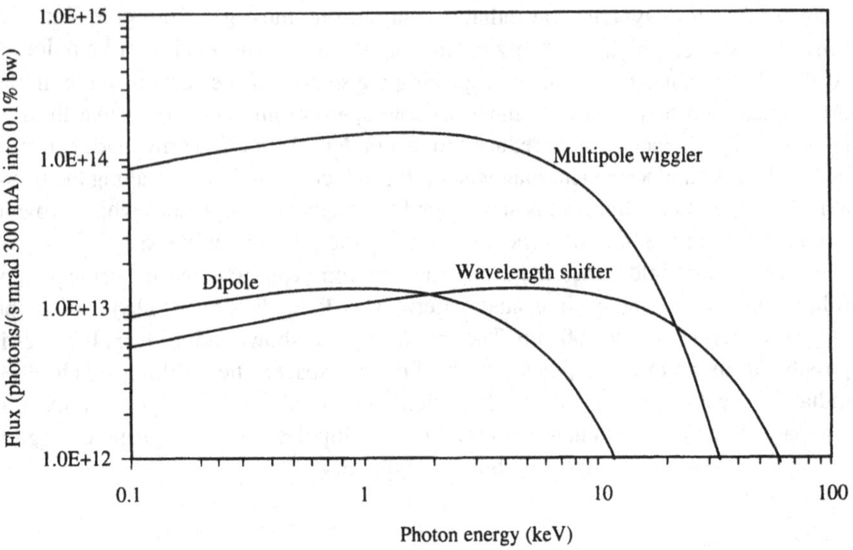

Fig. 13.4 Flux comparison.

wiggler and in a direction at right angles to the plane of the trajectory can be described approximately by

$$B_y(z) = B_0 \sin\left(\frac{2\pi z}{\lambda_0}\right), \tag{13.1}$$

where B_0 is the field amplitude on the axis and λ_0 is the magnetic period as shown in Fig. 13.5.

Two end poles are provided to match the beam trajectory inside and outside the multipole wiggler magnet. These poles are themselves a source of synchrotron radiation so the end pole field is arranged to provide a field of 1.8 T so that the additional spectrum is not too different from that from the nine 2 T poles. The magnetic length of the end pole is designed to provide the correct matching deflection angle.

As the electron travels through this field it follows a curved path along which an element of the path is given by $ds = R(s)\, d\theta$. The radius of curvature of the trajectory, $R(s)$, is related as usual to the magnetic field so that

$$d\theta = -\frac{ecB(s)}{E}\, ds$$

and the accumulated deflection angle at any point along the trajectory is

$$\theta(s) = -\frac{ec}{E}\int_0^s B(s)\, ds.$$

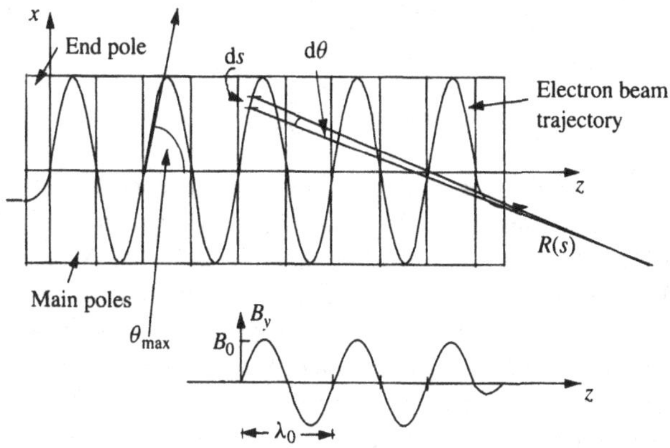

Fig. 13.5 Electron trajectory and magnetic field.

$B(s)$ can be replaced with its expression in eqn (13.1) and when the deflection angle is small we can write for the maximum deflection angle

$$\theta_{\max} = -\frac{ecB_0}{E}\int_0^{\lambda_0/4} \sin\left(\frac{2\pi z}{\lambda_0}\right) \mathrm{d}z.$$

Integrating and putting $E = \gamma m_0 c^2$ gives

$$\theta_{\max} = \frac{ecB_0\lambda_0}{2\pi\gamma m_0 c^2} = \frac{K}{\gamma}, \quad \text{where } K = 93.4 B_0\lambda_0.$$

The K value, which is proportional to the product of the maximum magnetic field on axis and the magnetic period, is an important parameter which depends only on the geometry and the magnetic field of the insertion device.

In fact, instead of θ we can write x', which is $\mathrm{d}x/\mathrm{d}z$ at the point s along the trajectory, so that

$$x' = -\frac{ec}{E}\int B_0 \sin\left(\frac{2\pi z}{\lambda_0}\right) \mathrm{d}z$$

and with the boundary condition that when $z = \lambda_0/4$, $x' = 0$, then

$$x'(z) = \frac{K}{\gamma}\cos\left(\frac{2\pi z}{\lambda_0}\right) \tag{13.2}$$

and further integration gives

$$x(z) = \frac{K}{\gamma}\frac{\lambda_0}{2\pi}\sin\left(\frac{2\pi z}{\lambda_0}\right) \tag{13.3}$$

for the equation of the trajectory of the electron in the horizontal plane, so that the maximum excursion is

$$x_{\max} = \frac{K}{\gamma} \frac{\lambda_0}{2\pi}. \tag{13.4}$$

It is instructive to calculate the maximum transverse excursion and the maximum deflection angle for the multipole wiggler of Fig. 13.3. This wiggler, with a field of 2 T and a period of 200 mm, has a K value of 37.4; so for 2 GeV electrons, $x_{\max} = 0.3$ mm and $\theta_{\max} = 9.5$ mrad. The deflection angles are small, which justifies the approximation made above, but large enough to provide for two independent beam lines from the device. The transverse motion of the electron beam is observed by the experimenter as an effective increase in the beam width. The maximum transverse velocity of the electrons is equal to cK/γ, so $\beta_{x\max}$ is about one-hundredth of the velocity of light.

References

1. V. P. Suller, *Synchrotron radiation. Sources and applications* (eds G. N. Greaves and I. H. Munro), Chapter 2. p. 39. Published by SUSSP, Edinburgh University Physics Department, Edinburgh, Scotland (1989).
2. G. N. Greaves, R. Bennett, P. J. Duke, R. Holt, and V. P. Suller, *Nucl. Instr. Methods*, **208**, 139–42 (1983).
3. N. Marks, G. N. Greaves, M. W. Poole, V. P. Suller, and R. P. Walker, *Nucl. Instr. Methods*, **208**, 97–103 (1983).
4. J. A. Clarke, N. Bliss, D. Bradshaw, C. Dawson, B. Fell, N. Harris, G. Hayes, M. W. Poole, and R. Reid, *J. Synchrotron Radiation*, **5**, 434–436 (1998).

14

Insertion devices—undulators

Elementary theory of undulators

Multipole wigglers have, in general, high magnetic fields and large K values. They are used to augment or enhance the synchrotron radiation spectrum at high photon energy. Undulators represent a logical extension of the multipole wiggler towards lower magnetic fields, more magnetic poles, and shorter magnetic wavelengths (see Fig. 13.1). These devices are designed with K values of the order of unity and operate in a completely different way to wigglers with large K. When $K \approx 1$ the maximum deflection angle of the electrons is of the same order as the angle of production of the radiation and the photons are emitted in a narrow forward cone. There is now a high degree of coherence between the radiation produced from each magnetic pole so that, in the forward direction, the maximum intensity is generated at a wavelength directly related to the magnetic wavelength. The maximum excursion of the electrons is of the order of $\lambda_0/2\pi\gamma$ and the insertion device is called an undulator. Undulators have a long history. Their properties were first examined by Motz[1] and the radiation spectra were first calculated in detail by Alferov *et al.*[2] but their practical realization was not implemented until much later. For a further study of these devices, the reader is also referred to Wiedemann.[3] Let us look at their mode of operation from an elementary viewpoint.

As the electron traverses the undulator, it follows a similar sinusoidal trajectory to that in the multipole wiggler. However, in this case, when $K \approx 1$, the maximum transverse velocity is of the order of c/γ, the transverse motion of the electron is non-relativistic and the behaviour of the electron is the same as that of an electron undergoing simple harmonic motion with one single frequency ω^*. To understand this, imagine a coordinate frame in which the longitudinal velocity of the electron is zero so that the electron has no forward motion. In this reference frame, the magnetic poles move past the electron with velocity close to the speed of light. The electron experiences an alternating magnetic field as the poles move past. In this field it behaves as though it were a dipole oscillator with frequency ω^* and radiating at that frequency. What is the value of ω^*? In the rest frame of the electron the length of the undulator, particularly the magnetic wavelength, appears contracted by a factor γ (see Chapter 3) so that the magnetic wavelength in the rest frame of the electron is $\lambda^* = \lambda_0/\gamma$, and the magnetic field drives the dipole oscillations with frequency $\omega^* = 2\pi\gamma c/\lambda_0$. This oscillation frequency is already much higher (by a factor of γ) than the natural frequency $2\pi c/\lambda_0$ corresponding to the wavelength of the undulator magnetic field. The observed frequency is higher still because an observer, at an angle θ in the laboratory, sees this dipole radiation shifted to an even shorter wavelength λ, through the relativistic Doppler effect, where [from eqn (3.19)]

$$\lambda = \lambda^{*}\gamma(1-\beta\cos\theta) = \lambda_0(1-\beta\cos\theta). \tag{14.1}$$

In the forward direction ($\theta = 0$), the observed wavelength of the radiation is $\lambda(1-\beta) = \lambda_0/2\gamma^2$.

What numerical value can we expect for the output wavelength? A 30 mm magnetic wavelength and an electron energy of 2 GeV would produce an output wavelength of about 1 nm. The same undulator used with 6 GeV electrons would yield a photon wavelength of about one-tenth of that, i.e. 0.1 nm (1 Å) because the output wavelength varies as $1/\gamma^2$. The observation of radiation at nanometre wavelengths, from a device whose natural wavelength is millimetres, is a remarkable verification of the special theory of relativity.

What effect will the undulator have on the electron beam trajectory? The maximum deflection angle of the electron beam would be just over $\frac{1}{4}$ mrad at 2 GeV and the corresponding maximum horizontal excursion would be 1.2 μm. For 6 GeV electrons these would be one-third of their 2 GeV values because the trajectory parameters vary as $1/\gamma$.

The above expression for the Doppler shifted frequency depends on a measurement of the difference between the arrival times of successive crests of the wave emitted by the electron in its rest frame and in the laboratory. This makes the Doppler shifted frequency very sensitive to the precise difference between β and unity. We have assumed that the frame in which the electron is at rest moves with a constant velocity β relative to the undulator magnet. This is not even true to a first approximation because the electron follows a sinusoidal trajectory in the magnet and the forward component of the velocity of the electron, β_z, is not equal to the value of β outside the undulator. In fact, β_z is a function of z and of K, and is always $\leq\beta$, which means the approximation above has overestimated the strength of the Doppler effect and made the observed wavelength appear shorter than it really is. An improvement to the approximation will be to calculate the average value of β_z and to use that in the formula for the Doppler shift.

The total velocity of the electron does not change as the electron travels through the undulator so that the transverse and longitudinal components are related by

$$\left(\frac{\mathrm{d}z}{\mathrm{d}t}\right)^2 = \beta^2c^2 - \left(\frac{\mathrm{d}x}{\mathrm{d}t}\right)^2$$

and, from eqn (13.2), in the small angle approximation,

$$\frac{\mathrm{d}x}{\mathrm{d}t} = \beta c\frac{K}{\gamma}\cos\frac{2\pi z}{\lambda_0},$$

which gives

$$\frac{\mathrm{d}z}{\mathrm{d}t} = \beta c\sqrt{1-\frac{K^2}{\gamma^2}\cos^2\frac{2\pi z}{\lambda_0}}$$

or, with an obvious change of notation,

$$\beta_z^2 = \beta^2 \left(1 - \frac{K^2}{\gamma^2} \cos^2 \zeta\right).$$

Now we obtain the rms average velocity

$$\begin{aligned}\beta_{z,\mathrm{av}}^2 &= \beta^2 \frac{\int_0^{\pi/2} \left(1 - (K^2/\gamma^2) \cos^2 \zeta\right) \mathrm{d}\theta}{\pi/2} \\ &= \beta^2 \left(1 - \frac{K^2}{2\gamma^2}\right)\end{aligned}$$

and

$$\beta_{z,\mathrm{av}} = \beta \left(1 - \frac{K^2}{4\gamma^2}\right). \tag{14.2}$$

If we now use this expression for the average β of the electron in eqn (14.1) for the Doppler shift, set $\cos\theta = 1 - \theta^2/2$, and neglect all terms such as θ/γ, K^2/γ^2, and higher orders, we obtain a good approximation for the wavelength of the output radiation from an undulator as seen by an observer at an angle θ to the beam direction:

$$\lambda \approx \frac{\lambda_0}{2\gamma^2} \left(1 + \frac{K^2}{2} + \gamma^2\theta^2\right). \tag{14.3}$$

Another way of imagining the output from an undulator is to consider it as a diffraction grating. The poles correspond to the lines on the grating. Consider Fig. 14.1. An electron emits a photon at A and after a time equal to $\lambda_0/c\beta_{z,\mathrm{av}}$ it emits a second photon at the next pole, A′. In the meantime, the photon has reached B after a time

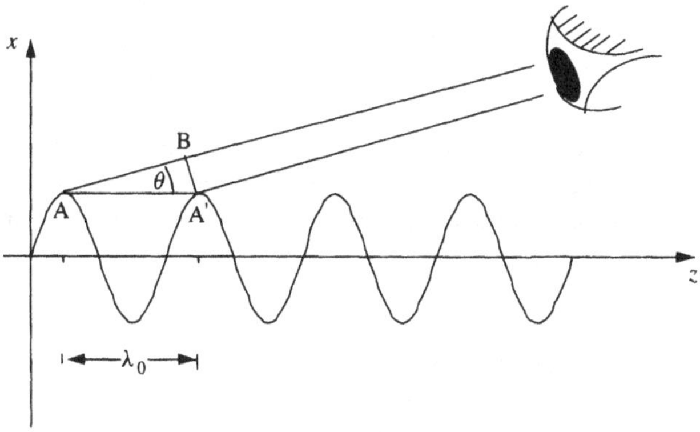

Fig. 14.1 Constructive interference of radiation from successive poles.

equal to $\lambda_0 \cos\theta/c$. The time difference, Δt, between the arrival times of the two photons at the observer is

$$\Delta t = \frac{\lambda_0}{c}\left(\frac{1}{\beta_{z,\mathrm{av}}} - \cos\theta\right).$$

The radiation detected by an observer is maximum when there is constructive interference between the two photons. That is, if the time difference between them is an integral number of photon wavelengths, $k\lambda$, divided by the velocity of light. Thus,

$$k\lambda = \lambda_0\left(\frac{1}{\beta_{z,\mathrm{av}}} - \cos\theta\right).$$

In the small angle approximation, as before, and using the expression for $\beta_{z,\mathrm{av}}$ in eqn (14.2) we obtain an expression for the fundamental wavelength, λ_k (with $k = 1$), and a series of harmonics with shorter wavelengths having values of $k = 2, 3, 5$, etc., so that

$$\lambda_k = \frac{\lambda_0}{2k\gamma^2}\left(1 + \frac{K^2}{2} + \gamma^2\theta^2\right). \tag{14.4}$$

These harmonics are to be expected even in the oscillating dipole model which was used to derive eqn (14.3). The expression for the average value of β_z [eqn (14.2)] is used to define a reference frame in which the radiating electron is at rest. In fact the electron velocity is a periodic function of z, so in the average rest frame the electron is only stationary in the limit when $K = 0$. The motion of the electron for three different values of K is shown in Fig. 14.2.

The trajectories have been normalized to the same transverse excursion in order to emphasize the relative importance of the transverse and longitudinal motion of the electron in the average rest frame. When K is small this motion is almost entirely in the x-direction (at right angles to the magnetic field vector) and the output radiation in the forward direction ($\theta = 0$) has a wavelength given by eqn (14.2) or by eqn (14.3) with $k = 1$. The angular distribution of the radiation corresponds to that emitted by a simple dipole aerial with a maximum in the forward direction. As K increases, the proportion of backward and forward motion in the direction of the observer increases and this motion is at twice the frequency of the fundamental. In the forward direction this produces no radiation but off-axis an observer sees the second harmonic at a wavelength given by eqn (14.3) with $k = 2$. As K increases further, higher order harmonics become more and more prominent, but always the even orders are suppressed in the forward direction.

Calculation of the photon flux

To set this on a numerical footing we must calculate the radiation intensity and, by implication, the photon flux produced by an electron traversing the undulator.

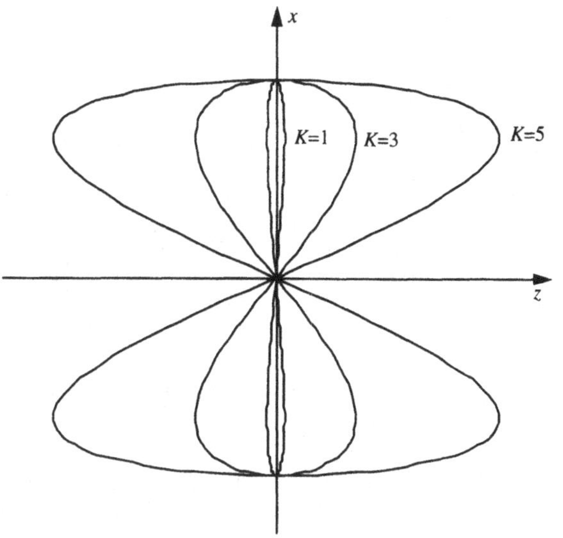

Fig. 14.2 Electron trajectory in the average rest frame as a function of K.

Our starting point is the same as that for the flux calculation from a dipole magnet in Chapter 5. We begin by calculating the strength of the electric field produced by the moving electron at the point of observation and then add up the contributions from each dipole in the undulator with due regard to the phase difference between each dipole source, as was described in the previous section. The energy received in one second at the point of observation, in a given frequency interval, is then given by the square of the electric field vector at that point.

As in Chapter 5, our starting point is the expression of eqn (5.38):

$$\frac{\mathrm{d}^2 I(\omega, \boldsymbol{n})}{\mathrm{d}\Omega\, \mathrm{d}\omega} = \frac{e^2\omega^2}{16\pi^3\varepsilon_0 c} \left| \int_{-\infty}^{+\infty} (\boldsymbol{n} \times (\boldsymbol{n} \times \boldsymbol{\beta})) \exp\left(+\mathrm{i}\omega\left(t' - \frac{\boldsymbol{n} \cdot \boldsymbol{R}(t')}{c}\right)\right) \mathrm{d}t' \right|^2 \tag{14.5}$$

This expression, given here in MKS units, is otherwise identical to that obtained by Jackson[4] and represents the energy emitted, within the frequency interval between ω and $\omega + \Delta\omega$, by one electron, as it passes through the undulator, into a solid angle $\Delta\Omega$ in the direction defined by the unit vector $\boldsymbol{n}$. The integral is to be evaluated over all time but can, in practice, be restricted to the time during which the electron is passing through the undulator. The variable of integration t' is the time of emission of the radiation. During the emission period between t' and $t' + \mathrm{d}t'$ the electron is located at a point (in the x–z plane) defined by the position vector $\boldsymbol{R}$ and the velocity vector $\boldsymbol{\beta}c$ which lies in the same plane. These vectors are shown in Fig. 14.3.

The observer views the undulator output from a location whose direction from the origin is defined by the unit vector $\boldsymbol{n}$ and whose distance from the undulator is large compared with the length of the undulator. The unit vector $\boldsymbol{n}$ lies in a plane which

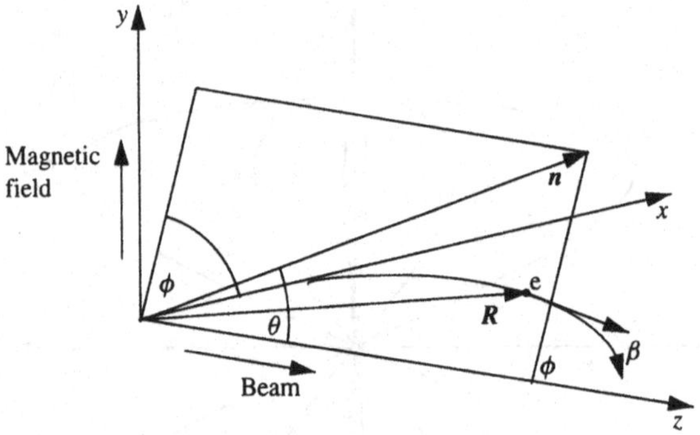

Fig. 14.3 Radiating electron—angles and vectors.

makes an angle ϕ with the x–z plane and can point in any direction θ within this plane. In practice, θ is a very small angle and the radiation is viewed close to the z-direction.

The vectors $\boldsymbol{R}$ and $\boldsymbol{\beta}$ which describe the motion of the electron in the x–z plane can be obtained from the previous treatment. We denote the average velocity of the electron along the z-axis as $\bar{\beta}c$ so that, from eqns (13.2) and (13.3),

$$\dot{x}(z) = \bar{\beta}c\frac{K}{\gamma}\cos\left(\frac{2\pi\bar{\beta}ct}{\lambda_0}\right)$$

and

$$x(t) = \frac{K}{\gamma}\frac{\lambda_0}{2\pi}\sin\left(\frac{2\pi\bar{\beta}ct}{\lambda_0}\right);$$

as before [eqn (14.2)]

$$\bar{\beta}c = \beta c\left(1 - \frac{K^2}{4\gamma^2}\right).$$

Similarly,

$$\begin{aligned}\dot{z} &= \beta c\sqrt{1 - \left(\frac{\bar{\beta}c}{\beta c}\right)^2 \frac{K^2}{\gamma^2}\cos^2\frac{2\pi z}{\lambda_0}} \\ &= \beta c\left(1 - \frac{K^2}{2\gamma^2}\cos^2\frac{2\pi z}{\lambda_0}\right)\end{aligned}$$

and, by integration,

$$z(t) = \bar{\beta}ct - \frac{K^2\lambda_0}{16\pi\gamma^2}\sin\left(\frac{4\pi}{\lambda_0}\bar{\beta}ct\right).$$

We note that the modulation frequency of the electron motion along the z-axis is twice the modulation frequency along the x-axis as pointed out in Fig. 14.2.

It is convenient to write the quantity $2\pi\bar{\beta}c/\lambda_0$ as a frequency which we denote by ω_0. Then the vector $\boldsymbol{R}$ becomes (using unit vectors $\boldsymbol{i}$, $\boldsymbol{j}$ and $\boldsymbol{k}$ along the x, y and z axes respectively)

$$\boldsymbol{R}(t') = \boldsymbol{i}\frac{K}{\gamma}\frac{\lambda_0}{2\pi}\sin(\omega_0 t') + \boldsymbol{k}\left(\bar{\beta}ct' - \frac{K^2}{8\gamma^2}\frac{\lambda_0}{2\pi}\sin(2\omega_0 t')\right); \tag{14.6}$$

the velocity vector becomes

$$\begin{aligned}\beta c(t') &= \frac{\mathrm{d}\boldsymbol{R}(t')}{\mathrm{d}t'} \\ &= \boldsymbol{i}\frac{K}{\gamma}\bar{\beta}c\cos(\omega_0 t') + \boldsymbol{k}\bar{\beta}c\left(1 - \frac{K^2}{4\gamma^2}\cos(2\omega_0 t')\right);\end{aligned} \tag{14.7}$$

and, from Fig. 14.3,

$$\boldsymbol{n} = \boldsymbol{i}\sin\theta\cos\phi + \boldsymbol{j}\sin\theta\sin\phi + \boldsymbol{k}\cos\phi. \tag{14.8}$$

Next, we compute the quantities required for eqn (14.5). The scalar product $\boldsymbol{n}\cdot\boldsymbol{\beta}(t')$ is

$$\boldsymbol{n}\cdot\boldsymbol{\beta} = \bar{\beta}\frac{K}{\gamma}\sin\theta\cos\phi\cos(\omega_0 t) + \bar{\beta}\cos\theta\left(1 - \frac{K^2}{\gamma^2}\cos(2\omega_0 t)\right), \tag{14.9}$$

which can be used to give the triple vector product

$$(\boldsymbol{n}\times\boldsymbol{\beta})\times\boldsymbol{\beta} = (\boldsymbol{n}\cdot\boldsymbol{\beta})\boldsymbol{n} - \boldsymbol{\beta}. \tag{14.10}$$

From eqns (14.6) and (14.8) the component of the position vector of the electron along the observation direction, expressed in units of time, is

$$\begin{aligned}\frac{\boldsymbol{n}\cdot\boldsymbol{R}(t')}{c} &= \frac{K}{\gamma}\frac{\lambda_0}{2\pi c}\sin(\omega_0 t')\sin\theta\cos\phi + \left(\bar{\beta}t' - \frac{1}{8}\frac{K^2}{\gamma^2}\frac{\lambda_0}{2\pi c}\sin(2\omega_0 t')\right)\cos\theta \\ &= \frac{K}{\gamma}\frac{\bar{\beta}}{\omega_0}\sin(\omega_0 t')\sin\theta\cos\phi + \left(\bar{\beta}t' - \frac{1}{8}\frac{K^2}{\gamma^2}\frac{\bar{\beta}}{\omega_0}\sin(2\omega_0 t')\right)\cos\theta\end{aligned}$$

and

$$\frac{\omega_0}{2\pi}\lambda_0 = \bar{\beta}c$$

as before. We can write

$$1 - \bar{\beta}\cos\theta = \frac{1}{2\gamma^2}\left(1 + \frac{K^2}{2} + \gamma^2\theta^2\right) = \frac{\omega_0}{\omega_1}$$

in which ω_1 is the frequency of the fundamental and we make the approximation that $K \ll \gamma$ and $\theta \ll 1$, so that the exponent $\mathrm{i}\omega(t' - \boldsymbol{n} \cdot \boldsymbol{R}(t')/c)$ becomes

$$\mathrm{i}\omega\left(t' - \frac{\boldsymbol{n}\cdot\boldsymbol{R}(t')}{c}\right)$$
$$= \mathrm{i}\frac{\omega}{\omega_1}\left(\omega_0 t' - \bar{\beta}\frac{K}{\gamma}\frac{\omega_1}{\omega_0}\theta\cos\phi\sin(\omega_0 t') + \frac{\bar{\beta}}{8}\frac{K^2}{\gamma^2}\frac{\omega_1}{\omega_0}\sin(2\omega_0 t')\right). \quad (14.11)$$

Now we return to the triple vector product, insert the scalar and vector quantities of eqns (14.7)–(14.9) explicitly into the identity (14.10) and make the same approximations to obtain

$$(\boldsymbol{n}\cdot\boldsymbol{\beta})\boldsymbol{n} - \boldsymbol{\beta} = \boldsymbol{i}\bar{\beta}\left(\theta\cos\phi - \frac{K}{\gamma}\cos(\omega_0 t')\right) + \boldsymbol{j}\bar{\beta}\theta\sin\phi. \quad (14.12)$$

We combine eqns (14.11) and (14.12) to yield an expression for the integral in eqn (14.5), which becomes

$$\bar{\beta}\int_{-\infty}^{\infty}\left((\boldsymbol{i}\cos\phi + \boldsymbol{j}\sin\phi)\theta - \boldsymbol{i}\frac{K}{\gamma}\cos(\omega_0 t')\right)$$
$$\times \exp\left(\mathrm{i}\frac{\omega}{\omega_1}\left(\omega_0 t' - \bar{\beta}\frac{K}{\gamma}\frac{\omega_1}{\omega_0}\theta\cos\phi\sin(\omega_0 t') + \frac{\bar{\beta}}{8}\frac{K^2}{\gamma^2}\frac{\omega_1}{\omega_0}\sin(2\omega_0 t')\right)\right)\mathrm{d}t'.$$

In order to evaluate this integral we expand the exponential function into a product of three exponential factors. We simplify the expressions by writing

$$S_1 = \bar{\beta}\frac{K}{\gamma}\frac{\omega_1}{\omega_0}\theta\cos\phi = \frac{2K\bar{\beta}\gamma\theta\cos\phi}{1 + K^2/2 + \gamma^2\theta^2},$$
$$S_2 = \frac{\bar{\beta}}{8}\frac{K^2}{\gamma^2}\frac{\omega_1}{\omega_0} = \frac{K^2\bar{\beta}}{4\left(1 + K^2/2 + \gamma^2\theta^2\right)} \quad (14.13)$$

so that, if $u = (\omega/\omega_1)S_1$ and $v = (\omega/\omega_1)S_2$ then the product of the exponential factors becomes

$$\exp\left(\mathrm{i}\frac{\omega}{\omega_1}\omega_0 t'\right)\exp\left(-\mathrm{i}u\sin(\omega_0 t')\right)\exp\left(\mathrm{i}v\sin(2\omega_0 t')\right).$$

Now we simplify the time dependence by expanding the second and third factors in terms of Bessel functions with integer order j according to the identity[5]

$$\exp(\mathrm{i}a\sin(\omega t)) = \sum_{j=0}^{\infty} J_j(a)\exp(\mathrm{i}j\omega t) \quad (14.14)$$

and we also place the origin of the coordinates at the centre of the undulator and take the limits of integration over the length L of the undulator. We put $L = N\lambda_0$, so that the integration limits run from $-\pi N/\omega_0$ to $+\pi N/\omega_0$. With these changes the integral becomes

$$\int_{-\pi N/\omega_0}^{\pi N/\omega_0} \left((\boldsymbol{i}\cos\phi + \boldsymbol{j}\sin\phi)\theta - \boldsymbol{i}\frac{K}{\gamma}\cos(\omega_0 t') \right) \times \sum_m J_m(u) \sum_n J_n(v) \exp(\mathrm{i}R_\omega \omega_0 t')\,\mathrm{d}t' \tag{14.15}$$

in which

$$R_\omega = \frac{\omega}{\omega_1} - m + 2n. \tag{14.16}$$

This integral divides into two terms. The first term is just the exponential function multiplied by a time independent term which we can leave outside the integral sign. The second term is the same exponential multiplied by $\boldsymbol{i}(K/\gamma)\cos(\omega_0 t')$. These integrals can be expressed in the form

$$\int_{-b}^{b} \exp(\mathrm{i}a\omega t)\,\mathrm{d}t = \int_{-b}^{b} (\cos(a\omega t) + \mathrm{i}\sin(a\omega t))\,\mathrm{d}t = \frac{2\sin(ab\omega)}{\omega}$$

and

$$\begin{aligned}\int_{-b}^{b} \cos(\omega t)\exp(\mathrm{i}a\omega t)\,\mathrm{d}t &= \frac{1}{2}\int_{-b}^{b} \exp(\mathrm{i}\omega t)\exp(\mathrm{i}a\omega t)\,\mathrm{d}t \\ &\quad + \frac{1}{2}\int_{-b}^{b} \exp(-\mathrm{i}\omega t)\exp(\mathrm{i}a\omega t)\,\mathrm{d}t \\ &= \frac{1}{2}\int_{-b}^{b} \exp(\mathrm{i}(a+1)\omega t)\,\mathrm{d}t + \frac{1}{2}\int_{-b}^{b} \exp(\mathrm{i}(a-1)\omega t)\,\mathrm{d}t \\ &= \frac{\sin((a+1)b\omega)}{\omega} + \frac{\sin((a-1)b\omega)}{\omega}.\end{aligned}$$

We make use of these expressions to evaluate the integral [eqn (14.15)] and obtain

$$\begin{aligned}\frac{\mathrm{d}^2 I(\omega)}{\mathrm{d}\Omega\,\mathrm{d}\omega} = \frac{e^2\omega^2 N^2 \bar{\beta}^2}{4\pi\varepsilon_0 c\omega_0^2} \Bigg| &\boldsymbol{i}\left(A_0 \sum_{m=-\infty}^{+\infty} J_m(u) \sum_{n=-\infty}^{+\infty} J_n(v) \frac{\sin(\pi N R_\omega)}{\pi N R_\omega} \right) \\ &+ \boldsymbol{i}\left(A_1 \sum_{m=-\infty}^{+\infty} J_m(u) \sum_{n=-\infty}^{+\infty} J_n(v) \frac{1}{2}\left(\frac{\sin(\pi N(R_\omega+1))}{\pi N(R_\omega+1)} + \frac{\sin(\pi N(R_\omega-1))}{\pi N(R_\omega-1)} \right)\right) \\ &+ \boldsymbol{j}\left(B_0 \sum_{m=-\infty}^{+\infty} J_m(u) \sum_{n=-\infty}^{+\infty} J_n(v) \frac{\sin(\pi N R_\omega)}{\pi N R_\omega} \right) \Bigg|^2.\end{aligned}$$

The variables u and v, which contain no time dependence, have been defined in eqn (14.13) and

$$A_0 = \theta \cos\phi, \quad A_1 = \frac{K}{\gamma}, \quad B_0 = \theta \sin\phi.$$

Let us examine this rather formidable expression. The factor of the form $|\sin(\pi N R_\omega)/\pi N R_\omega|^2$ or $|\sin(\pi N(R_\omega+1))/\pi N(R_\omega+1)|^2$ expresses the linewidth of the radiation. Its physical origin is the interference which occurs when the radiation from the individual sources (the poles of the undulator) is summed by the detector at the observation point. The function has a maximum value when R_ω is an integer and decreases rapidly as soon as R_ω moves away from an integral value in either direction, as the electromagnetic waves move further and further out of phase at the observation point as their frequency changes. The width of the line decreases rapidly as N, the number of periods, increases, because the effect of a small frequency shift on the total amplitude becomes more and more pronounced as the number of overlapping sources becomes larger.

This same linewidth factor, $|\sin(\pi N R_\omega)/\pi N R_\omega|^2$ can also be regarded as the result of a Fourier analysis of the output waveform into its component frequencies as described in Chapter 1. The result of this analysis, which has the form $(\sin \pi N\theta)/\pi N\theta$, must be squared to yield the power transported by the wave to the observer.

These effects are shown in Fig. 14.4 in which the calculated linewidth of the radiation is plotted for three different values of N. It is clear that in a typical undulator with ten or more periods all the radiation in a typical line is produced very close indeed to the wavelength for which R_ω (or $R_\omega \pm 1$) is close to zero. From eqn (14.16), $R_\omega = (\omega/\omega_1) - m + 2n$, so the effect of the interference function is to select only those frequency values which, together with the choice of the indices m and n, satisfy this restriction on R_ω. These frequencies correspond to the wavelengths defined by eqn (14.4) and we can write

$$\omega = \omega_k = k\omega_1 \quad \text{and} \quad R_\omega = k - m + 2n.$$

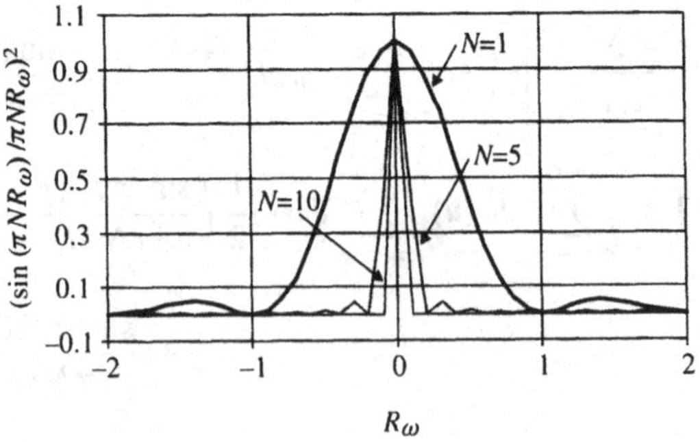

Fig. 14.4 Linewidth of radiation from an undulator with N periods.

We can now pick out from eqn (14.16) the photon energy spectrum for the harmonic number k. On tune, when the frequency is $\omega_k = k\omega_1$ we set $R_\omega = 0$, then, for terms containing R_ω the index $m = k+2n$ and for those containing $R_\omega \pm 1$, $m = k+2n\pm 1$. Off tune, we can write for the linewidth of the kth harmonic, $\Delta\omega_k = \omega - \omega_k = \omega - k\omega_1$ so that R_ω (or $R_\omega + 1$) becomes $\Delta\omega_k/\omega_1$ for the choice of indices above and the energy spectrum of the kth harmonic becomes

$$\frac{\mathrm{d}^2 I(\omega_k)}{\mathrm{d}\Omega\,\mathrm{d}\omega} = \frac{e^2\omega_k^2 N^2 \bar{\beta}^2}{4\pi\varepsilon_0 c\omega_0^2} \times \left(\frac{\sin(\pi N(\Delta\omega_k/\omega_1))}{\pi N(\Delta\omega_k/\omega_1)}\right)^2$$
$$\times \left| \boldsymbol{i}\left(A_0 \sum_{n=-\infty}^{+\infty} J_n(v) J_{k+2n}(u)\right)\right.$$
$$+\boldsymbol{i}\left(\frac{A_1}{2}6 \sum_{n=-\infty}^{+\infty} J_n(v) J_{k+2n+1}(u)\right)$$
$$\left. +\boldsymbol{j}\left(B_0 \sum_{n=-\infty}^{+\infty} J_n(v) J_{k+2n}(u)\right)\right|^2. \qquad (14.17)$$

This expression for the energy spectrum contains a component which corresponds to radiation polarized in the horizontal plane, in the direction denoted by the unit vector $\boldsymbol{i}$ and a component polarized in the vertical direction indicated by the unit vector $\boldsymbol{j}$. These two unit vectors are, by definition, orthogonal, so that when the modulus squared of this expression is taken, there are no cross terms. Let us concentrate on the radiation produced in the forward direction. In this case $\theta = 0$ so that A_0 and B_0 are both zero and there is, therefore, no vertically polarized radiation in the forward direction, as was noted previously. We examine the only remaining term, which is the middle one: S_1 and, therefore, u is equal to zero and

$$J_{k+2n\pm 1}(0) = \begin{cases} 1 & k+2n\pm 1 = 0, \\ 0 & \text{otherwise,} \end{cases}$$

so that the expression is non-zero only when

$$n = -\frac{k\pm 1}{2}.$$

We know, from eqn (14.14), that only integral orders appear in the expansion of the exponential functions in terms of Bessel functions, which means that the index k is restricted to odd values. There are no even order harmonics in the forward direction. We deal with the negative order Bessel functions by remembering that for integer orders

$$J_{-m}(x) = (m)^{-1} J_m(x)$$

and we reduce eqn (14.17) to

$$\left.\frac{\mathrm{d}^2 I(\omega_k)}{\mathrm{d}\Omega\,\mathrm{d}\omega}\right|_{\theta=0} = \frac{e^2\omega_k^2 N^2\bar{\beta}^2}{4\pi\varepsilon_0 c\omega_0^2} \times \left(\frac{\sin(\pi N(\Delta\omega_k/\omega_1))}{\pi N(\Delta\omega_k/\omega_1)}\right)^2 \times \left(\frac{K}{2\gamma}\right)^2 \left|J_{(k-1)/2}(v) - J_{(k+1)/2}(v)\right|^2 .$$

We can rewrite this expression as the peak number of photons generated into unit solid angle per second into a bandwidth $\Delta\omega/\omega$ after making the usual approximation for small K and large γ:

$$\left.\frac{\mathrm{d}N(\omega_k)}{\mathrm{d}\Omega\,\mathrm{d}t}\right|_{\theta=0} = \begin{cases} \alpha N^2\gamma^2\dfrac{\Delta\omega}{\omega} \times \dfrac{I}{e}\dfrac{k^2K^2}{(1+K^2/2)^2} \\ \quad \times \left|J_{(k-1)/2}\left(\dfrac{kK^2}{4(1+K^2/2)}\right) - J_{(k+1)/2}\left(\dfrac{kK^2}{4(1+K^2/2)}\right)\right|^2 \\ \quad k \text{ odd,} \\ 0 \quad k \text{ even.} \end{cases} \tag{14.18}$$

This very useful expression is identical to that derived by Kim.[6]

When k is odd, we can write eqn (14.18) in the form

$$\left.\frac{\mathrm{d}N(\omega_k)}{\mathrm{d}\Omega\,\mathrm{d}t}\right|_{\theta=0} = \alpha N^2\gamma^2\frac{\Delta\omega}{\omega} \times \frac{I}{e}\frac{k^2K^2}{(1+K^2/2)^2} \times F_k(K) \quad k \text{ odd},$$

where

$$F_k(K) = \frac{k^2K^2}{(1+K^2/2)^2}\left|J_{(k-1)/2}\left(\frac{kK^2}{4(1+K^2/2)}\right) - J_{(k+1)/2}\left(\frac{kK^2}{4(1+K^2/2)}\right)\right|^2$$

and $F_k(K)$ is plotted in Fig. 14.5 to show the relative contribution from the first three odd-numbered harmonics.

A detector placed on axis at a large distance from the undulator collects a number of photons, given by eqn (14.18), which depends on the subtended solid angle and the detection bandwidth. We have already shown that the undulator produces a series of discrete frequencies. The radiation at each frequency is spread over a bandwidth and is contained within an angular interval defined by the interference function

$$\left(\frac{\sin(\pi N(\Delta\omega_k/\omega_1))}{\pi N(\Delta\omega_k/\omega_1)}\right)^2 .$$

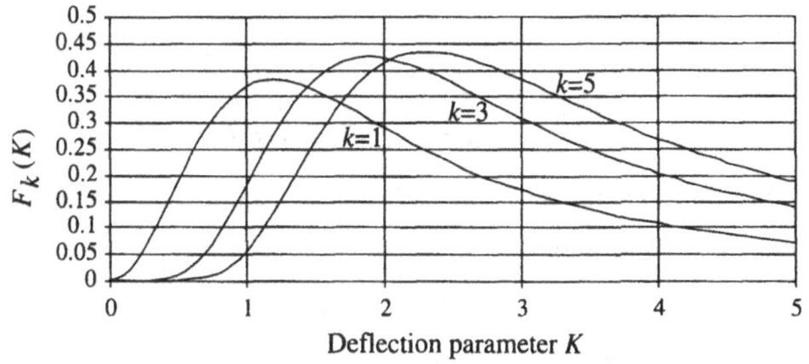

Fig. 14.5 Undulator output function $F_k(K)$ at $\theta = 0°$.

This function (Fig. 14.4) has a maximum value of unity and falls to one-half when $N \times (\Delta\omega_k/\omega_1) \approx 1/2$ so we can take the bandwidth of the kth harmonic of the radiation at frequency ω_k as

$$\frac{\Delta\omega_k}{\omega_k} = \frac{1}{kN}. \tag{14.19}$$

This is what would be expected from a physical argument. The electron generates a wave train of finite length, whose overall phase change is equal to 2π times the number of periods in the undulator. For a wave with frequency ω_k the phase difference between the ends of the wave train must be $2\pi kN$ radians. In order to reduce the amplitude of the wave train to zero at each end, there must be a wave present with longer (or shorter) wavelength so that at each end of the wave train the two waves are out of phase by π radians, i.e. a phase shift of 2π radians in $2\pi kN$ radians so that the fractional bandwidth must be of order $1/kN$.

What is the angular width of this kth harmonic radiation? We know from eqn (14.4) that an observer located exactly on axis sees radiation with frequency ω_k. An angular displacement off-axis, through an angle θ, changes the frequency by an amount $\Delta\omega_k = \omega(\theta) - \omega_k = \omega(\theta) - k\omega_1$, where

$$\begin{aligned} \omega(\theta) &= \frac{2k\gamma^2\omega_0}{1 + K^2/2 + \gamma^2\theta^2}, \\ \omega_k &= \frac{2k\gamma^2\omega_0}{1 + K^2/2}. \end{aligned} \tag{14.20}$$

As the angle θ increases, the interference function (Fig. 14.4) reaches zero when $kN \times (\Delta\omega_k/\omega_k) = 1$ and an aperture of this radius includes essentially all the radiation produced in the kth harmonic. This point is reached when

$$\omega(\theta) = \omega_k \left(1 + \frac{1}{kN}\right)$$

which, from eqn (14.20), gives

$$\theta^2 = \frac{1+K^2/2}{\gamma^2}\left(\frac{1}{1+kN}\right).$$

If we approximate the distribution in angle by a Gaussian with standard deviation σ_θ then, for large N, $kN \gg 1$, and

$$\sigma_\theta = \frac{1}{\gamma}\sqrt{\frac{1+K^2/2}{2kN}}.$$

The opening angle of the radiation from an undulator is reduced compared with that from a dipole magnet by a factor $\approx 1/\sqrt{kN}$. From eqn (14.4) the expression for σ_θ can be rewritten as

$$\sigma_\theta = \sqrt{\frac{\lambda_k}{N\lambda_0}} = \sqrt{\frac{\lambda_k}{L}}, \tag{14.21}$$

where N is the number of periods and L is the length of the undulator.

There is a simple physical argument for this result. Consider Fig. 14.6. A cone of radiation from within the undulator, spreading out at an angle $\Delta\theta$, has a diameter $\Delta r = \Delta\theta \times L/2$ at the exit. The emerging beam spreads out in such a way that the

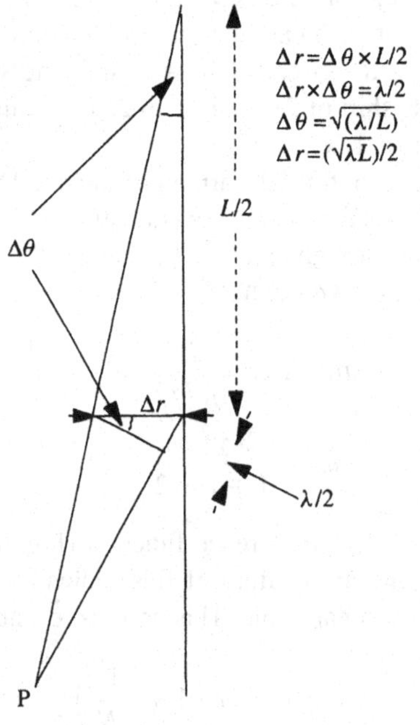

Fig. 14.6 Undulator radiation opening angle.

angular limits of the beam are determined by diffraction from a source of diameter Δr so that the waves reaching the point P are out of phase by $\lambda/2$. This corresponds to the first minimum of the Rayleigh distribution in diffraction theory. For this to be so, $\Delta r \times \Delta\theta = \lambda/2$ and $\Delta\theta \approx \sqrt{\lambda/L}$. By the same argument, $\Delta r = \frac{1}{2}\sqrt{\lambda L}$.

This width about the central axis is very narrow. For example, if $L = 1$ m and $\lambda_k = 10$ nm then $\sigma_\theta = 100\,\mu$rad, so that at 20 m from the undulator, a 4 mm diameter pinhole would be sufficiently wide to accept all the radiation at that wavelength (ignoring the diameter of the electron beam).

We can calculate the total number of photons passing through this pinhole if we assume that the angular distribution is Gaussian with a standard deviation σ_θ given by eqn (14.21). A rectangle with width $\sigma_\theta\sqrt{2\pi}$ will have the same area as the Gaussian distribution integrated from 0 to ∞ so we can write, for the photon flux in the forward cone,

$$\begin{aligned}\frac{\mathrm{d}N(\omega_k)}{\mathrm{d}t} &= 2\pi\sigma_\theta^2 \left.\frac{\mathrm{d}N(\omega_k)}{\mathrm{d}\Omega\mathrm{d}t}\right|_{\theta=0} \\ &= \pi\alpha N\frac{\Delta\omega}{\omega} \times \frac{I}{e} \times Q_k(K) \quad k \text{ odd}, \\ \text{where } Q_k(K) &= F_k(K)\left(1+\frac{K^2}{2}\right)\left(\frac{1}{k}\right). \end{aligned} \tag{14.22}$$

The function $Q_k(K)$ is plotted in Fig. 14.7.

In general, the output of the undulator as seen by a small detector centred on the axis of the undulator and subtending an angle $\sqrt{\lambda/N}$ is a central cone of radiation consisting of a fundamental and a series of harmonics whose wavelengths are given by eqn (14.4) with $\theta = 0$. If we increase the diameter of the detector so that the subtended angle eventually becomes of the order of $1/\gamma$, we intercept an increasing number of new series of harmonic, defined by the indices k and m and having a wavelength

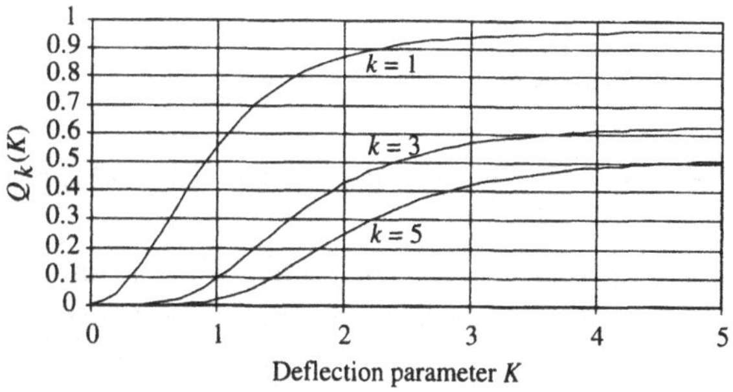

Fig. 14.7 Undulator output function $Q_k(K)$ at $\theta = 0°$.

$\lambda(\theta)_{k+m}$ equal to $\lambda(0)_k$. We can calculate these angles by observing that

$$\lambda(0)_k = \frac{\lambda_0}{2k\gamma^2}\left(1 + \frac{K^2}{2}\right),$$

$$\lambda(\theta)_{k+m} = \frac{\lambda_0}{2(k+m)\gamma^2}\left(1 + \frac{K^2}{2} + \gamma^2\theta^2\right),$$

so that

$$\lambda(\theta)_{k+m} = \frac{k}{(k+m)}\left(1 + \frac{\gamma^2\theta^2}{1+K^2/2}\right)\lambda(0)_k$$

and

$$\theta = \frac{1}{\gamma}\sqrt{\frac{m}{k}\left(1 + \frac{K^2}{2}\right)}$$

which is of the order of $1/\gamma$.

Total power output

The total power output from an undulator can be calculated using the same procedure as in Chapter 5. The expression for the power output from one electron with velocity $\boldsymbol{\beta}c$ and acceleration $\dot{\boldsymbol{\beta}}c$ is given by the Liénard formula [from eqn (5.23)]

$$\frac{\mathrm{d}U}{\mathrm{d}t} = \frac{2}{3}\frac{e^2}{4\pi\varepsilon_0 c}\gamma^6\left(\dot{\boldsymbol{\beta}}^2 - (\boldsymbol{\beta}\times\dot{\boldsymbol{\beta}})^2\right). \tag{14.23}$$

The electron velocity vector is given by eqn (14.7):

$$\boldsymbol{\beta}(t') = \boldsymbol{i}\frac{K}{\gamma}\bar{\beta}\cos(\omega_0 t') + \boldsymbol{k}\bar{\beta}\left(1 - \frac{K^2}{4\gamma^2}\cos(2\omega_0 t')\right)$$

and the acceleration vector is the derivative of this expression

$$\dot{\boldsymbol{\beta}}(t') = -\boldsymbol{i}\frac{K}{\gamma}\bar{\beta}\omega_0\sin(\omega_0 t') + \boldsymbol{k}\bar{\beta}\omega_0\frac{K^2}{2\gamma^2}\sin(2\omega_0 t').$$

Insertion of these expressions into eqn (14.23) following the rules for vector manipulation and neglecting terms which are small compared with unity

$$\frac{\mathrm{d}U}{\mathrm{d}t} = \frac{2}{3}\frac{e^2}{4\pi\varepsilon_0 c}\gamma^2 K^2\omega_0^2\sin^2(\omega_0 t).$$

The rms power radiated by one electron, passing through one period of the undulator, obtained by averaging over the factor $\sin^2\theta(\omega_0 t)$ is

$$\left.\frac{\mathrm{d}U}{\mathrm{d}t}\right|_{\mathrm{rms}} = \frac{1}{3}\frac{e^2}{4\pi\varepsilon_0 c}\gamma^2 K^2\omega_0^2$$
$$= \frac{4\pi^2}{3} r_e m_e c^3 \gamma^2 \frac{K^2}{\lambda_0^2} \quad (\mathrm{J/s}).$$

The energy lost by one electron as it traverses the whole length of the undulator $L = N\lambda_0$ in a time L/c is

$$\Delta E = \frac{4\pi^2}{3} r_e m_e c^2 \gamma^2 \frac{K^2}{\lambda_0^2} L$$
$$= \frac{4\pi^2}{3} r_e m_e c^2 \gamma^2 \frac{K^2}{\lambda_0} N \quad (\mathrm{J}).$$

When the electron beam current is I amperes the number of electrons passing through the undulator in one second, $\mathrm{d}n/\mathrm{d}t$, is equal to I/e so that the power output from the undulator is

$$\frac{\mathrm{d}U}{\mathrm{d}t} = \frac{4\pi^2}{3} r_e m_e c^2 \gamma^2 \frac{I}{e}\frac{K^2}{\lambda_0^2} L$$
$$= \frac{4\pi^2}{3} r_e m_e c^2 \gamma^2 \frac{I}{e}\frac{K^2}{\lambda_0} N \quad (\mathrm{W}). \tag{14.24}$$

Insertion of the constants into eqn (14.24) gives a useful practical expression for the total power produced by an undulator

$$\frac{\mathrm{d}U}{\mathrm{d}t} = 0.634 B_0^2 E^2 I L K \quad (\mathrm{W}).$$

B_0 is measured in tesla, E in GeV, I in amperes and L in metres.

Equation (14.24) can be compared with the corresponding expression for the power output from the electron current in a dipole magnet [from eqn (5.20)] in the form

$$\frac{\mathrm{d}U}{\mathrm{d}t} = \frac{2}{3} r_e m_e c^2 \gamma^4 \frac{1}{R^2} N_{\mathrm{rad}},$$

which, if we relate the orbit radius R_0 to the electron energy E using the formula $E = pc = BeRc$ and, if L is the length of the orbit in the dipole magnet, so that $N_{\mathrm{rad}} = (I/e) \times (L/c)$ can be written as

$$\frac{\mathrm{d}U}{\mathrm{d}t} = \frac{2}{3} r_e m_e c^2 \gamma^2 \frac{B^2 e^2 c^2}{(m_e c^2)^2}\frac{I}{e} L.$$

If the mean square field in the undulator is $B_0{}^2 = B^2/2$, then inserting the definition of K, this reduces to eqn (14.24) for the power loss from an undulator with length L.

Realization of undulators

Finally, let us calculate some order of magnitude values for the output from an undulator. A storage ring such as the ESRF at Grenoble operates at 6 GeV electron beam energy. If the stored beam current is 100 mA then an undulator 1.5 m long, $\lambda_0 = 5$ mm, with a field of 0.5 T, which is designed to generate a fundamental wavelength of 0.65 nm (1.9 keV)[7] produces an output power of 0.85 kW. This is not particularly large. It is less than the output of the kind of domestic heater used in the winter in countries with a cold climate.

However, this power output is far more concentrated than that from an electric fire. Kim[8] has calculated the following expression for the power radiated into unit solid angle:

$$\frac{\mathrm{d}^2U}{\mathrm{d}\Omega\,\mathrm{d}t} = \frac{7\pi}{4} r_e m_e c^2 \gamma^4 \frac{I}{e} \frac{G(K)}{\lambda_0^2} L$$
$$= \frac{7\pi}{4} r_e m_e c^2 \gamma^4 \frac{I}{e} \frac{G(K)}{\lambda_0} N \quad (\mathrm{W/rad^2}).$$

As we would expect, the power into unit solid angle is proportional to γ^4 because the radiation is produced into an opening angle which is proportional to $1/\gamma$ and the solid angle is proportional to $1/\gamma^2$. The function $G(K)$ tends to a constant value of unity once $K \gtrsim 1$ so in practical units we can write

$$\frac{\mathrm{d}^2U}{\mathrm{d}\Omega\,\mathrm{d}t} = 10.84\, B_0 E^4 I N \quad (\mathrm{W/mrad^2})$$

and the undulator referred to above would produce 21 kW/mrad^2. This power, in the X-ray region, is comparable to that generated by a high power laser in the visible region of the spectrum.

References

1. H. Motz, *J. Appl. Phys.*, **22**, 527 (1951).
2. D. F. Alferov, Y. A. Bashmakov, and E. G. Bessonov, *Soviet Physics—Technical Physics*, **18**, 1336 (1974).
3. H. Wiedemann, *Particle accelerator physics*, Vol II. Ch. 11, Springer-Verlag, Berlin, Germany (1996).
4. J. D. Jackson, *Classical electrodynamics* (2nd edn), equation 14.67. P. 671. John Wiley & Sons, New York (1975).
5. A. Gray and G. B. Mathews, *A treatise on Bessel functions*. Macmillan, London (1895); A. Erdélyi (ed.) *Higher transcendental functions*. McGraw-Hill, New York (1953).
6. K.-J. Kim, American Institute of Physics. Conference Proceedings 184 (eds M. Month and M. Deines) (1987 and 1988).
7. P. Ellaume, Theory of Undulators and Wigglers. CERN Accelerator School, 1989. CERN 90-03. Geneva, 1990, pp. 142–57.
8. K.-J. Kim, *Nucl. Instr. Methods*, **A246**, 71–6 (1986).

15

Recent developments and future prospects

Introduction

Synchrotron radiation is a highly versatile research tool which provides electromagnetic radiation over a wide range of wavelengths. It is highly collimated. When an undulator is the source, the radiation, especially in the X-ray region, is of unparalleled brightness.

In this concluding chapter we indicate how synchrotron radiation sources are being developed at the present time. These developments are being driven by the requirements of the synchrotron radiation users themselves.

High brightness sources

The most obvious direction of synchrotron radiation source development is in the provision of radiation sources with high brightness (photons/mm^2 mrad^2) and, by implication, with low electron beam emittance. The attraction of such sources is that radiation which emanates from a small source size can be focused to provide an extremely bright irradiated area on a target material. The ability to operate in this way is particularly important when the material for study is available in only a small quantity as is often the case with crystals of enzyme, protein, or virus.

The minimization of electron beam emittance was discussed briefly in Chapter 12. It was shown there that the horizontal electron beam emittance reduces to a simple approximate expression,

$$\varepsilon_x = 3.83 \times 10^{-13} \gamma^2 \theta_\mathrm{d}^3 F(\alpha_0, \beta_0, \gamma_0, s_\theta). \tag{15.1}$$

The value of F depends on the choice of magnetic lattice and from Table 12.1 is clear that the double bend or triple bend achromat gives the lowest emittance of the sources tabulated. The emittance also depends on the choice of storage ring energy. The demand for hard X-ray sources means that the electron energy must be increased compared to sources constructed 10–15 years ago. The ESRF (Grenoble, France), the Advanced Photon Source (APS, at the Argonne National Laboratory, near Chicago, USA), and SPring-8 (Nishi-Harima, Japan) have electron beam energies of 6, 7, and 8 GeV, respectively. All other things being equal, higher energy means higher emittance [eqn (15.1)]. The expectation of synchrotron radiation users that these sources are at least as bright as those constructed at lower energy places additional requirements on the basic source design. For example, a factor four increase in energy from the Berkeley ALS to the Grenoble ESRF (see Table 12.1) has been offset, in part, by increasing the number of bending magnets from 36 to 64 which decreases the bend angle in each dipole from 10° to 5.625°. The large circumference of these sources is determined partly by the high energy but predominantly by the requirement for long straight sections (drift spaces) for multipole wigglers and undulators.

Radiation from insertion devices

Synchrotron radiation with the highest possible brightness can be obtained by the use of insertion devices. Synchrotron radiation sources, proposed, designed, and constructed in recent years place strong emphasis on the incorporation of these devices (multipole wigglers and undulators) as an integral part of the storage ring from its first days of operation. This implies that to make the best possible use of these devices, the electron beam parameters must be matched as well as possible to the properties of the insertion device.

The radiation from an undulator at a particular wavelength is produced over a very narrow angular interval and the appropriate measure of its performance is photon brilliance (photons/mrad^2). Low electron beam emittance is a necessary but not a sufficient criterion for the achievement of high undulator brilliance. If we are to maximize the output brilliance, the angular divergence of the electron beam must be at least of the same order as the angular interval of the radiation. At any point around the storage ring the angular divergence of the electron beam in the horizontal and vertical planes, $\sigma'_{x,y}$, is related, approximately, to the electron beam emittance in the two planes by (Chapter 11)

$$\sigma'_{x,y}(s) \approx \sqrt{\frac{\varepsilon_{x,y}}{\beta_{x,y}(s)}}$$

so that the beta functions in the straight sections where undulators are to be installed must not be too low. On the other hand, to achieve high brightness (as opposed to brilliance alone), the same beta functions must not be too high because the beam sizes $\sigma_{x,y}$ in the straight section are given by

$$\sigma_{x,y}(s) = \sqrt{\varepsilon_{x,y}\beta_{x,y}(s)}.$$

As if this were not enough, there is a further condition which arises from the energy lost by the electrons themselves as they pass through the device. This can be comparable with the energy lost in one revolution around the storage ring. Taking the ESRF as an example, the electrons lose 4.75 MeV in the storage ring and an additional energy loss of 1.5 MeV in the insertion devices. Not only must this additional energy loss (which is about 25% of the total) be compensated by the radio-frequency system but also this energy loss, through the dispersion coefficient in the region of the insertion device leads to an increase (or, conceivably a decrease) in the size of the electron beam throughout the storage ring. This change in the beam dimensions (which affects the brightness of the beam for all users of the storage ring) can only be minimized by ensuring that the dispersion function has a low value in the drift spaces where the insertion devices are being located.

Clearly, a compromise must be reached between these conflicting requirements. The simple FODO lattice (such as that employed at the SRS) is not able to fulfil all these conditions and for this reason there needs to be recourse to more complex magnetic lattice designs involving additional dipole and quadrupole magnets. These ensure that the beta functions are matched as well as possible to the insertion device requirements in the drift spaces and that the dispersion function is low (or zero) in these regions.[2]

There are also practical considerations which affect the realization of both undulators and multipole wigglers. Short wavelength radiation from an undulator implies a short magnetic wavelength, which can only be achieved by closing down the undulator gap, in order to preserve the sinusoidal magnetic field variation. In the case of the multipole wiggler, high field again implies low gap so that the beta function in the vertical plane must be kept low as well. It is essential to preserve the regular magnetic field variation over the whole length of the undulator so that precision construction and location of the magnetic components is of great importance. Recent developments include the writing of computer codes for the field calculations which are optimized for these devices. Nevertheless the final adjustments of the magnetic field must be made empirically, using shims in the form of thin (0.1 mm) iron sheets attached to the magnetic poles.[3]

The importance of matching undulator and storage ring performance can be illustrated by reference to the ESRF.[4] By the middle of 1997, 50 insertion device segments were in operation in the storage ring serving 26 beamlines. New ring optics have been implemented which reduce the vertical beta function in the straight sections where insertion devices are installed. This has made it possible to reduce the vertical gap in the permanent magnet undulators, leading to an improved magnetic field configuration. The result is a factor of two increase in beam brightness through better matching of the photon beam emittance to the electron beam emittance. Although the lower vertical beta generates an increase in the vertical beam divergence, this was found to have a negligible effect on undulator performance.

Figure 15.1 shows the brightness routinely delivered in 1997. High and low beta refer to the type of straight section in which the device is installed. The brightness in the ID3 beamline (equipped with two undulator segments) reaches nearly 10^{20} photons/(s mm^2 mrad^2) into 0.1% bandwidth at around 5 keV(2.5 nm). Similar plots could be presented for most other operating synchrotron radiation sources.

Medium energy sources

The provision of high energy synchrotron radiation sources such as those mentioned above has highlighted the need for sources of medium energy radiation, built to modern specifications. An example of such a source is the Diamond project at the Daresbury Laboratory (although similar sources are planned for other places,

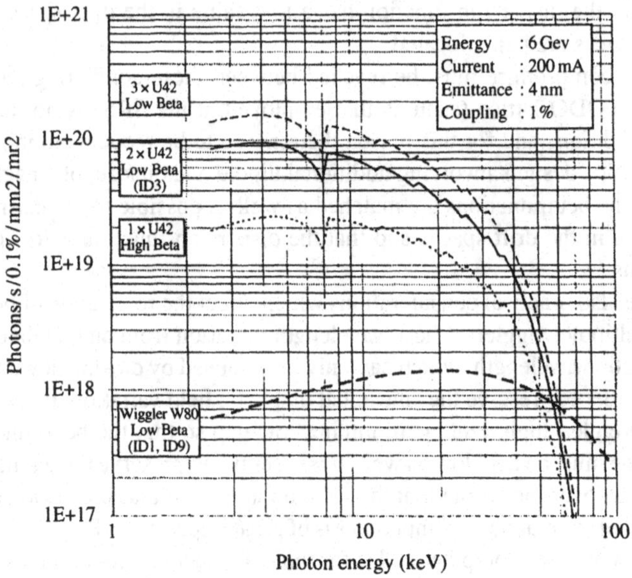

Fig. 15.1 Brightness routinely delivered at the ESRF.

e.g. Soleil, in France). The final specification of Diamond is not yet fixed but a 3 GeV ring is envisaged. The magnetic lattice would contain 16 double-bend achromat cells and the electron beam emittance would be less than 15 nm rad. Undulators would be used to provide high flux X-rays within a photon energy range of 3–30 keV and multi-pole wigglers would provide soft X-ray beam lines covering an energy range from 100 eV to 3 keV. Typical beam dimensions would be 0.2 mm horizontal × 0.04 mm vertical. The circumference of the ring would be about 340 m. Two straight sections 20 m in length would be included in the magnetic lattice to give scope for the accommodation of insertion devices with advanced or novel characteristics.

Coherence of undulator radiation

In general radiation sources are incoherent and cannot, without further conditioning, be used to form an interference pattern. For example, in the simple two-slit experiment, where it is desired to generate an interference pattern from two parallel slits, it is not sufficient just to illuminate the slits with radiation from a bright, extended source. The source must be followed by a pinhole aperture, so that the radiation falling on the two slits is coherent, i.e. there is a definite phase correlation between the wave trains from the source falling on each slit. The pinhole establishes a degree of transverse coherence across the beam in both the x- and y-directions.

Quite separate means must be employed (such as a monochromator) in order to establish longitudinal coherence along the direction of the beam (the z-coordinate).

The visibility of the interference pattern depends directly on the degree of coherence as described by Born and Wolf.[5]

The application of these beam conditioners leads, in most cases, to a significant and even intolerable loss of intensity, but, in the case of radiation from the undulator, the pinhole aperture is already present, being used to define the bandwidth of the radiation so it is legitimate to ask what degree of coherence can be expected.

Consider first the longitudinal coherence. Two parallel wave trains, with wavelengths λ and $\lambda + \Delta\lambda$, starting out in phase, become increasingly out of phase, as seen by an observer further down the beam line. After they have travelled a distance $\lambda^2/\Delta\lambda$ they are out of phase by 2π radians. It is usual to call this distance the longitudinal coherence length, l_c, so that, from eqn (14.19),

$$l_c = \frac{\lambda^2}{\Delta\lambda} = Nk\lambda.$$

The undulator described in the previous chapter would provide radiation at the fundamental wavelength of 0.65 nm ($k = 1$) with a natural longitudinal coherence length of 0.2 μm without any further beam conditioning. The natural bandwidth of the radiation is about 0.33%, so a coherence length of 2 μm could be achieved by the use of a monochromator if the loss of intensity corresponding to a reduced bandwidth of 0.033%, were acceptable.

In the transverse direction, we are interested in the degree of coherence between two points P_1 and P_2 separated by a distance d which subtends an angle θ at a point S in the source (see Fig. 15.2).

When P_1 and P_2 coincide, the degree of coherence is unity and reduces as the two points move further and further apart, except in the special case where the source itself is a point. More generally, according to the Van-Cittert–Zernike theorem, the intensity (and phase) of the radiation at P_2 is determined by the diffraction pattern of the source, centred on P_1. More precisely, the coherence function μ_{12} is given by the Fourier transform of the source intensity distribution function $I(\xi, \eta)$, normalized to

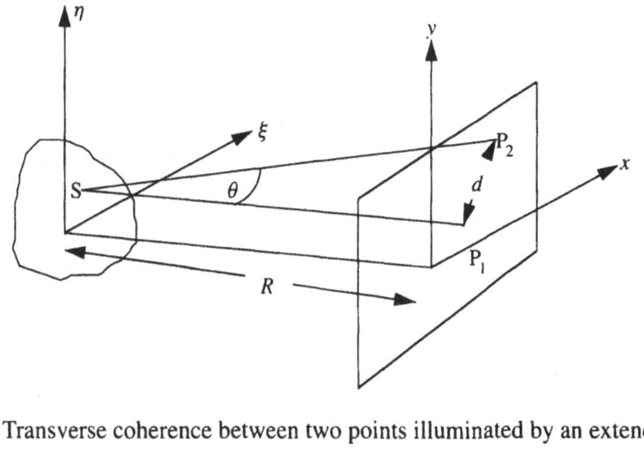

Fig. 15.2 Transverse coherence between two points illuminated by an extended source.

the total intensity of the source:[5]

$$\mu_{12} = \exp(\mathrm{i}\psi)\frac{\iint_\sigma I(\xi,\eta)\exp(-\mathrm{i}k(p\xi + q\eta))\,\mathrm{d}\xi\,\mathrm{d}\eta}{\iint_\sigma I(\xi,\eta)\,\mathrm{d}\xi\,\mathrm{d}\eta}, \tag{15.2}$$

where p and q are the components of the angle θ projected onto the horizontal (ξX) and vertical (ηY) planes (see Fig. 15.2). The angle ψ is the phase difference between the rays reaching P_1 and P_2 from O. Equation (15.2) is a precise definition but its application depends on the assumptions made about the intensity distribution across the undulator aperture and also the degree of partial coherence which is acceptable. For the latter, it is usual to suppose that the radiation is acceptably coherent if P_2 lies within the central region of the diffraction pattern of the source. At the first diffraction minimum, no interference pattern is visible so $\mu_{12} = 0$ by definition. For a uniform Gaussian intensity distribution with radius r, $\mu_{12} = 0$ when $\theta = 0.61\lambda/r$. For a rectangular slit, width a, the corresponding value of θ is $0.5\lambda/a$ and it is convenient to take this as the diffraction limit. This choice has been justified by Kim,[6] who describes the distribution in phase space of the radiation from a single electron by the convolution of two Gaussian functions with standard deviations σ_x and $\sigma_{x'}$ so that the beam brightness at the point (x, x') is

$$\begin{aligned} B(x,x') &= \left.\frac{\mathrm{d}N}{\mathrm{d}t}\right|_{(0,0)} \frac{1}{\sigma_x\sqrt{2\pi}}\frac{1}{\sigma_{x'}\sqrt{2\pi}}\exp\left(-\frac{x^2}{2\sigma_x^2}\right)\exp\left(-\frac{x^2}{2\sigma_{x'}^2}\right) \\ &= \left.\frac{\mathrm{d}N}{\mathrm{d}t}\right|_{(0,0)} \frac{1}{2\pi\sigma_x\sigma_{x'}}\exp\left(-\frac{1}{2}\left(\frac{x^2}{\sigma_x^2}+\frac{x^2}{\sigma_{x'}^2}\right)\right) \end{aligned}$$

and the on-axis brightness therefore is

$$B(0,0) = \left.\frac{\mathrm{d}N}{\mathrm{d}t}\right|_{(0,0)} \frac{1}{2\pi\sigma_x\sigma_{x'}}$$

so that for this to be coherent within the criterion developed above, the product $\sigma_x\sigma_{x'}$ must be $\lesssim\lambda/4\pi$. This imposes a further limit on the beam size in addition to those discussed in the previous section. If we suppose that $\sigma_{x'} \approx \sqrt{\lambda/L}$ then the transverse beam size must be $\approx\sqrt{\lambda L}/4\pi$. In the case of the ESRF the beam emittance itself is routinely 4 nm rad whereas the $\lambda/4\pi$ value at 0.1 nm is $\approx 10^{-2}$ nm rad. This means that the raw output from the undulator is a long way from transverse coherence.

The desire to achieve a high-power coherent beam in the X-ray region is leading to new directions for synchrotron radiation sources. The first is the construction and operation of Free Electron Lasers (FELs) and the second is the use of linear accelerators (linacs) to provide electron beams with diffraction-limited emittance in the transverse direction. Briefly, the low synchrotron radiation output from an electron (or positron) undergoing linear acceleration (which has been discussed in Chapter 5) means that the beam emittance is not determined by the excitation effects which determine the emittance in a storage ring. In the case of a linac the beam emittance is determined by the emittance at injection (from the electron gun) and is then reduced in inverse proportion to the increase in energy down the linac. This beam can then be transported to an undulator (or switched between several undulators) to provide radiation to experiments in a single pass through the device.

The FEL effect, referred to above, arises because the intense, diffraction-limited electron beam, in its passage down the undulator, can produce an electromagnetic field at the output wavelength of the undulator which is sufficiently dense that the electron beam current density is modulated at the output wavelength. This always occurs in an undulator but, in the situations described in Chapter 14, the effect is so weak as to be negligible. If the undulator is very long, this spontaneous emission generates further modulation so that the undulator acts as an amplifier producing coherent emission at a wavelength given by eqn (14.4). Such SASE (Self-Amplified Spontaneous Emission) devices are currently in the design stage with test experiments envisaged in the near future. A project underway at the Stanford Linear Accelerator Centre (SLAC) is looking towards a SASE FEL operating in the wavelength range from 0.15 to 1.5 nm. The brightness envisaged would be up to 10^{23} average and up to 10^{34} peak, in the usual units.[7] Such a device creates quite new experimental opportunities as well as operating in a regime where there has been no practical experience so far.

The project mentioned above is only one of a large number of FEL developments covering a wide range of the electromagnetic spectrum.[8]

High quality coherent output becomes easier to achieve as the wavelength of the radiation gets longer. In the infrared region of the spectrum there are many examples of beam lines from storage ring dipole magnets. The output from these sources compares very favorably with that from thermal sources, not only in brightness (some 1000 times brighter) but also in terms of signal/noise ratio. Of course, FEL sources in the infrared region would be even brighter but, at the present time, bending magnet sources have the distinct advantage of continuous and predictable operation.[9] They follow a routine operating pattern compared with most FEL. The latter are still experiments in themselves rather than research tools.

Beam current and beam lifetime considerations

The ultimate limitation on electron beam lifetime in low emittance sources such as those mentioned above is the Touschek effect (see Chapter 8) which is proportional to the particle density in the electron bunch. The longest lifetimes will be obtained when the electrons are distributed equally between all the radio-frequency buckets. However, this does not necessarily mean the best conditions for all the experiments because users of the radiation source who, for example, are making lifetime measurements of photon-induced excited states may require a longer time between bunches than would be provided if all the bunches were present. Multibunch operation can also result in current limitation because of resonant effects in the radio-frequency system induced by the regular passage of the bunched electron beam through the radio-frequency cavities. These so-called higher order oscillatory modes (HOMs) also reduce the beam lifetime. By filling two-thirds of the available radio-frequency buckets the ESRF has achieved routinely a beam lifetime (April 1997) of 50 h and a current of 200 mA.[10]

Conclusion

Although the synchrotron radiation production processes are well understood, and have been for many years, this is not quite true for the sources themselves. The latter, especially those constructed recently and operating at high energy are electromagnetic devices with considerable complexity. Experience in operating these sources is already leading to major improvements in the characteristics of the photon beams offered to users. Experience indicates that although the outline properties of the storage ring are specified at the design stage it is essential that the design contains sufficient flexibility to incorporate new and unforeseen requirements.

Future developments which could be foreseen are:

- A magnetic lattice which allows storage ring operation under several beta function modes.
- Incorporation of high order multipolar magnets into the design so as to provide flexible control of the energy acceptance of the electron bunches.
- Radio-frequency systems which allow for the investigation and elimination of beam induced instabilities which would limit the beam lifetime and current.
- Provision of short pulse radiation ($\approx$1 ps) for timing experiments.
- The ability to insert new magnetic devices into the straight sections to produce radiation with specific spectral and/or polarization properties.
- Precise control of individual magnet coordinates within the lattice in order to control the coupling between horizontal and vertical betatron oscillation. One of the aims should be to operate with vertical beam emittances in the order of 10^{-3} nm rad with coupling well below 1% but another might be to increase the vertical emittance with a corresponding reduction in the horizontal plane.
- Continuous monitoring and control not only of electron beam position and angle but also of other beam parameters such as beam emittance and low amplitude-induced oscillations. Such procedures, applied routinely, would ensure that users obtain photon beams of the highest possible brightness on a routine basis.
- Careful examination and monitoring of all sources of storage ring failure in order to eliminate disruption to the experimental programme.

Epilogue

The last few years have seen a substantial increase in the number of synchrotron radiation sources operating world-wide as well as proposals for the construction of new facilities. At the present time the requirements of the users are making greater and greater demands on source operation. It is still the case that the limits on the exploitation of new ideas are financial rather than practical. This should ensure that the upgrading of existing sources and the eventual creation of new ones are driven by the research aims of the synchrotron radiation users.

References

1. A. Ropert, High brilliance lattices and the effects of insertion devices. CERN Accelerator School. 1989. CERN 90-03. Geneva 1990, pp. 158–94.
2. R. Chasman and K. Green, Preliminary design of a dedicated synchrotron radiation facility, *IEEE Trans. Nucl. Sci.,* **NS-22**, 1765 (1975).
3. J. Chavanne, P. Elleaume, and P. Van Vaerenbergh, The ESRF insertion devices, *J. Synchrotron Radiation*, **5**, 196–201 (1998).
4. Quoted from Highlights 1996/1997 European Synchrotron Radiation Facility BP220, F-38043 Grenoble cedex, France, November 1997.
5. M. Born and E. Wolf, *Principles of optics* (5th edn) Chapter 10, p. 491 *et seq.* Pergamon Press, Oxford, England (1975).
6. K.-J. Kim, Brightness, coherence and propagation characteristics of synchrotron radiation, *Nucl. Instr. Methods*, **A246**, 71–6 (1986).
7. H. Winick, Synchrotron radiation sources—present capabilities and future directions, *J. Synchrotron Radiation,* **5**, 168–75 (1998); K.-J. Kim, Advanced capabilities for future light sources, *J. Synchrotron Radiation*, **5**, 202–7 (1998).
8. W. B. Colson, Short-wavelength free-electron lasers in 1997. In *Free Electron Lasers 1997. Proceedings of the 19th International FEL Conference.* Beijing, China, August 1997. Published in *Nucl. Instr. Methods in Physics Research*, **A407**, 26–9 (1998); H. P. Freund and V. L. Granatstein. Long-wavelength free electron lasers in 1997, *ibid.*, 30–3.
9. See for example *Il Nuovo Cimento* 20D, 375 *et seq.* (1998) for a series of articles listing infrared sources and their uses.
10. Highlights 1996/97. European Synchrotron Radiation Facility, p. 98. ESRF November 1997.

APPENDIX A

Vector algebra

Prelude

This appendix contains a sufficient introduction to the properties of vectors required for the understanding of the material in this book. For a complete mathematical treatment of these and other topics the reader should consult one or more of the many excellent publications which are available.[1]

Definitions—scalars and vectors

A scalar quantity has only one component, magnitude. Examples are energy (E), density (ρ), electric charge (q or e), phase (ϕ), etc. Scalar quantities obey all the laws of arithmetic and in particular, addition, subtraction, multiplication, and division.

A vector quantity has more than one component; it possesses both magnitude and direction. Examples are: location in space ($\boldsymbol{r}$), velocity ($\boldsymbol{v}$), momentum ($\boldsymbol{p}$), electric field strength ($\boldsymbol{E}$), magnetic field strength ($\boldsymbol{B}$), etc.

The magnitude of a vector quantity $\boldsymbol{a}$ is denoted by $|\boldsymbol{a}|$ or simply, a. For example, $|\boldsymbol{r}|$ or r is the length of the position vector $\boldsymbol{r}$.

A unit vector is denoted by $\hat{\boldsymbol{a}}$. It has unit length so that $\hat{\boldsymbol{a}} = \boldsymbol{a}/|\boldsymbol{a}|$.

Vectors in a right-handed coordinate system

The laws of vector algebra are independent of what the vectors represent. In order to relate a vector quantity to observations in the real world we must be able to express the vector in terms of measurable quantities.

Figure A.1 shows a right-handed coordinate system or reference frame in which the point P with position vector $\boldsymbol{r}$ has coordinates x, y, and z. These coordinates are related to $\boldsymbol{r}$ by $x = |\boldsymbol{r}| \cos\theta_x$, $y = |\boldsymbol{r}| \cos\theta_y$, and $z = |\boldsymbol{r}| \cos\theta_z$, where the angles θ_x, θ_y, and θ_z are the angles between the direction of $\boldsymbol{r}$ and the coordinate axes.

The coordinates x, y, and z are scalar quantities but can be expressed as vectors using the unit vectors $\hat{\boldsymbol{\imath}}$, $\hat{\boldsymbol{\jmath}}$, and $\hat{\boldsymbol{k}}$ which point along the direction of the coordinate axes. We can write $\boldsymbol{r}$ as the sum of these vectors:

$$\boldsymbol{r} = x\hat{\boldsymbol{\imath}} + y\hat{\boldsymbol{\jmath}} + z\hat{\boldsymbol{k}}.$$

We read this definition by saying that we can reach the point P, the end point of $\boldsymbol{r}$, either by going a distance r along the direction OP from the origin O or by going a

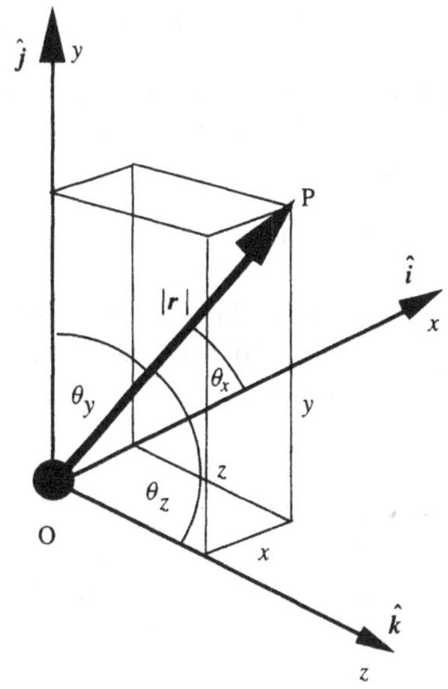

Fig. A.1 Unit vectors in a Cartesian coordinate system.

distance x along the direction $\hat{\boldsymbol{i}}$, followed by distances y and z along the directions $\hat{\boldsymbol{j}}$ and $\hat{\boldsymbol{k}}$, respectively. This defines what we mean by the addition of vectors and also the multiplication of a vector by a scalar quantity. Note that, because $\boldsymbol{i}$, $\boldsymbol{j}$, and $\boldsymbol{k}$ are always unit vectors the caret sign will be omitted when they are used later.

Vector multiplication—scalar product

The square of the position vector, $|\boldsymbol{r}|^2$, occurs in the expression for the electric and magnetic field strength. In order to find the value of $|\boldsymbol{r}|^2$ we can define the *scalar* product (sometimes call the dot product) of two vectors. The scalar product of two vectors $\boldsymbol{a}$ and $\boldsymbol{b}$ is written as

$$\boldsymbol{a} \cdot \boldsymbol{b} = |\boldsymbol{a}||\boldsymbol{b}| \cos \theta,$$

where θ is the angle between them. It is obvious that the scalar product is just the component of one vector along the direction of the other and is zero when the two vectors are at right angles so that $\cos\theta = 0$ and a maximum when the vectors are parallel. If we apply the definition of the scalar product to the expression for $\boldsymbol{r}$ then $\boldsymbol{r} \cdot \boldsymbol{r}$ can be expanded as

$$\boldsymbol{r} \cdot \boldsymbol{r} = (x\boldsymbol{i} + y\boldsymbol{j} + z\boldsymbol{k})^2$$

which is just $|\boldsymbol{r}|^2 = x^2 + y^2 + z^2$, as would be expected from the Pythagoras theorem (sixth century BC). To evaluate this expression we have made use of the fact that

$\boldsymbol{i} \cdot \boldsymbol{i} = \boldsymbol{j} \cdot \boldsymbol{j} = \boldsymbol{k} \cdot \boldsymbol{k} = 1$ and, because $\boldsymbol{i}$, $\boldsymbol{j}$, and $\boldsymbol{k}$ are mutually perpendicular, $\boldsymbol{i} \cdot \boldsymbol{j} = \boldsymbol{j} \cdot \boldsymbol{k} = \boldsymbol{k} \cdot \boldsymbol{i} = 0$. We would have obtained the same result for $\boldsymbol{r} \cdot \boldsymbol{r}$ if we had evaluated $\boldsymbol{r} \cdot (x\boldsymbol{i} + y\boldsymbol{j} + z\boldsymbol{k})$ because $\boldsymbol{r} \cdot \boldsymbol{i}$ is the component of $\boldsymbol{r}$ along the x-axis and similarly for $\boldsymbol{r} \cdot \boldsymbol{j}$ and $\boldsymbol{r} \cdot \boldsymbol{k}$. In general the scalar product $\boldsymbol{a} \cdot \boldsymbol{b}$ can be expanded, in the Cartesian coordinate system to

$$\boldsymbol{a} \cdot \boldsymbol{b} = a_x b_x + a_y b_y + a_z b_z.$$

Here a_x, a_y, and a_z are the components of $\boldsymbol{a}$ along the three axes and likewise for $\boldsymbol{b}$. These components are scalar quantities so that $\boldsymbol{a} \cdot \boldsymbol{b} = \boldsymbol{b} \cdot \boldsymbol{a}$. When the vectors $\boldsymbol{a}$ and $\boldsymbol{b}$ are identical, $\boldsymbol{a} = \boldsymbol{b}$ and $\boldsymbol{a} \cdot \boldsymbol{a} = |\boldsymbol{a}|^2 = a^2$.

Vector multiplication—vector product

In order to write down the expression for the magnetic field generated by a moving charge we require the vector product which is written as

$$\boldsymbol{c} = \boldsymbol{a} \times \boldsymbol{b} = |\boldsymbol{a}||\boldsymbol{b}| \sin\theta.$$

The new vector $\boldsymbol{c}$ generated by the product of $\boldsymbol{a}$ and $\boldsymbol{b}$ points in a direction at right angles to both $\boldsymbol{a}$ and $\boldsymbol{b}$. The angle θ between the two vectors $\boldsymbol{a}$ and $\boldsymbol{b}$ is positive when measured in the anticlockwise direction in a right-handed coordinate system. If $\boldsymbol{a}$ and $\boldsymbol{b}$ are interchanged then θ is negative and $\boldsymbol{a} \times \boldsymbol{b} = -\boldsymbol{b} \times \boldsymbol{a}$. When the angle between the vectors is zero then $\boldsymbol{a} \times \boldsymbol{b} = 0$.

The vector products of the unit vectors shown in Fig. A.1 are:

$$\boldsymbol{i} \times \boldsymbol{j} = \boldsymbol{k}, \quad \boldsymbol{j} \times \boldsymbol{k} = \boldsymbol{i}, \quad \boldsymbol{k} \times \boldsymbol{i} = \boldsymbol{j}$$

and

$$\boldsymbol{i} \times \boldsymbol{i} = \boldsymbol{j} \times \boldsymbol{j} = \boldsymbol{k} \times \boldsymbol{k} = 0.$$

The vector product can be expanded into its components in terms of these unit vectors in the following way:

$$\begin{aligned} \boldsymbol{a} \times \boldsymbol{b} &= \left(a_x\boldsymbol{i} + a_y\boldsymbol{j} + a_z\boldsymbol{k}\right) \times \left(b_x\boldsymbol{i} + b_y\boldsymbol{j} + b_z\boldsymbol{k}\right) \\ &= \left(a_y b_z - a_z b_y\right)\boldsymbol{i} + \left(a_z b_x - a_x b_z\right)\boldsymbol{j} + \left(a_x b_y - a_y b_x\right)\boldsymbol{k} \end{aligned}$$

which can be written in the form of a determinant as

$$\begin{aligned} \boldsymbol{a} \times \boldsymbol{b} &= \begin{vmatrix} \boldsymbol{i} & \boldsymbol{j} & \boldsymbol{k} \\ a_x & a_y & a_z \\ b_x & b_y & b_z \end{vmatrix} \\ &= \left(a_y b_z - a_z b_y\right)\boldsymbol{i} - \left(a_x b_z - a_z b_x\right)\boldsymbol{j} + \left(a_x b_y - a_y b_x\right)\boldsymbol{k}. \end{aligned}$$

We follow the usual rules for the expansion of a determinant so that the determinant for $\boldsymbol{b} \times \boldsymbol{a}$ is written as

$$\begin{aligned}\boldsymbol{b} \times \boldsymbol{a} &= \begin{vmatrix} \boldsymbol{i} & \boldsymbol{j} & \boldsymbol{k} \\ b_x & b_y & b_z \\ a_x & a_y & a_z \end{vmatrix} \\ &= \left(b_y a_z - b_z a_y\right)\boldsymbol{i} - \left(b_x a_z - b_z a_x\right)\boldsymbol{j} + \left(b_x a_y - b_y a_x\right)\boldsymbol{k}\end{aligned}$$

so that

$$\boldsymbol{a} \times \boldsymbol{b} = -\boldsymbol{b} \times \boldsymbol{a}$$

as before.

Vector multiplication—scalar triple product

We can write the result, s, of any scalar product of three vectors as

$$\begin{aligned}s &= \boldsymbol{a} \cdot \boldsymbol{b} \times \boldsymbol{c} \\ &= \left[a_x\boldsymbol{i} + a_y\boldsymbol{j} + a_z\boldsymbol{k}\right] \cdot \left[\left(b_y c_z - b_z c_y\right)\boldsymbol{i} - \left(b_x c_z - b_z c_x\right)\boldsymbol{j} + \left(b_x c_y - b_y c_x\right)\boldsymbol{k}\right] \\ &= a_x\left(b_y c_z - b_z c_y\right) - a_y\left(b_x c_z - b_z c_x\right) + a_z\left(b_x c_y - b_y c_x\right) \\ &= \begin{vmatrix} a_x & a_y & a_z \\ b_x & b_y & b_z \\ c_x & c_y & c_z \end{vmatrix}.\end{aligned}$$

In fact s is the volume of a rectangular solid whose sides are a, b, and c and the resulting product is the same (except for a sign) whatever the order of the vectors so that $\boldsymbol{a} \cdot \boldsymbol{b} \times \boldsymbol{c} = \boldsymbol{b} \cdot \boldsymbol{c} \times \boldsymbol{a} = \boldsymbol{c} \cdot \boldsymbol{a} \times \boldsymbol{b}$ and $\boldsymbol{a} \cdot \boldsymbol{b} \times \boldsymbol{c} = -\boldsymbol{a} \cdot \boldsymbol{c} \times \boldsymbol{b}$, etc. If any two vectors are equal or if the vectors all lie in the same plane the result is zero.

Vector multiplication—vector triple product

The vector product u of any three vectors is written as

$$\boldsymbol{u} = \boldsymbol{a} \times \boldsymbol{b} \times \boldsymbol{c}.$$

We can expand this product by setting

$$\begin{aligned}\boldsymbol{d} &= \boldsymbol{b} \times \boldsymbol{c} \\ &= \begin{vmatrix} \boldsymbol{i} & \boldsymbol{j} & \boldsymbol{k} \\ b_x & b_y & b_z \\ c_x & c_y & c_z \end{vmatrix} \\ &= \begin{vmatrix} b_y & b_z \\ c_y & c_z \end{vmatrix}\boldsymbol{i} - \begin{vmatrix} b_x & b_z \\ c_x & c_z \end{vmatrix}\boldsymbol{j} + \begin{vmatrix} b_x & b_y \\ c_x & c_y \end{vmatrix}\boldsymbol{k} \\ &= d_x\boldsymbol{i} + d_y\boldsymbol{j} + d_z\boldsymbol{k},\end{aligned}$$

then

$$\begin{aligned} \boldsymbol{u} &= \boldsymbol{a} \times \boldsymbol{d} \\ &= \begin{vmatrix} \boldsymbol{i} & \boldsymbol{j} & \boldsymbol{k} \\ a_x & a_y & a_z \\ d_x & d_y & d_z \end{vmatrix} \\ &= \begin{vmatrix} \boldsymbol{i} & \boldsymbol{j} & \boldsymbol{k} \\ a_x & a_y & a_z \\ \begin{vmatrix} b_y & b_z \\ c_y & c_z \end{vmatrix} & -\begin{vmatrix} b_x & b_z \\ c_x & c_z \end{vmatrix} & \begin{vmatrix} b_x & b_y \\ c_x & c_y \end{vmatrix} \end{vmatrix}. \end{aligned}$$

We can expand this expression and collect terms in $\boldsymbol{i}$, $\boldsymbol{j}$, and $\boldsymbol{k}$ to reach the useful result that

$$\boldsymbol{u} = (\boldsymbol{a} \cdot \boldsymbol{c})\boldsymbol{b} - (\boldsymbol{a} \cdot \boldsymbol{b})\boldsymbol{c}.$$

This result shows that $\boldsymbol{a} \times \boldsymbol{b} \times \boldsymbol{c}$ is a vector in the plane defined by $\boldsymbol{b}$ and $\boldsymbol{c}$. This is to be expected because $\boldsymbol{d} = \boldsymbol{b} \times \boldsymbol{c}$ is perpendicular to both $\boldsymbol{b}$ and $\boldsymbol{c}$ and $\boldsymbol{a} \times \boldsymbol{d}$ is perpendicular to $\boldsymbol{d}$ and must therefore lie in the same plane as $\boldsymbol{b}$ and $\boldsymbol{c}$.

Square of a vector product

One useful application of the properties of the triple scalar and vector products is the determination of the square of the vector product $(\boldsymbol{a} \times \boldsymbol{b}) \cdot (\boldsymbol{a} \times \boldsymbol{b})$. We write

$$\begin{aligned} (\boldsymbol{a} \times \boldsymbol{b}) \cdot (\boldsymbol{a} \times \boldsymbol{b}) &= (\boldsymbol{a} \times \boldsymbol{b}) \cdot \boldsymbol{d} \\ &= (\boldsymbol{b} \times \boldsymbol{d}) \cdot \boldsymbol{a} \\ &= (\boldsymbol{b} \times \boldsymbol{a} \times \boldsymbol{b}) \cdot \boldsymbol{a}. \end{aligned}$$

Now we expand the triple vector product to give

$$(\boldsymbol{b} \times \boldsymbol{a} \times \boldsymbol{b}) = (\boldsymbol{b} \cdot \boldsymbol{b})\boldsymbol{a} - (\boldsymbol{b} \cdot \boldsymbol{a})\boldsymbol{b}$$

with the result

$$\begin{aligned} (\boldsymbol{a} \times \boldsymbol{b}) \cdot (\boldsymbol{a} \times \boldsymbol{b}) &= ((\boldsymbol{b} \cdot \boldsymbol{b})\boldsymbol{a} - (\boldsymbol{b} \cdot \boldsymbol{a})\boldsymbol{b}) \cdot \boldsymbol{a} \\ &= a^2 b^2 - (\boldsymbol{a} \cdot \boldsymbol{b})^2. \end{aligned}$$

Differentiation of vectors

A common problem is the determination of the velocity of an object whose position is given by a vector $\boldsymbol{r}(t)$.

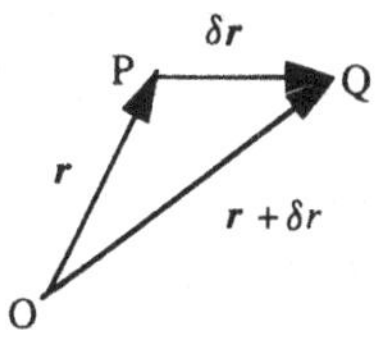

Fig. A.2 Infinitesimal change of a vector.

In Fig. A.2, a particle located at a point P moves by a small amount $\delta\boldsymbol{r}$ to a new position Q denoted by $\boldsymbol{r} + \delta\boldsymbol{r}$ in a time δt. The velocity $\boldsymbol{v}$ is defined in the usual way as the limiting value of $\delta\boldsymbol{r}/\delta t$ as δt tends to zero so that

$$\boldsymbol{v}(t) = \frac{\mathrm{d}\boldsymbol{r}(t)}{\mathrm{d}t} = \lim_{\delta t \to 0}\left\{\frac{(\boldsymbol{r}(t+\delta t) - \boldsymbol{r}(t))}{\delta t}\right\}.$$

The displacement from P to Q can be in any direction. In particular, OPQ can form an isosceles triangle so that the magnitude $|\boldsymbol{r}|$ remains the same and only its direction changes.

The gradient operator

In Chapter 2 we show how the electric field could be described in terms of a potential function $V(\boldsymbol{r})$ so that

$$\boldsymbol{E}(\boldsymbol{r}) = -\,\mathrm{grad}\,V(\boldsymbol{r}) = -\nabla V(\boldsymbol{r}).$$

In this expression, ∇ is the gradient operator, grad, defined as

$$\nabla = \frac{\partial}{\partial x}\boldsymbol{i} + \frac{\partial}{\partial y}\boldsymbol{j} + \frac{\partial}{\partial z}\boldsymbol{k}.$$

so that when grad operates on a scalar function of position, such as the potential function,

$$\nabla V(\boldsymbol{r}) = \frac{\partial V(\boldsymbol{r})}{\partial x}\boldsymbol{i} + \frac{\partial V(\boldsymbol{r})}{\partial y}\boldsymbol{j} + \frac{\partial V(\boldsymbol{r})}{\partial z}\boldsymbol{k}. \qquad \text{(A.1)}$$

The symbol for the partial derivative, $\partial/\partial x$, means that the operation of differentiation is to be carried out for x only, assuming that y and z remain constant and similarly for $\partial/\partial y$ and $\partial/\partial z$. When we carry out this operation, we are calculating the gradient of the function along each axis in turn. The final result is the vector sum of the three gradients which is a vector whose direction is that of the maximum rate of change of the function with position and the length of the vector is the rate of change itself.

The result of the gradient operator acting on itself, $\nabla \cdot \nabla$, is written ∇^2 and from eqn (A.1), together with the rules for the scalar product

$$\nabla^2 V(\boldsymbol{r}) = \frac{\partial^2 V(\boldsymbol{r})}{\partial x^2} + \frac{\partial^2 V(\boldsymbol{r})}{\partial y^2} + \frac{\partial^2 V(\boldsymbol{r})}{\partial z^2};$$

∇^2 is often call the Laplacian.

The gradient operator—divergence and rotation

The gradient operator is itself a vector so it can operate on vector functions such as electric and magnetic fields in two ways corresponding to the scalar and vector products of two vectors. These are, for the scalar product or divergence:

$$\nabla \cdot \boldsymbol{A} = \frac{\partial A_x}{\partial x} + \frac{\partial A_y}{\partial y} + \frac{\partial A_z}{\partial z}$$

and for the vector product or curl:

$$\nabla \times \boldsymbol{A} = \left(\frac{\partial B_z}{\partial y} - \frac{\partial B_y}{\partial z}\right)\boldsymbol{i} + \left(\frac{\partial B_x}{\partial z} - \frac{\partial B_z}{\partial x}\right)\boldsymbol{j} + \left(\frac{\partial B_y}{\partial x} - \frac{\partial B_x}{\partial y}\right)\boldsymbol{k}$$

$$= \begin{vmatrix} \boldsymbol{i} & \boldsymbol{j} & \boldsymbol{k} \\ \dfrac{\partial}{\partial x} & \dfrac{\partial}{\partial y} & \dfrac{\partial}{\partial z} \\ B_x & B_y & B_z \end{vmatrix}.$$

As before $\nabla \times \nabla = 0$.

Reference

1. Mary L. Boas, *Mathematical methods in the physical sciences* (2nd edn). John Wiley & Sons, New York, USA (1966 and 1983).

Index

Printed in the United States
By Bookmasters